Particulates in Water

Particulates in Water

Characterization, Fate, Effects, and Removal

Michael C. Kavanaugh, EDITOR

James M. Montgomery,
Consulting Engineers, Inc.

James O. Leckie, EDITOR

Stanford University

Based on a symposium

sponsored by the Division of

Environmental Chemistry

at the 175th Meeting of the

American Chemical Society,

Anaheim, California,

March 13–15, 1978.

ADVANCES IN CHEMISTRY SERIES **189**

AMERICAN CHEMICAL SOCIETY

WASHINGTON, D. C.   1980

Library of Congress CIP Data

Particulates in water.
 (Advances in chemistry series; 189 ISSN 0065-2393)
 "Based on a symposium sponsored by the Division
of Environmental Chemistry at the 175th meeting of
the American Chemical Society, Anaheim, California,
March 13-15, 1978."

 Includes bibliographies and index.

 1. Water—Purification—Congresses. 2. Particles—
Environmental aspects—Congresses.
 I. Kavanaugh, Michael C., 1940- . II. Leckie, Jim,
1939- . III. American Chemical Society. Division of
Environmental Chemistry. IV. Series.

QD1.A355 no. 189   [TD433]   540s [628.3'5]
ISBN 0-8412-0499-3   ADCSAJ   189 1–401 1980

Advances in Chemistry Series

M. Joan Comstock, *Series Editor*

FOREWORD

ADVANCES IN CHEMISTRY SERIES was founded in 1949 by the American Chemical Society as an outlet for symposia and collections of data in special areas of topical interest that could not be accommodated in the Society's journals. It provides a medium for symposia that would otherwise be fragmented, their papers distributed among several journals or not published at all. Papers are reviewed critically according to ACS editorial standards and receive the careful attention and processing characteristic of ACS publications. Volumes in the ADVANCES IN CHEMISTRY SERIES maintain the integrity of the symposia on which they are based; however, verbatim reproductions of previously published papers are not accepted. Papers may include reports of research as well as reviews since symposia may embrace both types of presentation.

CONTENTS

PREFACE

In the scientific and engineering community, concern for protection of aquatic environments from chemical pollution is shared by individuals from diverse disciplines. An area of overlapping interests is the study of particulates in water, their characterization, their interaction with the solution phase, and their control. Because many chemical and microbiological water contaminants are associated with colloidal or suspended particulate matter, a symposium on this subject seemed timely to provide a forum for various specialists to exchange ideas, methods, and models used to investigate the fate and effects of particulates and their associated materials in various aqueous environments.

The similarities and analogies observed between the research investigations of particulate specialists are intriguing. The mechanism of particulate capture by marine zooplankton mimics particulate removal by microscreens and granular-media filters used in water treatment. Modeling and analytical techniques now being used to investigate the dynamics of particulate suspensions in rivers, lakes, and estuaries parallel similar efforts conducted recently by air pollution scientists. These and other similarities were the major impetus for the organization of this symposium.

This book contains many of the papers presented at the symposium which addressed a variety of questions on the characterization, removal, fate, and effects of particulates in water, including:

1. What are the sampling and measurement problems associated with the physical and chemical characterization of particulates in water?
2. Are the particulates discharged into receiving waters a sink or a source of contaminants in aquatic environments?
3. How are adsorbed or incorporated pollutants distributed in the size classes of the particulate size spectrum?
4. How do particle dynamics in aquatic environments influence the fate of contaminants associated with the particulate phase?
5. Can the characterization of the physical, chemical, and microbiological properties of the particulates as a function of size be used to improve the accuracy of management and design decisions for water quality control?

Answers to these questions have been developed slowly in comparison to rapid advances in our knowledge of air particulates based on studies completed in the late 1960's. This is partly a result of superior sampling devices for size and chemical characterization of air particulates, and partly a result of the more heterogeneous nature of water particulates. The size spectrum of particulates in water extends from colloidal humic substances 1 nm in size, to large aggregates such as fecal pellets or marine snow with sizes up to 10^{-2} m. The distribution of shapes, densities, surface chemical properties, and chemical composition may vary widely with size. Some fractions of the size spectrum may be living, and all particulates are subject to diverse physical–chemical and biological processes that can alter size distributions, shape, or chemical composition.

Given this heterogeneity, it is not surprising that characterization of aqueous particulates is difficult and complex. Sizing of particulates smaller than 1 μm requires electron microscopic techniques. Some colloidal particulates have a nonrigid structure that necessitates special preparatory methods for sizing. Particulates larger than about 50 μm are usually present in low number concentrations, requiring large volume sampling. Fractionation techniques used to prepare samples for physical and chemical characterization of different size classes have proved to be time-consuming and of questionable accuracy. Finally, all sampling and size measurement techniques are subject to errors due to possible changes in the size distribution during sample collection, storage, and instrumental analysis.

The participants in this symposium addressed many of these problems. Methods for characterizing size and chemical composition by size were discussed, and several models of metal and virus adsorption were presented. Modeling particulate dynamics in rivers and the ocean provided new insights into the fate of contaminants associated with particulates. Papers on applications of size distribution measurements for selection, process modeling, and control of solid/liquid separation processes demonstrated the analytical value of particle counting compared to cumulative measurements of particulate concentration.

It is hoped that these papers will stimulate renewed interest in the impact of particulates on the fate and effects of materials deleterious to aquatic environments. Research in this area should provide improved modeling tools for water quality managers and designers of pollution control and water treatment facilities.

The organization of this symposium was aided by numerous individuals who provided suggestions on topics and speakers. We would particularly like to thank J. J. Morgan, W. Stumm, C. R. O'Melia, and

R. R. Trussell. We wish to acknowledge the essential financial support of James M. Montgomery, Consulting Engineers, Inc., and the secretarial skills of Judi Burle.

James M. Montgomery, Consulting
  Engineers, Inc.
11800 Sunrise Valley Drive
Reston, Virginia 22091

MICHAEL C. KAVANAUGH

Stanford University
Stanford, California 94305

JAMES O. LECKIE

April 15, 1980

Characterization of Surface Chemical Properties of Oxides in Natural Waters

The Role of Specific Adsorption in Determining the Surface Charge

HERBERT HOHL, LAURA SIGG, and WERNER STUMM

Institute for Water Resources and Water Pollution Control (EAWAG), Swiss Federal Institute of Technology, Zürich, Switzerland

The specific adsorption of H^+, OH^-, cations, and anions on hydrous oxides and the concomitant establishment of surface charge can be interpreted in terms of the formation of surface complexes at the oxide–water interface. The fixed charge of the solid surface and the pH of its isoelectric point can be measured experimentally by determining the proton balance at the surface (from alkalimetric titration curve) and by the analytical determination of the extent of adsorbate adsorption. Equilibrium constants established for the surface coordination reactions can be used to predict pH_{IEP}, to calculate adsorption isotherms, and to estimate concentration–pH regions for which the hydrous oxide dispersions are stable from a colloid-chemical point of view.

Oxides, especially those of silicon, aluminum, and iron, are abundant components of the earth's crust; they participate in geochemical reactions and in many chemical processes in natural waters, and often occur as colloids in water and waste treatment systems. The properties of the phase boundary between a hydrous oxide surface and an electrolyte solution depend on the forces operating on ions and water molecules by the solid surface and on those of the electrolyte upon the solid surface. The presence of an electric charge on the surface of particles often is essential for their existence as colloids; the electric double layer on their surface hinders the attachment of colloidal particles to each other, to other surfaces, and to filter grains.

0-8412-0499-3/80/33-189-001$07.75/0

It is the purpose of this chapter:

1. to discuss the effects of specific adsorption on the surface charge of hydrous oxides; specifically, to interpret the specific adsorption of H^+, OH^-, cations, and anions, and the concomitant establishment of surface charge in terms of the formation of surface complexes at the oxide–water interface;

2. to document experimentally with the help of simple model systems how the specific adsorption of solutes (surface coordination of cations and anions) affects the fixed charge and the isoelectric point of the surface, and to illustrate that the fixed charge can be measured experimentally by determining the proton balance at the surface (alkalimetric titration curve) and by the analytical determination of adsorbate adsorption; and

3. to illustrate that equilibrium constants can be used to predict pH_{IEP}, to calculate adsorption isotherms and the surface charge as a function of pH and other solution variables, and to estimate the pH-solute domains of the colloidal stability of oxide dispersions.

Specific Chemical Interaction with H^+, OH^-, Cations, and Anions at the Oxide Surface

The surface chemical properties of the oxide surface are sensitive to the composition of the aqueous phase because adsorption or binding of solutes to the surface may increase, decrease, or reverse the effective surface charge on the solid (1–9). One speaks of specific chemical interactions if binding mechanisms other than electrostatic interactions are significantly involved in the adsorption process.

Specific adsorption of cations and anions on hydrous oxide surfaces may be interpreted as a surface coordination reaction (10) (Figure 1).

The surfaces of metal or metalloid hydrous oxides are generally covered with OH groups which exhibit amphoteric properties; the pH-dependent charge of an oxide results from proton transfers at the surface:

$$\equiv MeOH_2^+ \rightleftarrows \equiv MeOH + H^+;$$

$$K^s_{a1} = \{\equiv MeOH\}[H^+]/\{\equiv MeOH_2^+\} \quad (1)$$

$$\equiv MeOH \rightleftarrows \equiv MeO^- + H^+;$$

$$K^s_{a2} = \{\equiv MeO^-\}[H^+]/\{\equiv MeOH\} \quad (2)$$

where [ ] and { } indicate concentrations of species in the aqueous phase (mol dm^{-3}) and concentrations of surface species (mol kg^{-1}), respec-

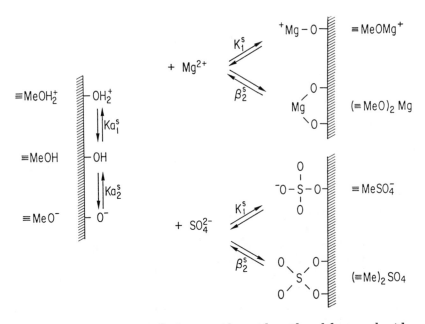

Figure 1. Interaction of hydrous oxides with acid and base and with cations and anions, exemplified here for the specific adsorption of Mg^{2+} and SO_4^{2-}, is interpreted in terms of surface complex formation and ligand-exchange equilibria.

tively. That portion of the charge attributable to specific interactions with H^+ and OH^- ions corresponds to the difference in protonated and deprotonated $\equiv$MeOH groups.

Operationally, there is a similarity between H^+ and metal ions (Lewis acids) and OH^- and bases (Lewis bases). The OH group on a hydrous oxide surface has a complex-forming oxygen-donor group like OH^- or OH groups attached to other elements (phosphate, silicate, polysilicate). Proton and metal ions compete with each other for the available coordinating sites on the surface:

$$\equiv MeOH + M^{z+} \rightleftharpoons \equiv MeOM^{(z-1)} + H^+;$$
$$*K^s_1 = \{\equiv MeOM^{(z-1)}\}[H^+]/\{\equiv MeOH\}[M^{z+}] \quad (3)$$

$$\begin{aligned}\equiv Me-OH \\ \equiv Me-OH\end{aligned} + M^{z+} \rightleftharpoons (\equiv MeO)_2M^{(z-2)} + 2H^+;$$
$$*\beta^s_2 = \{(\equiv MeO)_2M^{(z-2)}\}[H^+]^2/\{\equiv MeOH\}^2[M^{z+}] \quad (4)$$

The extent of coordination is related to the exchange (or displacement) of H^+ by M^{z+} ions. Similarly, ligand exchange with coordinating anions leads to a release of OH^- from the surface:

$$\equiv Me\text{-}OH + A^{z-} \rightleftarrows \equiv Me\text{-}A^{(z-1)} + OH^-; \quad K^s_1 \qquad (5)$$

$$\begin{array}{l} \equiv Me\text{-}OH \\ \equiv Me\text{-}OH \end{array} + A^{z-} \rightleftarrows (\equiv Me)_2 A^{(z-2)} + 2OH^-; \quad \beta^s_2 \qquad (6)$$

For protonated anions the ligand exchange may be accompanied by a deprotonation of the ligand at the surface. For example, in the case of HPO_4^{2-}:

$$\equiv MeOH + HPO_4^{2-} \rightleftarrows \equiv MeHPO_4^- + OH^- \rightleftarrows \equiv MePO_4^{2-} + H_2O \quad (7)$$

It is also conceivable that in a high pH condition, adsorption of a metal ion may be accompanied by hydrolysis (8).

As Figure 1 illustrates qualitatively, the specific adsorption of H^+ and cations increases the net charge while the specific adsorption of OH^- and anions decreases the net charge of the particle surface. As we shall see, the equilibrium constants that characterize the processes described (1–7) can be used to quantify the extent of adsorption and the resultant net charge of the oxide particle surface as a function of pH and solute activity.

Experimental Measurement of Surface Charges. If the fixed charge of an oxide particle $\bar{\sigma}$ ($C\ m^{-2}$) arises from the specific adsorption of H^+, OH^-, cations, and anions by the hydrous oxide surface, it is possible to determine (in principle) its value by determining experimentally (analytically) the extent of adsorption of charged species:

1. in the absence of specifically adsorbable cations and anions:

$$\bar{\sigma} = F\ (\Gamma_H - \Gamma_{OH}); \qquad (8)$$

2. in the presence of a specifically adsorbable cation M^{z+}:

$$\bar{\sigma} = F\ (\Gamma_H - \Gamma_{OH} + z\ \Gamma_{M^{z+}}); \qquad (9)$$

3. in the presence of a specifically adsorbable anion $H_nA^{(z-n)-}$:

$$\bar{\sigma} = F\ (\Gamma_H - \Gamma_{OH} - z\ \Gamma_{A^{z-}}); \qquad (10)$$

or, generally,

$$\bar{\sigma} = F\ (\Gamma_H - \Gamma_{OH} + \sum_{ji} z_j\ \Gamma_{M_i^{z+}} - \sum_{ji} z_j\ \Gamma_{A_i^{z-}}) \quad (11)$$

where F is Faraday's constant ($C\ mol^{-1}$), z is the magnitude of the charge of the nonhydrolyzed cation or deprotonated anion, and Γ_H, Γ_{OH}, $\Gamma_{M^{z+}}$, and $\Gamma_{A^{z-}}$ are the adsorption densities (mol m^{-2}) of H$^+$ (and its complexes), of OH$^-$ (and its complexes), of cations, and of the deprotonated anion, respectively. (If a mole of an anion like H$_2$PO$_4^-$ is adsorbed, the surface gains the equivalent of 2 mol of protons (to be included in the Γ_H) and 1 mol of PO$_4^{3-}$.) Before considering the experimental methods available we should reconsider some definitions.

Definitions

The electric state of a surface depends on the spatial distribution of free (electronic or ionic) charges in its neighborhood. This distribution is usually idealized as an electrochemical double layer; one layer of the double layer is envisaged as a fixed charge or surface charge attached to the particle or solid surface, while the other layer is distributed more or less diffusely in the liquid in contact. (The description of the double layer given here corresponds essentially to that given by the International Union of Pure and Applied Chemistry (IUPAC) (11).) This layer contains an excess of counterions, opposite in sign to the fixed charge, and usually (but not always) a deficit of coions of the same sign as the fixed charge. Counterions and coions in immediate contact with the surface are said to be located in the Stern layer; ions further away from the surface form the diffuse or Gouy layer.

Charged or Electrified Interface. This reflects an unequal distribution of charges (usually ions) at the phase boundary; that is, it results from a localized disturbance of the electroneutrality. The interfacial system as a whole (the region from one bulk phase to the other) is electrically neutral; that is, the sum of the charges is zero:

$$\bar{\sigma} + \sigma_G = 0 \qquad (12)$$

where σ is the fixed charge density; it may comprise the surface charge σ_o, or σ_s, any charge density in the Stern layer. The location of the dividing plane between σ_o and σ_G will determine how high the surface charge is.

Potential-Determining Ions. These are by definition (11) those species of ions which by virtue of their equilibrium distribution between the two phases (or by their equilibrium with electrons in one of the phases) determine the interfacial potential difference, that is, the difference in galvanic potential between these phases.

It is readily seen that the electrode potential (or the potential difference between the electrode and the solution) of a silver electrode depends on the Ag$^+$ ion activity of the solution (Nernst equation)

$$E = E_o + \frac{RT}{F} \ln \frac{\{Ag^+\}}{\{Ag(s)\}} \qquad (13)$$

or, the potential of an Ag/AgCl electrode is related to the activity of Ag^+ or Cl^- ions. Accepting the establishment of potential-determining equilibria between two phases, we may also apply the concept of potential-determining species to dispersed particles. It has been shown that the Nernst equation holds for silver–silver halogenide colloids, for example, for $AgBr(s)$

$$\psi_o = \frac{RT}{F} \ln \frac{\{Ag^+\}}{\{Ag_o^+\}} = -\frac{RT}{F} \ln \frac{\{I^-\}}{\{I_o^-\}} \qquad (14)$$

where ψ_o is called the surface potential (that is, the difference in potential between the surface and the solution), and $\{Ag_o^+\}$ and $\{I_o^-\}$ are the activities of Ag^+ and I^-, respectively, at a reference point, that is, at a point of zero potential. While a standard reference electrode (for example, H_2 electrode) is a standard reference state for electrode potentials, the point of zero potential or the zero point of charge may serve as a reference state in surface chemistry. Since there are oxide electrodes, for example, the antimony oxide electrode, which show a potental dependence on $\{H^+\}$ or the interrelated $\{OH^-\}$ in accordance with the Nernst equation

$$E = E_o + \frac{RT}{F} \ln \frac{\{H^+\}}{\{H_o^+\}} \qquad (15)$$

one expects that H^+ and OH^- are determining the potential of hydrous oxide surfaces:

$$\psi_o = \frac{RT}{F} \ln \frac{\{H^+\}}{\{H_o^+\}} \quad \text{or}$$

$$\psi_o = \frac{2.3RT}{F} (pH_o - pH) = \frac{2.3RT}{F} (pOH - pOH_o) \qquad (16)$$

where $\{H^+\}$ and $\{OH^-\}$ are the H^+ and OH^- activities, respectively, at the point of zero potential or at the point of zero charge, and pH_o or pH_{ZPC} is the pH of zero point of charge. The experimental observations with most oxides investigated are not in accord with the Nernst equation (Equation 16) (Figure 2). Presumably, the electrochemical potential of the solid is not independent of how many H^+ or OH^- ions are adsorbed (12). As Figure 2 illustrates, H^+ and OH^- nevertheless have a significant influence on the surface potential and surface charge of oxides, and even if the Nernst equation is not fulfilled we may speak of H^+ and OH^- ions as potential-determining ions. It is also appropriate to speak of a pH_{ZPC}. (The concept of potential-determining ions has been restricted by Lyklema (9) to constituent ions of the sorbent. However, O_2^- ions (which are

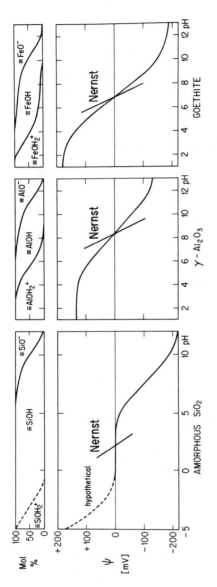

Figure 2. Dependence of surface potential on pH.

The acid–base characteristics of the surface groups (relative speciation of surface groups as a function of pH in upper figure) determine the pH of zero potential (point of zero proton condition $\{\equiv MeOH_2^+\} = \{\equiv MeO^-\}$). The Nernst equation—a surface potential dependence on pH of $(RT/F) \ln 10 (= 59 \; mvolt \; at \; 25°C)$—is not fulfilled. The lines in the lower figure were calculated from alkalimetric and acidimetric titration curves using Equation 26 (15).

components of the hydrous oxides) are in reversible equilibrium with OH^- ($O_2^- + H_2O \rightleftarrows 2OH^-$); hence OH^- is in equilibrium with both phases. For a further discussion on the invalidity of the Nernst equation see Ref. 15.) The pH_{ZPC} is different for different metal oxides; it is a measure of the acidity and basicity of the hydrated surface-oxide groups (Figure 2). It is constant for a given oxide and is independent of the concentration of indifferent (nonspecifically adsorbable) solutes in the solution.

Species that interact with potential-determining ions and alter their activity must also influence the surface potential; they have been called potential-determining species of the second kind (14). For example, NH_3 alters the activity of $\{Ag^+\}$ and thus affects the surface potential of silver halogenides. Similarly, metal ions that coordinate with surface OH^- groups

$$\equiv MeOH + M^{z+} \rightleftarrows \equiv MeOM^{(z-1)+} + H^+ \qquad (17)$$

or ligands that can exchange with surface OH^- groups

$$\equiv MeOH + A^{z-} \rightleftarrows \equiv MeA^{(z-1)-} + OH^- \qquad (18)$$

influence the surface potential (see Equations 3–7).

Thus, in general, from a coordination-chemical point of view, the following type of chemical equilibria between the solid and the solution or between the surface phase and the solution may be considered potential-determining equilibria:

$$Ag(s) = \mathbf{Ag^+} + \mathbf{e} \qquad (a)$$

$$AgBr(s) + \mathbf{e} = Ag(s) + \mathbf{Br^-} \qquad (b)$$

$$AgBr(s) + \mathbf{Ag^+} = Ag_2Br^+(surface) \qquad (c)$$

$$CuS(s) + \mathbf{H_2S} = \mathbf{Cu}(SH)_2(surface) \qquad (d)$$

$$FePO_4(s) + \mathbf{H_2PO_4^-} = Fe(HPO_4)_2^-(surface) \qquad (e)$$

$$FeOOH(s) + \mathbf{H^+} = Fe(OH)_2^+(surface) \qquad (f)$$

$$FeOOH(s) + \mathbf{OH^-} + H_2O = Fe(OH)_4^-(surface) \qquad (g)$$

$$SiO_2(s) + \mathbf{OH^-} = SiO_2(OH)^-(surface) \qquad (h)$$

$$SiO_2(OH)(surface) + \mathbf{Pb^{2+}} = SiO_2(OPb)^+(surface) + H^+ \qquad (i)$$

$$FeOOH(s) + \mathbf{HPO_4^{2-}} = FeOPO_4H^-(surface) + \mathbf{OH^-} \qquad (k)$$

$$S_n(s) + \mathbf{HS^-} = S_nS^-(surface) + \mathbf{H^+} \qquad (l)$$

$$(19)$$

In each case, the ion, species, or electron (boldface in Equation 19) that crosses or reacts at the interphase establishes the chemical equilibrium at the interface (equality of the electrochemical potential between solid or surface phase and solution) and thus determines (or influences) the interfacial potential difference. (The electron in Equations a and b may stand for a suitable reductant.) However, without further information the validity of the Nernst equation may not be inferred.

Charge. While the surface potential is not accessible by direct experimentation, the charge of the particle can be measured (at least in principle). However, the precise identification of charge density depends on the precise definition adopted for surface charge. (We ignore here charges caused by isomorphic substitution in the solid oxide as well as charges caused by adsorbing surface-active ions (soaps, fatty acids, polyelectrolytes, polymeric hydrolysis products).) The fixed charge attached to an oxide particle is given primarily by the proton transfer reactions that (1) render the surface positive and (2) render the surface negative; it is the charge given by the number of $\equiv MeOH_2^+$ groups minus the number of $\equiv MeO^-$ groups per unit surface. It is no more difficult to evaluate the surface charge in the case of surface-coordinating cations (Reactions 3 and 4) or anions (Reactions 5 and 6). Colloid chemists often place these specifically adsorbed cations and anions into the Stern layer. (Grahame (*16*) has placed specifically adsorbable ions at the mercury surface into a plane of closest approach (at a distance corresponding to the radius of a nonhydrated ion).) From a coordination-chemical point of view it does not appear very meaningful to assign a surface coordinating ion M^{z+} in a $\equiv MeOM^{(z-1)+}$ group or A^{z-} in a $\equiv MeA^{(z-1)-}$ group into a different layer than H^+ or OH^- in a $\equiv MeOH$ group. It is for this reason that we define the fixed charge $\bar{\sigma}$ of the oxide particle to include all the charges caused by bound (specifically adsorbed) cations and anions in addition to H^+ and OH^-, and consign all nonspecifically adsorbed counterions into the diffuse layer (Figure 3, Equation 11); that is, the fixed charge includes the charge caused by the binding of H^+ and/or OH^- (disturbance of surface proton balance) as well as that of bound metal ions and bound anions. Thus, in general, potential-determining species (Equation 19, c–l) are assumed to cross the interphase and are charge determining. Of course more elaborate models (for example, the basic and extended Stern or triple-layer models) can be used to describe the electric double layer. The adsorbed ions may be relegated to different mean planes of adsorption. Although these models may be expressed with mass-law and mass-balance equations, they differ merely in how the adsorption energy is separated (on the basis of nonthermodynamic assumptions) into electrostatic and chemical

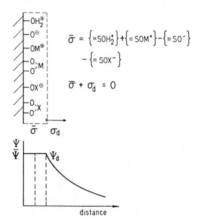

Oxide Interface and Charge Distribution
(schematically)

*Figure 3.   Juxtaposition of the thermodynamic model of the oxide sur-
face with the simplified electric double-layer model*

contributions (*27*). Westall and Hohl (*27*) have compared five electrostatic models for the oxide–solution interface, that is, the constant-capacitance model (*2, 4*), the diffuse-layer model (*1, 3*), the basic Stern-layer
model (*28*), the triple-layer model (*8*), and the Stern model, and con-
clude that all models represent the experimental data equally well,
although the values of corresponding parameters in different models are
not the same. Thus, all the models may be viewed as being of the correct
mathematical form to represent the data but are not necessarily accurate
physical descriptions of the interface.

The definition of the ZPC depends on the definition of the fixed
surface charge. If the latter is defined as, for example, by Lyklema (*12*)
to be caused by H^+ and OH^- only, ZPC corresponds to the pH where the
proton balance at the surface is zero ($\Gamma_H - \Gamma_{OH} = 0$); if we adopt our

definition for the fixed charge (Equation 11), that is, if we assume that the fixed charge includes ions attached by surface coordination, then the point of zero fixed surface charge is given by the condition

$$\Gamma_H - \Gamma_{OH} + \sum_{ji} z_j \Gamma_{M_i^{z+}} - \sum_{ji} z_j \Gamma_{A_i^{z-}} = 0 \qquad (20)$$

Since a particle surface that fulfills Equation 20 shows no electroosmosis (electrophoresis), this condition corresponds to the isoelectric point (IEP). Thus, pH_{IEP} is the pH where a particle carries no fixed charge. In addition, we can define that portion of the fixed surface charge caused by the proton balance of the surface only $- F(\Gamma_H - \Gamma_{OH})$ (cf. Equation 8). If this is zero, we can define a pH of zero proton condition $pH_{(\Gamma_H-\Gamma_{OH}=0)}$. This corresponds to the pH_{ZPC} defined by Lyklema (9, 12) for oxide surfaces. According to our picture, ZPC might stand for zero proton condition rather than for zero point of charge. (In the absence of specifically adsorbing species zero proton condition corresponds to zero point of charge, and also to IEP.)

Electrokinetic Potential (Zeta Potential ζ). ζ is the potential drop across the mobile part of the double layer that is responsible for electrokinetic phenomena, for example, electrophoresis (= motion of colloidal particles in an electric field) (11). It is assumed that the liquid adhering to the solid (particle) surface and the mobile liquid are separated by a shear plane (slipping plane). Unfortunately, the division between fixed surface charge of the particle and the diffuse charge of the solution does not necessarily coincide with the shear plane. The electrokinetic charge is the charge on the shear plane.

Experimentally Accessible Parameters.   Three quantities are readily determined experimentally:

1. The proton balance at the surface $\Gamma_H - \Gamma_{OH}$ obtained from alkalimetric or acidimetric titration permits the determination of that portion of the charge which is attributable to H^+ or OH^-. With the help of these curves, the acidity and basicity of the $\equiv$MeOH groups and the pH of zero proton condition can be determined (see Figures 9a, b, and c).
2. The adsorption of charge (potential)-determining ions other than H^+ or OH^- can also be determined analytically, for example, by measuring the extent of their adsorption by analyzing changes in the solution or on the solid. In principle, the extent of metal ion or anion adsorption can also be determined quantitatively from the shift in the alkalimetric–acidimetric titration curve caused by the presence of the coordinating cation (Equation 3) or anion (Equation 4) (1, 17) (see also Appendix I). Information gained from measurements (1) and (2) allows one to

estimate whether in a given pH range free or hydrolyzed metal ions or protonated or deprotonated ligands are adsorbed.

3. The electrophoretic mobility permits one to calculate the ζ potential. Often ψ_G, the potential drop across the diffuse part of the double layer, is taken to be identical to ζ. Hence, the electrokinetic charge may also be set approximately equal to the diffuse charge.

$$\sigma_{electrokinetic} \approx \sigma_G = -\bar{\sigma} = -(\sigma_o + \sigma_s) \qquad (21)$$

Thus, electrokinetics reflects the fixed surface charge (plus any specifically adsorbed ions in the Stern layer not yet accounted for in the fixed surface charge). If $\zeta = 0$, then $pH \approx pH_{IEP}$.

Experiments with Model Systems

The principles involved in the experimental determination of the fixed surface charge as a function of pH can be explained with the help of Figure 4. A model oxide with a large specific surface area (preferably a few $m^2 \ g^{-1}$) is dispersed in a solution of an inert (nonspecifically adsorbable cations and anions) salt solution, for example, $NaNO_3$. Aliquots of the dispersion are titrated with standard base (NaOH) and with standard acid (HNO_3). The resultant titration curve may be compared with that obtained in the absence of the oxide, and the quantity of H^+ or OH^- bound ($\Gamma_H - \Gamma_{OH}$) is calculated (see also Figures 9a, b, and c). The point of zero proton condition ($\{\equiv MeOH_2^+\} = \{\equiv MeO^-\}$) corresponds to the pH where the surface is uncharged and is identical to IEP. The dispersion may then be titrated in the presence of an adsorbable cation. The titration curve is shifted toward higher pH values (Figures 4 and 5) in such a way as to lower the pH of zero proton condition at the surface. At this point that portion of the charge that is attributable to H^+ and OH^- or their complexes becomes zero. (If a hydrolyzed metal ion is adsorbed, its OH^- will be included in the proton balance; similarly, in case of adsorption of protonated anions, their H^+ will be included in the proton balance.) Because of the binding of M^{z+} to the surface ($\Gamma_{M^{z+}}$), the fixed surface charge increases or becomes less negative; at the pH where the fixed surface charge becomes zero, the IEP is shifted to higher pH values. Correspondingly, specifically adsorbable anions (Figures 4 and 5) increase the pH of zero proton condition but lower the pH of IEP.

Surface Coordination of γ-Al_2O_3 with Mg^{2+} and SO_4^{2-}. Figures 5a and 9a illustrate the titration of γ-AlOOH with base and acid. In the absence of specifically adsorbable ions the point of zero proton condition

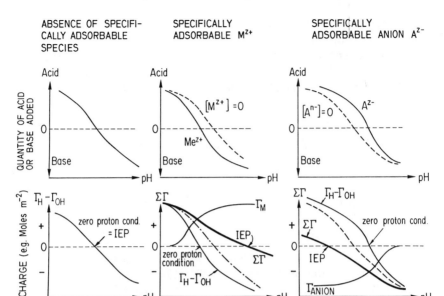

Figure 4. The net charge at the hydrous oxide surface is established by the proton balance (adsorption of H or OH⁻ and their complexes) at the interface and specifically bound cations or anions.

This charge can be determined from an alkalimetric–acidimetric titration curve and from a measurement of the extent of adsorption of specifically adsorbed ions. Specifically adsorbed cations (anions) increase (decrease) the pH of the IEP but lower (raise) the pH of the zero proton condition.

(= zero fixed charge) is at pH = 8.1; Mg^{2+} ions shift the titration curve in such a way that more base is required, and alternatively, SO_4^{2-} ions shift the titration curve in the opposite direction so as to increase the amount of acid consumed (Figure 5a). The adsorption of Mg^{2+} and SO_4^{2-} was determined analytically from aliquots for each point in the titration curve. From this information and the titration curve, the fixed charge was computed (Figures 5b and c). That the IEP is shifted to lower pH values by SO_4^{2-} and to higher pH values by Mg^{2+} is also seen from the pH shift observed for optimum sedimentation rate.

Mathematical equations for the titration curves shown in Figure 5a are given in Appendix I. These equations can be derived from charge-balance (or proton balance) conditions. Equations 3 and 5 of Appendix II illustrate that the shift in the titration curve at any pH is related quantitatively to the extent of specific adsorpion (Equations 17 and 18). The latter can be calculated independently of adsorption measurements of the shifts (17).

Silicic Acid. Figure 6 (from Ref. *18*) illustrates the effect of silicic acid on the titration curve of goethite (α-FeOOH). Dissolved silica is primarily present as H_4SiO_4 below pH $= 9$. (The acidity constants of H_4SiO_4 at $I = 0.5$, 25°C, are $pK_{a_1} = 9.5$, $pK_{a_2} = 12.6$.) The shift in the titration curve caused by silica reflects a release of protons which can be explained with the reactions

$$\equiv\!FeOH + H_4SiO_4 = \equiv\!FeOSi(OH)_3 + H_2O;$$

$$\log K^s_1 = 4.1 \quad (22)$$

$$\equiv\!FeOSi(OH)_3 = \equiv\!FeOSiO(OH)_2^- + H^+;$$

$$\log K^s_{a3}\ (\text{intr.}) = -7.4 \quad (23)$$

where K^s_1 is a charge-independent and K^s_{a3} a charge-dependent equilibrium constant. Because of the deprotonation reactions (Equations 22 and 23) the adsorption of dissolved silica decreases the fixed charge of the goethite surface (Figure 6b). With the help of calculations using the equilibrium constants, the extent of adsorption as a function of pH can be predicted (Figure 6c).

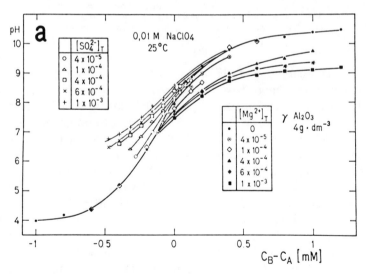

Figure 5.   Surface coordination of γ-Al$_2$O$_3$ with Mg^{2+} and SO$_4^{2-}$: (a) (above) titration of γ-Al$_2$O$_3$ in the presence of Mg^{2+} or SO$_4^{2-}$; (b) (facing page) surface charge of γ-Al$_2$O$_3$ in the presence of Mg^{2+} (I) $= 10^{-2}$M NaClO$_4$; T $= 25°$C); (c) (facing page) surface charge of γ-Al$_2$O$_3$ in the presence of SO$_4^{2-}$ (I $= 10^{-2}$M NaClO$_4$; T $= 25°$C)

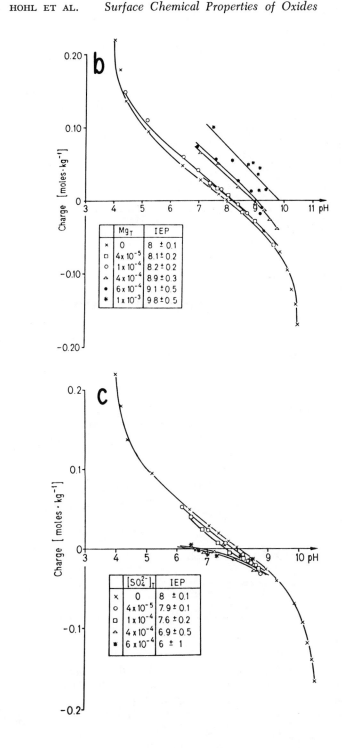

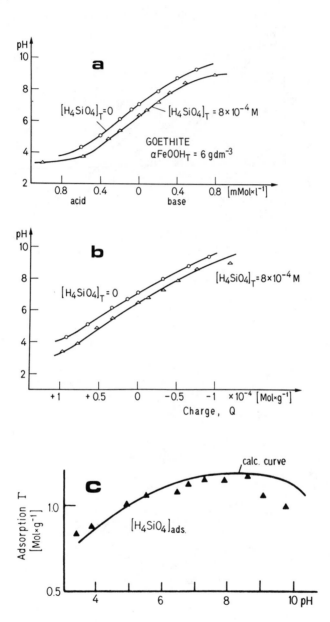

Figure 6.   Adsorption of aqueous silica on goethite (α-FeOOH) and its effect on the alkalimetric titration curve and surface charge.

The adsorption of silicic acid on α-FeOOH tends to release protons (a) and causes a decrease in surface charge (b). The extent of adsorption as a function of pH can be predicted by an equilibrium model that considers the equilibrium constants given in Equations 22 and 23 and the acidity constant of H_4SiO_4 and $\equiv FeOH_2^+$ (18).

Organic Acids. Adsorption of polar organic substances often gives results similar to those illustrated for silicate. The adsorption of the organic acid is accompanied by a deprotonation of the adsorbate. Figure 7 (from Ref. *19*) shows the extent of adsorption of phthalic acid (H_2X) on Al_2O_3 as a function of pH. This adsorption causes a charge reduction over most of the pH range. Adsorption and charge reduction can be explained by considering the reactions

$$\equiv\!AlOH + H_2X = \equiv\!AlXH + H_2O; \quad \log K^s_1 = 7.3 \qquad (24)$$

$$\equiv\!AlXH = H^+ + \equiv\!AlX^-; \quad \log K^s_{a_3} \text{ (intr.)} = -4.7 \qquad (25)$$

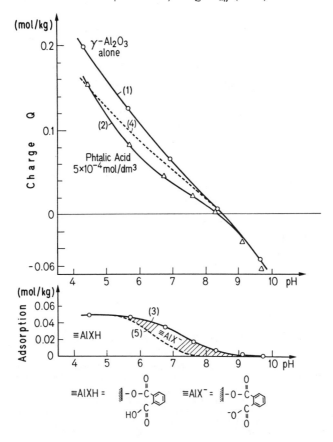

Figure 7.   Adsorption of phthalic acid (H_2X) on γ-Al_2O_3 and its effect on the surface charge.

The charge vs. pH Curves 1 and 2 were calculated from alkalimetric titration curves of Al_2O_3 in the absence and presence of phthalic acid, respectively. The adsorption of total phthalic acid (formation of $\equiv\!AlXH$ and $\equiv\!AlX^-$) was deter-mined analytically (Curve 3). Curve 4 was calculated from Curves 1 and 3 assuming that uncharged $\equiv\!AlXH$ only is formed. Curve 5 was calculated with the equilibrium constants given in Equation 24, the acidity constants of phthalic acid ($-\log K_{a_1} = 2.8$, $-\log K_{a_2} = 4.9$) (19).

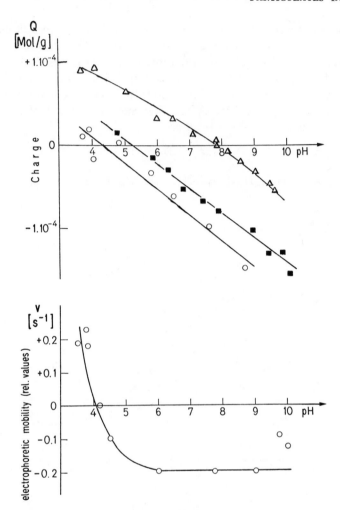

Figure 8.   Effect of adsorption of phosphate on goethite on its surface charge.

The fixed charge is computed from alkalimetric titration curves and from analytical determination of the quantity of P_T adsorbed. For the electrophoretic mobility measurement (bottom), a ratio of $P_T/FeOOH$ equal to that of the lower charge vs. pH curve (top) was used (equilibrium values of P_T in solution are approximately the same) (18). (top) (◯) $P_T = 2 \times 10^{-4}M$, 1.2 g FeOOH/L; (■) $P_T = 5 \times 10^{-4}M$, 6.0 g FeOOH/L; (△) $P_T = 0$, 6.0 g FeOOH/L; (bottom) $P_T = 2 \times 10^{-5}M$, 120 mg FeOOH/L.

The charge reduction observed at the lower pH values where Equation 24 predominates can be explained by considering that some of the protolyzable ≡AlOH groups have been replaced by ≡AlXH; that is, the fraction {≡AlOH$_2^+$}/({≡AlO$_T$}) has decreased as a result of the adsorption. The examples given in Figures 6 and 7 illustrate that a combined measurement of the proton balance and adsorption can be used to predict whether in a given pH region protonated or deprotonated ligands, or free or hydrolyzed metal ions are adsorbed. Supplementary electro-kinetic measurements are also of predictive value.

Phosphate. The binding of phosphate to hydrous oxides, especially Al_2O_3 and FeOOH, is also characterized by a proton release and a shift of IEP to lower pH values. With goethite (α-FeOOH) dispersed in phosphate solutions, the fixed charge was computed as a function of pH from titration curves (surface proton balance) and from analytic infor-mation (phosphate adsorbed) (18) (Figure 8). Reasonable agreement with electrokinetic data was obtained.

Computation of Adsorption and Surface Charge from Equilibrium Constants

The acid–base and adsorption equilibria can be characterized by equilibrium constants. Appendix II, for example, gives such constants for the proton transfer equilibria of γ-Al_2O_3 and its coordination with Mg^{2+} and $SO_4{}^{2-}$.

The values given are intrinsic constants, that is, constants valid for a hypothetically uncharged surface (for a given ionic medium at a given ionic strength). The acidity constants defined in Reactions 1 and 2 are experimentally accessible from the titration curves; they are microscopic acidity constants. Each loss of a proton reduces the charge on the surface polyacid and thus affects the acidity of the neighbor groups. Hence, the free energy of deprotonation may be interpreted to consist of a chemical contribution (dissociation) as measured by an intrinsic acidity constant $K^s{}_a$ (intr.), and of an electrostatic contribution (the removal of the proton from the site of the dissociation into the bulk of the solution) $\exp (F\psi / RT)$; thus

$$K^s{}_a = K^s{}_a \text{ (intr.) } \exp (F\psi / RT) \qquad (26)$$

where ψ is the potential difference between the surface site and the bulk solution. There is no direct way to obtain ψ. However, it is possible to determine the microscopic constants experimentally as a function of fixed charge and to extrapolate these constants to zero surface charge to obtain intrinsic constants (15). (For a more detailed discussion see Ref. 20. As Schindler (20) points out, measurements in ionic media reveal no infor-mation about the interaction between the reacting $\equiv$MeOH and the inert medium ions. Davis et al. (8) accounted for such interactions by esti-mating ion-pair formation constants for species such as $\equiv$MeOH$_2{}^+$ClO$_4{}^-$ and $\equiv$MeO$^-$Na$^+$.) Equation 26, then, allows one to estimate ψ. (This approach has been used to construct the ψ vs. pH curves of Figure 2.)

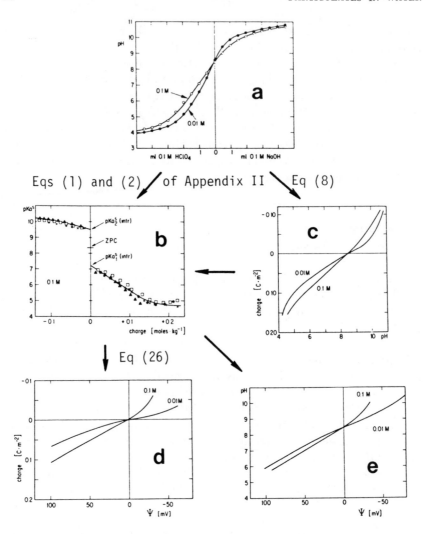

Figure 9. Exemplification of the method used in deriving the fixed charge of an oxide in absence of specifically adsorbable ions, surface acidity constants, and surface potential from alkalimetric–acidimetric titration curves.

The alkalimetric titration curve (a) permits the calculation of the surface charge as a function of pH (c) caused by the disturbance of the proton balance and the microscopic acidity constants as a function of the charge (b); intrinsic constants are obtained by extrapolation to zero surface charge (b). With the help of Equation 26, the surface potential either as a function of charge (d) or pH may be calculated. Experimental data are for γ-Al_2O_3 in $NaClO_4$ solutions (17).

The point of zero proton condition (the pH of fixed charge = 0 in the absence of specifically sorbed ions = IEP) is given by

$$\text{pH}_{\text{zero proton condition}} = \frac{1}{2}\left[\text{p}K^s_{a1}\,(\text{intr.}) + \text{p}K^s_{a2}\,(\text{intr.})\right] \qquad (27)$$

This pH should agree with the common intersection point of the titration curves obtained with different concentrations of an inert salt (Figure 9a). Equations similar to 26 can be written for all the constants given in Appendix II.

Shift in IEP.   The intrinsic constants may be used to estimate the composition of the oxide surface as a function of solution variables (pH, concentration of specifically adsorbable cations and anions). Without correcting for coulombic attraction or repulsion such calculations should give reasonable predictions only for surfaces that have a fixed surface charge of zero (or nearly zero). Hence, it should be possible to use intrinsic constants to predict shifts in IEP caused by specific cation and anion binding. Equation 20 gives the condition for zero fixed charge (IEP).

In our model system (γ-Al_2O_3, Mg^{2+}, SO_4^{2-}), the IEP in the presence of Mg^{2+} is characterized by

$$\text{IEP:} \quad \{\equiv AlOH_2^+\} + \{\equiv AlOMg^+\} = \{\equiv AlO^-\} \tag{28}$$

while the condition of zero proton condition is given by

$$\text{zero proton condition:} \quad \{\equiv AlOH_2^+\} =$$
$$\{\equiv AlO^-\} + \{\equiv AlOMg^+\} + 2\{(\equiv AlO)_2Mg\} \tag{29}$$

By using the intrinsic equilibrium constants given in Appendix II, and mass-balance equations for Mg_T and the available surface groups (a, and $\{\equiv AlO_T\}$), the shift in IEP as a function of Mg^{2+} or Mg_T can be predicted. Generally, for the specific binding of a cation M^{2+}, the following relationship can be derived:

$$[H^+_{IEP(M^{2+})}]^2 \cong [H^+_{IEP(0)}]^2 \left(1 - \frac{{}^*K^s_1}{K^s_{a2}} \cdot [M^{2+}]\right) \tag{30}$$

where $[M^{2+}]$ is the equilibrium concentration of M^{2+}.

In a similar way, the shift caused by SO_4^{2-} on IEP and zero proton condition can be derived from:

$$\text{IEP:} \{\equiv AlOH_2^+\} = \{\equiv AlSO_4^-\} + \{\equiv AlO^-\} \tag{31}$$

$$\text{zero proton condition:} \{\equiv AlOH_2^+\} =$$
$$\{\equiv AlSO_4^-\} + 2\{(\equiv Al)_2SO_4\} + \{\equiv AlO^-\} \tag{32}$$

Thus in the presence of a specifically adsorbable anion A^{2-}, the IEP of Al_2O_3 is given by

$$[H^+_{IEP(A^{2-})}]^2 \cong [H^+_{IEP(O)}]^2 \left[1 \Big/ \left(1 - \frac{K^s_1 \cdot K^s_{a1}}{K_w} [A^{2-}] \right) \right] \quad (33)$$

where K_w is the ion product of water. As Equations 30 and 33 suggest, the change in $[H^+_{IEP}]$ is related to the concentration of the specifically adsorbable cations and anions and their affinity to the surface groups, that is, the respective surface complex formation constants. In the presence of coordinating cations and anions the change in IEP's can be estimated by

$$[H^+_{IEP(M^{2+},A^{2-})}]^2 = [H^+_{IEP(O)}]^2 \frac{1 - (^*K^s_1/K^s_{a2})\,[M^{2+}]}{1 - \left(\dfrac{K^s_1 \cdot K^s_{a1}}{K_w} \right) [A^{2-}]}$$

$$(34)$$

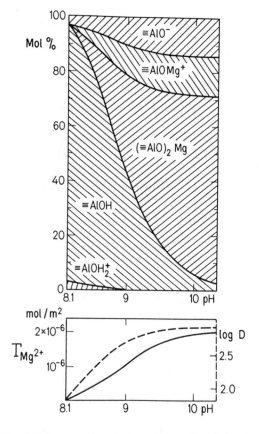

Figure 10. Composition of the surface phase of γ-Al_2O_3 at the IEP as a function of $[Mg^{2+}]$ in solution

The Surface Composition at the IEP. In Figure 10 the speciation at the Al_2O_3 surface in the presence of Mg^{2+} is calculated (with the equations of Appendix II) as a function of pH_{IEP}. According to the equilibrium conditions, every pH_{IEP} is related to a specific equilibrium concentration of free Mg^{2+}. (The calculation is based on the equations given in Appendix II and on Equation 28; furthermore $\{\equiv AlO_T\} = \{\equiv AlOH_2^+\} + \{\equiv AlOH\} + \{\equiv AlO^-\} + \{\equiv AlOMg^+\} + 2\{(\equiv AlO)_2-Mg\}$.) It is seen that the extent of Mg binding to the oxide surface (uncharged surface) increases with increasing pH; $(\equiv AlO_2)Mg$ and $\equiv AlOMg^+$ substitute for $\equiv AlOH$ groups.

Figure 11 represents the shift in pH_{IEP} caused by Mg^{2+} and SO_4^{2-}. The agreement between experimentally determined IEP's and calculated ones becomes rather poor as soon as $\Delta pH_{IEP} > 1$. As Figures 10a and b indicate, the surface coverage with Mg^{2+} reaches a "saturation" in higher pH regions (pH > 9, $\Delta pH_{IEP} > 1$). Under these conditions, the first term in Equation 28 becomes negligible; hence the equation reduces to $\{\equiv AlOMg^+\} = \{\equiv AlO^-\}$ which is equivalent to

$$*K^s_1 \{\equiv AlOH\} [M^{2+}]/[H^+] = K^s_{a2} \{\equiv AlOH\}/[H^+], \quad \text{or}$$

$$[M^{2+}] = K^s_{a2}/*K^s_1 \tag{35}$$

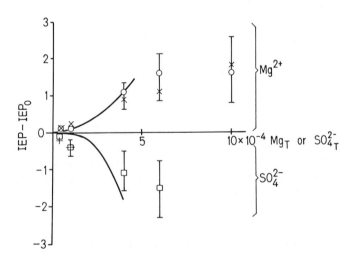

Figure 11.   Shift of the IEP of γ-Al_2O_3 (4 g/L γ-Al_2O_3) caused by Mg^{2+} or SO_4^{2-}.

Curves were calculated using Equations 28 and 31 and the equilibrium constants given in Appendix II. These calculations show that the Al_2O_3 surface can no longer remain uncharged at any pH above $[Mg_T] \sim 5 \times 10^{-4}M$ or $[SO_4^{2-}]_T \approx 4 \times 10^{-4}M$, respectively. Experimentally determined values for pH_{IEP} above these concentrations of Mg^{2+} and SO_4^{2-} are relatively insensitive to the relatively small residual surface charges that may prevail under these conditions. (top) Mg^{2+}: (○) from titration; (×) from settling velocity; (——) calculated curve; (bottom) SO_4^{2-}: (□) from titration; (+) from settling velocity.

For the Al_2O_3–Mg^{2+} system, Equation 35 gives $[Mg^{2+}] = 2.5 \times 10^{-4}M$. For this concentration of Mg^{2+}, $pH_{IEP} \cong 9.3$. Any attempt to further increase the surface coverage in Mg by increasing free Mg^{2+} in solution results in a positively charged surface, that is, with equilibrium concentrations of Mg^{2+} larger than $2.5 \times 10^{-4}M$. There is no pH value for which the Al_2O_3 surface remains uncharged. The increasing coverage of the surface with Mg^{2+} with increasing pH_{IEP} (Figure 10b) results from the availability of a larger number of coordination sites at the higher pH value.

Correcting for Coulombic Interaction. The surface speciation as a function of solution variables can be computed if we can correct our equibrium constants for electrostatic attraction or repulsion. Westall (21, 22, 23) has developed a computer program that permits one to compute iteratively the composition of the surface and its charge from a set of equilibrium constants. Figures 12 and 13 illustrate the application of this computation to the interaction of o-phosphate with goethite (α-FeOOH). This interaction is rather involved because various monodentate and bidentate surface species have to be assumed to account for the experimental observations (18, 24):

$\equiv FeOPO_3H_2, \equiv FeOPO_3H^-, \equiv FeOPO_3^{2-},$

A set of equilibrium constants has been determined (18) which is compatible with experimental observations. These constants can now be used to make generalizations and to predict adsorption isotherms and the fixed surface charge of FeOOH as a function of pH and other solution variables. A satisfactory agreement between model calculations and experimental results (adsorption data and measurements of electrophoretic mobility) is obtained.

Predicting Domains of Colloid Stability. With the help of the equilibrium constants, an isoelectric line in a pH vs. concentration of specifically adsorbing solute may be computed. An oxide dispersion is uncharged and will coagulate in the immediate proximity of this line. Colloid stability is assumed to occur on either side of the line at positive and negative threshold charges. Figure 13 shows the pH–log P_T ranges in which goethite is colloid-chemically stable as a negatively or posi-

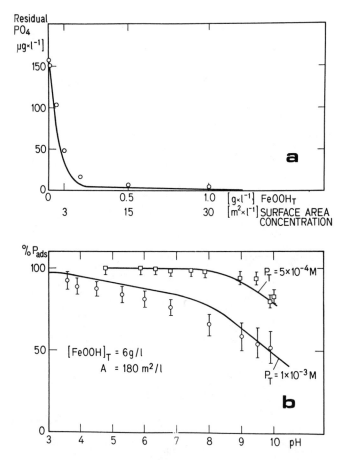

Figure 12.   Interaction of phosphate with goethite (α-FeOOH). Extent of adsorption is predicted with the help of the equilibrium constants given in Table I (18). (a): (———) calculated curve; (○) experimental measurements; pH = 7; (b): [FeOOH]$_T$ = 6 g/L; A = 180 m²/L.

Table I.   Equilibria on the Interaction of Phosphate with Goethite (α-FeOOH)

		log K_{intr}
$\equiv FeOH_2^+ \rightleftharpoons \equiv FeOH + H^+$	K^s_{a1}	−6.4
$\equiv FeOH \rightleftharpoons \equiv FeO^- + H^+$	K^s_{a2}	−9.25
$\equiv FeOH + H_3PO_4 \rightleftharpoons \equiv FePO_4H_2 + H_2O$	K^s_1	9.5
$\equiv FePO_4H_2 \rightleftharpoons \equiv FePO_4H^- + H^+$	K^s_{a4}	−4.4
$\equiv FePO_4H^- \rightleftharpoons \equiv FePO_4^{2-} + H^+$	K^s_{a5}	−6.6
$2 \equiv FeOH + H_3PO_4 \rightleftharpoons \equiv Fe_2PO_4H + 2H_2O$	β_2^s	8.5
$\equiv Fe_2PO_4H \rightleftharpoons \equiv Fe_2PO_4^- + H^+$	K^s_{a3}	−4.0

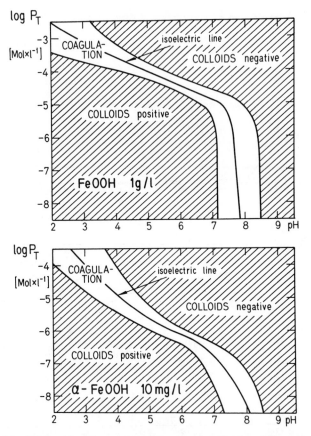

Figure 13.   Plot of log P_T vs. pH domain of colloid stability of goethite dispersions.

These domains were calculated using equilibrium constants (Table I) and corrections for coulombic attraction or repulsion. Rapid coagulation should occur in the proximity of the isoelectric line. Colloid stability is assumed to occur at positive and negative charges corresponding to a zeta potential of 30 mvolts (18).

tively charged colloid. The computed diagram is similar to a diagram for the α-Fe_2O_3–phosphate system experimentally determined by Breeuwsma and Lyklema (25).

The distinction between inert counterions and surface-coordinating (specifically adsorbable) ions is readily seen in coagulation experiments; the critical coagulation concentration of the specifically sorbable ion depends on the surface area concentration of the dispersed colloid (Figure 14).

James and Healy (26) have shown with measurements of electrophoretic mobility that in systems of a colloid oxide and metal ions two

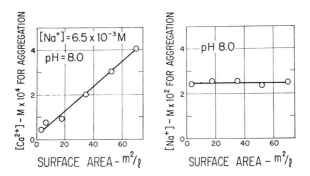

Figure 14. Relationship between MnO_2 colloid surface area concentration and critical coagulation concentration of Ca^{2+} and Na^+.

Because of the specific interaction of the oxide surface with Ca^{2+}, a stoichiometric relationship exists between the critical coagulation concentration and the surface area concentration; however, in the case of Na^+ this interaction is weaker, so primarily compaction of the diffuse part of the double layer causes destabilization (1).

charge reversals at different pH values may be observed. For example, in a system of SiO_2–M^{2+}, the silica is negatively charged at low pH. With increased pH the extent of adsorption increases, and a charge reversal from negative to positive occurs at the $pH_{IEP(A)}$. On further increase in pH, the charge may revert back to negative at $pH_{IEP(B)}$. Schindler (20) has shown that the coordination chemistry model can account fully for these observations. The first charge reversal is primarily caused by the increase in coordinative sites of the surface and the larger extent of M^{2+} adsorption with increasing pH. The second charge reversal results primarily from the lowering of free M^{2+} ions (increased consumption of M^{2+} by the surface and by hydrolysis with increased pH).

Appendix I. Alkalimetric and Acidimetric Titration of Hydrous Oxide

C_A, C_B are concentrations of strong acid and strong base, respectively, ΔC_B is the shift in the titration curve at constant pH caused by the presence of a specific adsorbable cation or anion; $\Delta C_B = (C_B - C_A) - (C_B^* - C_A^*)$ where concentrations with a * are those in the presence of specifically adsorbable ions. α_0, α_1, α_2 = degrees of protolysis of $\equiv MeOH_2^+$, $\equiv MeOH$, and $\equiv MeO^-$, respectively. $\alpha_0 = \{\equiv MeOH_2^+\}/\{\equiv Me_T\}$, $\alpha_1 = \{\equiv MeOH\}/\{\equiv Me_T\}$, $\alpha_2 = \{\equiv MeO^-\}/\{\equiv Me_T\}$; $\alpha_0 = (1 + K^s_{a_1}/[H^+] + K^s_{a_1} K^s_{a_2}/[H^+]^2)^{-1}$; $\alpha_1 = ([H^+]/K^s_{a_1} + 1 + K^s_{a_2}/[H^+])^{-1}$; $\alpha_2 = ([H^+]^2/K^s_{a_1} K^s_{a_2} + [H^+]/K^s_{a_2} + 1)^{-1}$. a = quantity of oxide used (kg/L); Q = charge in mol kg^{-1}. It is assumed that, at a given pH, the proportion of protonated to nonprotonated free surface groups is not affected by adsorption.

1. Absence of specific adsorption of ions:

$$(C_B - C_A + [H^+] - [OH^-])/a =$$
$$\{\equiv MeO^-\} - \{\equiv MeOH_2^+\} = -Q \quad (1)$$

2. In presence of specific adsorption of M^{2+}:

$$(C_B{}^* - C_A{}^* + [H^+] - [OH^-])/a =$$
$$\{\equiv Me\overset{*}{O}^-\} - \{\equiv Me\overset{*}{O}H_2^+\} + \{\equiv MeOM^+\} + 2\{(\equiv MeO)_2M\} \quad (2)$$
$$\Delta C_B/a = -(2\alpha_0 + \alpha_1)(\{\equiv MeOM^+\} + 2\{(\equiv MeO)_2M\}) \quad (3)$$

3. In presence of specific adsorption of A^{2-}:

$$(C_B{}^* - C_A{}^* + [H^+] - [OH^-])/a =$$
$$\{\equiv Me\overset{*}{O}^-\} - \{\equiv Me\overset{*}{O}H_2^+\} - \{\equiv MeA^-\} - 2\{(\equiv Me)_2A\} \quad (4)$$
$$\Delta C_B/a = +(\alpha_1 + 2\alpha_2)(\{MeA^-\} + 2\{(\equiv Me)_2A\}) \quad (5)$$

4. In presence of weak acid HA:

$$(C_B{}^* - C_A{}^* + [H^+] - [OH^-]/a = \{\equiv MeO^{-*}\} -$$
$$\{\equiv MeOH_2^{+*}\} - \{\equiv MeA\} - 2\{\equiv Me_2A^+\} +$$
$$(HA_T - [HA](1/a) \quad (6)$$

$$\Delta C_B/a = {}^*Q - Q = (\alpha_1 + 2\alpha_2)[\{\equiv MeA\} +$$
$$2\{\equiv Me_2A^+\}] - (HA_T - [HA])(1/a) \quad (7)$$

Appendix II. Equilibrium Constants for the Al_2O_3, Mg^{2+}, SO_4^{2-} Systems [a]

$\equiv AlOH_2^+ \rightleftarrows \equiv AlOH + H^+$

$$K^s_{a1} \text{ (intr.)} = \frac{\{\equiv AlOH\} \cdot [H^+]}{\{\equiv AlOH_2^+\}} = 10^{-6.6} \quad (1)$$

$\equiv AlOH \rightleftarrows \equiv AlO^- + H^+$

$$K^s_{a2} \text{ (intr.)} = \frac{\{\equiv AlO^-\} \cdot [H^+]}{\{\equiv AlOH\}} = 10^{-9.6} \quad (2)$$

$\equiv AlOH + Mg^{2+} \rightleftarrows \equiv AlOMg^+ + H^+$

$$*K^s_1 \text{ (intr.)} = \frac{\{\equiv AlOMg^+\} \cdot [H^+]}{\{\equiv AlOH\} \cdot [Mg^{2+}]} = 10^{-6} \quad (3)$$

$2\equiv AlOH + Mg^{2+} \rightleftarrows (\equiv AlO)_2Mg + 2H^+$

$$*\beta^s_2 \text{ (intr.)} = \frac{\{(\equiv AlO)_2Mg\} \cdot [H^+]^2}{\{\equiv AlOH\}^2 \cdot [Mg^{2+}]} = 10^{-13.8} \quad (4)$$

$\equiv AlOH + SO_4^{2-} \rightleftarrows \equiv AlOSO_3^- + OH^-$

$$K^s_1 \text{ (intr.)} = \frac{\{\equiv AlOSO_3^-\} \cdot [OH^-]}{\{\equiv AlOH\} \cdot [SO_4^{2-}]} = 10^{-3.6} \quad (5)$$

$2\equiv AlOH + SO_4^{2-} \rightleftarrows (\equiv AlO)_2SO_2 + 2OH^-$

$$\beta^s_2 \text{ (intr.)} = \frac{\{(\equiv AlO)_2SO_2\} \cdot [OH^-]^2}{\{\equiv AlOH\}^2 \cdot [SO_4^{2-}]} = 10^{-9.4} \quad (6)$$

[a] The experimental approach used is that described in Ref. 17.

Acknowledgment

Valuable advice given by J. A. Davis, Robert Kummert, Paul Schindler, and John Westall is acknowledged. This work was supported by the Swiss National Foundation.

Literature Cited

1. Stumm, W.; Huang, C. P.; Jenkins, S. R. "Specific Chemical Interaction Affecting the Stability of Dispersed Systems," *Croat. Chem. Acta* **1970**, *42*, 223.
2. Stumm, W.; Hohl, H.; Dalang, F. "Interaction of Metal Ions with Hydrous Oxide Surfaces," *Croat. Chem. Acta* **1976**, *48*, 491.
3. Huang, C. P.; Stumm, W. "Specific Adsorption of Cations on Hydrous γ-Al_2O_3," *J. Colloid Interface Sci.* **1973**, *43*, 409.
4. Schindler, P. W.; Wälti, E.; Fürst, B. "The Role of Surface Hydroxyl Groups in the Surface Chemistry of Metal Oxides," *Chimia* **1976**, *30*, 107.
5. Schindler, P. W.; Fürst, B.; Dick, R.; Wolf, P. U. "Ligand Properties of Surface Silanol Groups," *J. Colloid Interface Sci.* **1976**, *55*, 469.
6. Hingston, F. J.; Posner, A. M.; Quirk, J. P. "Anion Adsorption by Goethite and Gibbsite," *J. Soil Sci.* **1972**, *23*, 177.
7. Parfitt, R. L.; Atkinson, R. J.; Smart, R. S. C. "The Mechanism of Phosphate Fixation by Iron Oxides", *Soil Sci. Soc. Am., Proc.* **1975**, *39*, 837.
8. Davis, J. A.; James, R. O.; Leckie, J. O. "Surface Ionization and Complexation at the Oxide-Water Interface," *J. Colloid Interface Sci.* **1978**, *63*, 480.
9. Lyklema, J. "Interfacial Electrochemistry of Hydrophobic Colloids," In "The Nature of Seawater"; Goldberg, E. D., Ed.; Dahlem Konferenzen: Berlin, 1975.
10. Morgan, J. J.; Stumm, W. "The Role of Multivalent Metal Oxides in Limnological Transformations as Exemplified by Iron and Manganese," In "Proceedings 2nd International Water Pollution Research Conference"; Pergamon: New York, 1965; p. 163.
11. Everett, D. H. "International Union of Pure and Applied Chemistry (IUPAC): Defintions, Terminology and Symbols in Colloid and Surface Chemistry," *Pure Appl. Chem.* **1972**, *31*, 618.
12. Lyklema, J. "Electrochemistry of Reversible Electrodes and Colloidal Particles," In "Physical Chemistry: Enriching Topics from Colloids and Surface Science"; van Olphen, H., Mysels, K. J., Eds.; Theorex: La Jolla, CA, 1975.
13. Parsons, R. "Thermodynamics of Electrified Interphases," In "Physical Chemistry: Enriching Topics from Colloids and Surface Science"; van Olphen, H., Mysels, K. J., Eds.; Theorex: La Jolla, CA, 1975.
14. Parks, G. A. "Adsorption in the Marine Environment," In "Chemical Oceanography"; Riley, J. P., Skirrow, G., Eds.; Academic: New York, 1975.
15. Hohl, H.; Schindler, P.; Stumm, W. "Interactions of Metal Ions with Hydrous Oxides," *Adv. Colloid Sci.* (in press).
16. Grahame, D. C. *Chem. Rev.* **1947**, *41*, 441.
17. Hohl, H.; Stumm, W. "Interaction of Pb^{2+} with Hydrous γ-Al_2O_3," *J. Colloid Interface Sci.* **1976**, *55*, 281.
18. Sigg, L., Doctoral Thesis, Swiss Fed. Inst. Technol., Zurich, 1979.
19. Kummert, R., Doctoral Thesis, Swiss Fed. Inst. Technol., Zurich, 1979.
20. Schindler, P. "Surface Complexes at Oxide-Water Interfaces," In "Adsorption of Inorganics at the Solid-Liquid Interface"; Anderson, M. A., Rubin, A., Eds.; Ann Arbor Science: Ann Arbor, 1979.

21. Westall, J. C., personal communication, 1977.
22. Westall, J. C.; Zachary, J. L.; Morel, F. M. M. MINEQL, a Computer Program for the Calculation of Chemical Equilibrium Composition of Aqueous Systems. Technical Note No. 18, Water Quality Lab., Dept. Civ. Engineering, MIT, Cambridge, MA, 1976.
23. Westall, J. C., Chapter 2 in this book.
24. Gupta, S. K., Doctoral Thesis, University of Berne, 1976.
25. Breeuwsma, A.; Lyklema, J. "Physical and Chemical Adsorption of Ions in the Electrical Double Layer on Hematite (α-Fe$_2$O$_3$)," *J. Colloid Interface Sci.* **1973**, *43*, 437.
26. James, R. O.; Healy, T. W. "Adsorption of Hydrolyzable Metal Ions at the Oxide-Water Interface," *J. Colloid Interface Sci.* **1972**, *40*, 53.
27. Westall, J.; Hohl, H. *Adv. Colloid Sci.* **1980**, *12*, 265.
28. Bowden, J. W.; Posner, A. M.; Quirk, J. P. *Austr. J. Soil Res.* **1977**, *15*, 121.

RECEIVED October 16, 1978.

2

Chemical Equilibrium Including Adsorption on Charged Surfaces

JOHN WESTALL

Swiss Federal Institute of Technology, EAWAG,
CH-8600 Duebendorf, Switzerland

A general method for the formulation of chemical equilibrium problems allows modeling of the adsorption of ionic species on charged surfaces. The energy of adsorption is expressed by a constant chemical energy term and a charge-dependent electrostatic energy term. Various expressions can be used to define the electrostatic potential–charge relationship. The method was developed for adsorption at hydrolyzable oxide surfaces and is applicable to any number and type of adsorbed species (surface complexes) and any number of soluble species. Application of the method to a constant-capacitance double-layer and triple-layer model is demonstrated, and other applications are cited.

Interest in the fate of chemical substances in the aquatic environment has stimulated research in many areas, including the adsorption of materials on hydrolyzable metal oxide surfaces. Quantitative interpretation of adsorption on these surfaces is complicated because the electrostatic energy of adsorption is variable, and often a large number of chemical species are present in solution and adsorbed to the surface. In this chapter a chemical equilibrium approach is used to interpret adsorption on charged surfaces.

The basis for the discussion of adsorption on charged surfaces is the surface complexation model. The precept for this model is the use of the standard mass-action and mass-balance equations from solution chemistry to describe the formation of surface complexes. Use of these equations results in a Langmuir isotherm for the saturation of the surface with adsorbed species. There are of course other models that satisfy these precepts, but which are not generally referred to as surface complexation models, for example, the Stern model (1).

0-8412-0499-3/80/33-189-033$05.00/0

While there are similar mass-balance and mass-action equations in all surface complexation models, there are a great number of ways to formulate the electrostatic energy associated with adsorption on charged surfaces. Customarily the electrostatic energy of an adsorbed ion of formal charge z at a plane of potential ψ is taken by Coulomb's law to be $zF\psi$, but the relationships used to define surface potential ψ as a function of surface charge σ, or any other experimentally observable variable, are different. In addition, different descriptions of the surface/solution interface have been used, that is, division of the interface into different layers, or planes, to which different ions are assigned formally.

Because of the variable electrostatic energy term in the mass-action laws and the great number of species to be considered in a surface/ solution equilibrium problem, particularly when a multilayer interface is considered, traditional approaches to chemical equilibrium problems (ligand number $\bar{n}$ and ionization or neutralization fractions α) become complicated, and an intuitive feel for the problem is lost. Here the need for a general, systematic approach to chemical equilibrium, including species adsorbed at a charged surface, is indicated.

The mathematical formulation of chemical equilibria presented here is not only a method of solving a problem, but also a way of defining the problem. Concepts such as potential determining ions, fixed charge, etc. have absolutely unambiguous meanings. In addition, when different electrostatic models are formulated on the same mathematical basis, models that appear different by the authors' description may turn out to be mathematically degenerate. Systematic comparison of models also shows that parameters such as interlayer capacitances and intrinsic adsorption constants are model specific and not necessarily interchangeable from one model to another. As more and more adsorption experiments are made and adsorption constants are published, it is important that these constants be used only in models for which they are valid.

In this chapter we present a general method for solving surface/ solution equilibrium problems described by a surface complexation model, applicable for arbitrary surface layer charge/potential relationships and arbitrary surface/solution interface structures.

The method is based on the chemical equilibrium program MINEQL (2), modified to include the coulombic energy of adsorption caused by the charged surface. First the principles of MINEQL are presented through a short example from solution chemistry and then the extension of the method to the constant-capacitance double-layer model of the surface/solution interface used by Stumm, Schindler, and co-workers (3, 4) is demonstrated. Finally, the use of the method in the "triple-layer site-binding" model introduced by Yates, Levine, and Healy (5), and used by Davis, James, and Leckie (6), is shown. In each case the mathematics are described in sufficient detail to be reproducible.

Equilibria in Solution

To illustrate our approach to chemical equilibrium problems, we consider an example from solution chemistry: equilibria among a metal, hydrolysis species of the metal, a protonated ligand, the metal/ligand complex, hydrogen, and hydroxide ion. We define species as every chemical entity to be considered in the equilibrium problem; those for this example are listed in Table I. Then we define a set of components in such a way that every species can be written as the product of a reaction involving only the components, and no component can be written as the product of a reaction involving only the other components. This is similar to Gibbs' (7) definition of components. In algebraic terms, the components can be described as a linearly independent set spanning species space. The set of components is certainly not unique, but once it has been defined, the representation of the species in terms of this set of components is unique. For the example we have chosen, the set of components is shown in Table I.

Table I.   Chemical Equilibrium in Solution

Species and Mass-Action Equations

$$
\begin{array}{l}
M^{2+} \\
MOH^+ \quad [M^{2+}] \qquad\qquad\qquad [H^+]^{-1} \quad K_1 = [MOH^+] \\
HL \\
L^- \qquad\qquad\quad [HL] \qquad [H^+]^{-1} \quad K_2 = [L^-] \\
ML^+ \quad [M^{2+}] \quad [HL] \quad [H^+]^{-1} \quad K_3 = [ML^+] \\
H^+ \\
OH^- \qquad\qquad\qquad\qquad\quad [H^+]^{-1} \quad K_4 = [OH^-]
\end{array}
$$

Components $\quad M^{2+}, \quad HL, \quad H^+$

Algebraic Description of Problem[a]

$${}^t T^{[b]} = [\, T_M \qquad T_{HL} \qquad T_H \,]$$

$${}^t X^{[c]} = [\, [M^{2+}] \quad [HL] \quad [H^+] \,]$$

$$
A^{[d]} =
\begin{bmatrix}
1 & 0 & 0 \\
1 & 0 & -1 \\
0 & 1 & 0 \\
0 & 1 & -1 \\
1 & 1 & -1 \\
0 & 0 & 1 \\
0 & 0 & -1
\end{bmatrix}
\qquad
C^{[e]} =
\begin{bmatrix}
[M^{2+}] \\
[MOH^+] \\
[HL] \\
[L^-] \\
[ML^+] \\
[H^+] \\
[OH^-]
\end{bmatrix}
\qquad
K'^{[f]} =
\begin{bmatrix}
1.0 \\
K_1 \\
1.0 \\
K_2 \\
K_3 \\
1.0 \\
K_4
\end{bmatrix}
$$

[a] See Table II for equations in A, X, T, C, K.
[b] T is the vector of total concentrations of components.
[c] X is the vector of free concentrations of components.
[d] A is the stoichiometry matrix: a_{ij} is the stoichiometric coefficient of component j in species i. Thus each row pertains to an individual species and each column pertains to a component.
[e] C is the vector of concentrations of species.
[f] K is the vector of stability constants of species.

As chemical equilibrium problems are normally posed, we are given the total (analytical) concentrations of all components, the stoichiometry and stability constants of all species, and are asked to find the free equilibrium concentration of all components, from which we can easily compute the free concentrations of all species. This problem is solved as follows (see Tables I and II for description of symbols): from an initial guess for the concentration of components, the approximate concentration of the species can be computed:

$$\log C_i = \log K_i + \sum_j a_{ij} \log X_j \tag{1}$$

From the approximate concentration of species the error in the material-balance equation for each component can be computed:

$$Y_j = \sum_i a_{ij} C_i - T_j \tag{2}$$

By the multidimensional Newton–Raphson method the error in the material-balance equation and the computed derivative of the errors with respect to the concentration of components (Jacobian of Y with respect to X) can be used to compute an improved guess for the concentration of components:

$$Z \cdot \Delta X = Y \tag{3}$$

The elements of the Jacobian Z are given by:

$$z_{jk} = \frac{\partial Y_j}{\partial X_k} = \sum_i (a_{ij} \, a_{ik} \, C_i / X_k) \tag{4}$$

This iteration procedure is carried out until the error in the material-balance equation is small with respect to the terms in the equation. The method converges over a wide range of initial guesses.

The mathematical formulation given here does not include any explicit correction for activity coefficients of ionic species. In principle, these activity coefficients can be included as corrections to the stability constants. For ionic species in solution, individual ion activity coefficients can be estimated from one of the semiempirical expressions based on the Debye–Hückel theory. For charged surface species there is no established approach to activity coefficients. As a first approximation, we may consider that the primary effect of ionic strength on the activity of charged surface groups is accounted for in the ionic strength-dependent part of the double-layer theory, and ignore further activity coefficients for charged surface groups.

Table II. Description of Chemical Equilibrium Problem

Scalar[a]	Description	Matrix or Vector	Description
a_{ij}	Stoichiometric coefficient of component j in species i	A	Matrix of a_{ij}
C_i	Free concentration of species i	C	Vector of C_i
		C^*	Vector of log C_i
K_i	Stability constant of species i	K^*	Vector of log K_i
T_j	Total analytical concentration of component j	T	Vector of T_j
X_j	Free concentration of component j	X	Vector of X_j
		X^*	Vector of log X_j
Y_j	Residual in material balance equation for component j	Y	Vector of Y_j
z_{jk}	Partial derivative $(\partial Y_j / \partial X_k)$	Z	Jacobian of Y with respect to X

Mass-Law Equations

$$\log C_i = \log K_i + \sum_j a_{ij} \log X_j \qquad C^* = K^* + AX^*$$

Mass-Balance Equations

$$Y_j = \sum_i a_{ij} C_i - T_j \qquad Y = {}^tAC - T$$

Iteration Procedure (Newton–Raphson)

$$z_{jk} = \sum_i (a_{ij} a_{ik} C_i / X_k) = \frac{\partial Y_j}{\partial X_k} \qquad \begin{array}{l} Z \cdot \Delta X = Y \\ \Delta X = X_{\text{original}} - X_{\text{improved}} \end{array}$$

[a] Indices: i is used to denote any species; j and k are used to denote any component.

Constant-Capacitance Model

The modification of this method to include surface charge and coulombic energy is straightforward and is demonstrated with an equilibrium problem involving surface hydrolysis with the constant-capacitance double-layer model of Stumm, Schindler, and co-workers (3, 4). The species to be considered are the neutral, deprotonated and protonated surface species as well as the aqueous hydrogen and hydroxide ions, for which the mass-action equations are given in Table III. From inspection of these equations it is apparent that the electrostatic potential term $e^{-F\psi/RT}$ appears in the same form as the chemical concentration, or chemical potential term, of the hydrogen ion or of the free surface groups.

Table III.   Surface Hydrolysis—Constant-Capacitance Model

Species and Mass-Action Equations

H^+				
OH^-			$[H^+]^{-1}$	$K_1 = [OH^-]$
SOH^{2+}	$[SOH]$	$(e^{-F\psi/RT})$	$[H^+]$	$K_2 = [SOH_2^+]$
SOH				
SO^-	$[SOH]$	$(e^{-F\psi/RT})^{-1}$	$[H^+]^{-1}$	$K_3 = [SO^-]$

Components $SOH,$ $e^{-F\psi/RT},$ H^+

Algebraic Description of Problem

$${}^tT = [T_{SOH} \quad C \cdot \psi \cdot s \cdot a / F \quad T_H]$$

$${}^tX = [[SOH] \quad (e^{-F\psi/RT}) \quad [H^+]]$$

$$A = \begin{bmatrix} 0 & 0 & 1 \\ 0 & 0 & -1 \\ 1 & 1 & 1 \\ 1 & 0 & 0 \\ 1 & -1 & -1 \end{bmatrix} \qquad C = \begin{bmatrix} [H^+] \\ [OH^-] \\ [SOH_2^+] \\ [SOH] \\ [SO^-] \end{bmatrix} \qquad K = \begin{bmatrix} 1.0 \\ K_1 \\ K_2 \\ 1.0 \\ K_3 \end{bmatrix}$$

Jacobian

1. For all elements of the Jacobian except $z_{\psi,\psi}$; $z_{jk} = \sum_i (a_{ij}\, a_{ik}\, C_i / X_k)$

2. For $z_{\psi,\psi}$; $z_{\psi,\psi} = \sum_i (a_{i\psi}\, a_{i\psi}\, C_i / X_\psi) + C \dfrac{sa}{F} \dfrac{RT}{FX_\psi}$

This suggests that it would be appropriate to include electrostatic potential in the set of components. Then the concentrations of all the species can be written as functions of the concentrations of the components.

The next question that arises is, having included the electrostatic potential in the set of components, how do we write a "total concentration" for this component? For the other components, hydrogen ion and surface hydroxyl groups, the total concentration is determined simply by how much acid or base, or how much surface we have added to the system. In the case of the electrostatic component, we can use the independent electrostatic charge–potential relationship to define a total concentration or charge for the surface—in the case of the constant-capacitance model

$$\sigma = C\,\psi \qquad [C/m^2] \tag{5}$$

or converted to molar quantities:

$$T_\sigma = \sigma \cdot \frac{sa}{F} \qquad [mol/L] \tag{6}$$

The subscript σ or ψ is used to denote the electrostatic component (for example, T_σ, X_ψ). When the equilibrium problem is solved, the charge calculated by summing the charged surface species must be equal to the electrostatically calculated charge T_σ; in this way the electrostatic component is analogous to the other components.

We included the electrostatic term in the set of components and expressed the total concentration of this component; now it is necessary to modify certain elements of the Jacobian. Since the total concentration of the electrostatic component is not experimentally determined, but rather a function of the potential, the derivative of T_σ with respect to X_ψ is not zero and therefore must be attached to the normal expression given previously for a Jacobian element:

$$z_{\psi\psi} = \frac{\partial Y_\psi}{\partial X_\psi} = \sum_i (a_{i\psi}\, a_{i\psi}\, C_i/X_\psi) - \frac{\partial T_\sigma}{\partial X_\psi}$$

$$\frac{\partial T_\sigma}{\partial X_\psi} = -\frac{sa}{F}\, C\, \frac{RT}{FX_\psi}$$

(7)

This completes the modifications of the general chemical equilibrium method for surface adsorption involving a constant-capacitance model. The equations are summarized in Table III.

One should note that this formulation is completely general. Other soluble species and adsorbed species (for example, adsorbed anions and cations, and their soluble hydrolysis products) could be included in the problem simply by adding the new component and the new stoichiometries and stability constants in the general formulation of the problem.

Also important is the fact that certain mathematical simplifications (for example, the approximation that on the acidic side of the ZPC no deprotonated surface groups exist and vice versa) made by other authors (8) are no longer necessary with the current formulation of the problem.

Triple-Layer Model

Finally, we demonstrate the application of our general formulation of surface/solution equilibria to a more involved model of the surface/solution interface, the triple-layer site-binding model of Yates, Levine, and Healy. Again we discuss the principles of the method using simple hydrolysis equilibria, but the extension to more complicated equilibria is straightforward.

A schematic description of the surface/solution interface is given in Figure 1. Immediately at the surface, in the 0 plane, are specifically adsorbed hydrogen and hydroxide ions that experience the potential ψ_0 and contribute to the charge σ_0. At the inner Helmholtz plane, or β

layer, are electrolyte counterions bound pairwise to oppositely charged surface groups by both specific chemical energy and electrostatic energy. These electrolyte counterions experience the potential ψ_β and contribute to the charge σ_β. The outer Helmholtz plane or d layer is the innermost plane of the Gouy–Chapman diffuse region. The potential at the d plane is ψ_d and the charge contained in the entire diffuse region is given by the Gouy–Chapman (Poisson–Boltzmann) theory for a monovalent electrolyte:

$$\sigma_d = -(8\epsilon\epsilon_0 RTI)^{1/2} \sinh (F\psi_d/2RT) \qquad (8)$$

The 0 and β planes are separated by a region of capacitance C_1, and the β and d planes are separated by a region of capacitance C_2.

This description of the surface/solution interface requires us to define the species given in Table IV, and to add three electrostatic components ψ_β, ψ_0, and ψ_d to the normal set of chemical components SOH, Na$^+$, Cl$^-$, and H$^+$.

Now we examine the electrostatic description of the interface in order to determine total concentrations for the three electrostatic components. The charge in the 0 plane is given electrostatically by:

$$\sigma_0 = C_1 (\psi_0 - \psi_\beta) \qquad (9)$$

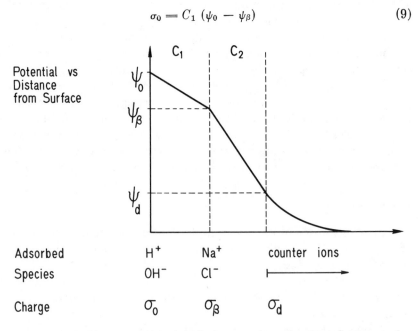

Figure 1.   Schematic of surface/solution interface showing placement of ions, the potential, and the charge in the two inner planes and the diffuse layer (the positions of the planes are not to scale)

and in the β plane by

$$\sigma_\beta = C_1 \, (\psi_\beta - \psi_0) + C_2 \, (\psi_\beta - \psi_d) \tag{10}$$

and in the d region by

$$\sigma_d = C_2 \, (\psi_d - \psi_\beta) \tag{11}$$

These last three equations are used to define the total concentrations of the three electrostatic components (see also Table IV). When the equilibrium problem is solved, the charge calculated by summation of charged species in the 0 and β planes must be equal to the electrostatically calculated charge given previously (Equations 9–11).

$$Y_{\sigma_0} = \sum \text{(charged species in } 0) - T_{\sigma_0} = 0 \tag{12}$$

$$Y_{\sigma_\beta} = \sum \text{(charged species in } \beta) - T_{\sigma_\beta} = 0 \tag{13}$$

No species are designated for the diffuse region; it is necessary to use the Gouy–Chapman theory to calculate the sum of charged species in the diffuse region, and this value must equal the previous electrostatic value.

$$Y_{\sigma_d} = -(8\epsilon\epsilon_0 RTI)^{1/2} \sinh (F\psi_d/2RT) - T_{\sigma_d} = 0 \tag{14}$$

Finally, modifications of the elements of the Jacobian are necessary because of the dependence of the T_σ's on the X_ψ's. The principle is the same as discussed for the constant-capacitance model; the modified elements are given in Table IV.

The triple-layer site-binding model now fits within the scheme of the general equilibrium problem given in Equations 1–3. Other adsorbed cations and anions can be included in the equilibria simply by adding the appropriate components and species.

Applications

The method developed here for the description of chemical equilibria including adsorption on charged surfaces was applied to interpret phosphate adsorption on iron oxide (9), and to study electrical double-layer properties in simple electrolytes (6), and adsorption of metal ions on iron oxide (10). The mathematical formulation was combined with a procedure for determining constants from experimental data in a comparison of four different models for the surface/solution interface: a constant-capacitance double-layer model, a diffuse double-layer model, the triple-layer model described here, and the Stern model (11). The reader is referred to the Literature Cited for an elaboration on the applications.

Table IV. Surface Hydrolysis—Triple-Layer Model

Species and Mass–Action Equations

Na$^+$
Cl$^-$
H$^+$
OH$^-$
SOH
SOH$_2^+$
SO$^-$
SOH$_2^+$Cl$^-$
SO$^-$Na$^+$

$$[\text{H}^+] \qquad\qquad K_1 = [\text{OH}^-]$$

$$[\text{SOH}]\,[\text{H}^+]\qquad\qquad\qquad\qquad (e^{-F\psi_0/RT}) \qquad\qquad K_2 = [\text{SOH}_2^+]$$
$$[\text{SOH}]\,[\text{H}^+]^{-1}\qquad\qquad\qquad\quad (e^{-F\psi_0/RT})^{-1} \qquad\qquad K_3 = [\text{SO}^-]$$
$$[\text{SOH}]\,[\text{H}^+]\,[\text{Cl}^-]\,(e^{-F\psi_\beta/RT})^{-1}\,(e^{-F\psi_0/RT}) \qquad K_4 = [\text{SOH}_2^+\text{Cl}^-]$$
$$[\text{SOH}]\,[\text{H}^+]^{-1}\,[\text{Na}^+]\,(e^{-F\psi_\beta/RT})\,(e^{-F\psi_0/RT})^{-1} \qquad K_5 = [\text{SO}^-\,\text{Na}^+]$$

Components

SOH Na$^+$ Cl$^-$ ψ_β ψ_0 H$^+$ ψ_d

Algebraic Description of Problem

$${}^tX = [\,[\text{SOH}]\ \ [\text{Na}^+]\ \ [\text{Cl}^-]\ \ (e^{-F\psi_\beta/RT})\ \ (e^{-F\psi_0/RT})\ \ [\text{H}^+]\ \ (e^{-F\psi_d/RT})\,]$$

$$A = \begin{bmatrix}
0 & 1 & 0 & 0 & 0 & 0 & 0 \\
0 & 0 & 1 & 0 & 0 & 0 & 0 \\
0 & 0 & 0 & 0 & 0 & 1 & 0 \\
0 & 0 & 0 & 0 & 0 & -1 & 0 \\
1 & 0 & 0 & 0 & 0 & 0 & 0 \\
1 & 0 & 0 & 0 & 1 & 1 & 0 \\
1 & 0 & 0 & 0 & -1 & -1 & 0 \\
1 & 0 & 1 & -1 & 1 & 1 & 0 \\
1 & 1 & 0 & 1 & -1 & -1 & 0
\end{bmatrix}$$

$$C = \begin{bmatrix}
[\text{Na}^+] \\
[\text{Cl}^-] \\
[\text{H}^+] \\
[\text{OH}^-] \\
[\text{SOH}] \\
[\text{SOH}_2^+] \\
[\text{SO}^-] \\
[\text{SOH}_2^+\text{Cl}^-] \\
[\text{SO}^-\,\text{Na}^+]
\end{bmatrix}$$

$${}^tT = [\,T_{\text{SOH}}\quad T_{\text{Na}^+}\quad T_{\text{Cl}^-}\quad T_{\sigma_\beta}{}^{[a]}\quad T_{\sigma_0}{}^{[b]}\quad T_{\sigma_d}{}^{[e]}\quad T_{\text{H}}\,]$$

Modification of Jacobian (Triple-Layer Model) [a]

$$z_{\psi_0\psi_0} = \Sigma(\psi_0,\psi_0) + C_1 \frac{sa}{F}\frac{RT}{FX_{\psi_0}}$$

$$z_{\psi_0\psi_\beta} = \Sigma(\psi_0,\psi_\beta) - C_1 \frac{sa}{F}\frac{RT}{FX_{\psi_\beta}}$$

$$z_{\psi_0\psi_a} = 0$$

$$z_{\psi_\beta\psi_0} = \Sigma(\psi_\beta,\psi_0) - C_1 \frac{sa}{F}\frac{RT}{FX_{\psi_0}}$$

$$z_{\psi_\beta\psi_\beta} = \Sigma(\psi_\beta,\psi_\beta) + (C_1 + C_2)\frac{sa}{F}\frac{RT}{FX_{\psi_\beta}}$$

$$z_{\psi_\beta\psi_a} = \Sigma(\psi_\beta,\psi_a) - C_2 \frac{sa}{F}\frac{RT}{FX_{\psi_a}}$$

$$z_{\psi_a\psi_0} = 0$$

$$z_{\psi_a\psi_\beta} = \Sigma(\psi_a,\psi_\beta) - C_2 \frac{sa}{F}\frac{RT}{FX_{\psi_\beta}}$$

$$z_{\psi_a\psi_a} = \frac{-F}{2RT}(8\epsilon\epsilon_0 RTI)^{1/2}\cosh\left(\frac{F\psi_a}{2RT}\right) + C_2 \frac{sa}{F}\frac{RT}{FX_{\psi_a}}$$

[a] $T_{\sigma\beta} = s \cdot a/F\,(C_1(\psi_\beta - \psi_0) + C_2(\psi_\beta - \psi_a))$.

[b] $T_{\sigma 0} = s \cdot a/F\,(C_1(\psi_0 - \psi_\beta))$.

[c] $T_{\sigma a} = s \cdot a/F\,(C_2(\psi_a - \psi_0))$.

[a] The symbol $\Sigma(j,k)$ stands for the usual Jacobian element computed in solution chemistry problems: $\Sigma(j,k) = \sum_i (a_{ij}\, a_{ik}\, C_i/X_k)$.

Acknowledgment

This work was supported in part by the Swiss National Foundation.

Glossary of Symbols (see also Table II)

a = concentration of suspended solid [g/L]
C = specific capacitance [F/m^2]
F = Faraday [C/mol]
I = ionic strength [mol/L]
R = gas constant [J/($^\circ$K mol)]
s = specific surface area [m^2/g]
T = temperature [$^\circ$K]
ϵ = dielectric constant
ϵ_0 = permitivity of free space [C^2/(J m)]
σ = specific surface charge [C/m^2]
ψ = electrostatic potential [V]

Literature Cited

1. Stern, O. Z. Elektrochem. Angew. Phys. Chem. **1924**, *30*, 508.
2. Westall, J.; Zachary, J. L.; Morel, F. "MINEQL, a Computer Program for the Calculation of Chemical Equilibrium Composition of Aqueous Systems"; Technical Note No. 18, Ralph M. Parsons Laboratory, MIT: Cambridge, 1976.
3. Huang, C. P.; Stumm, W. *J. Colloid Interface Sci.* **1973**, *43*, 409.
4. Schindler, P. W.; Furst, B.; Dick, R.; Wolf, P. U. *J. Colloid Interface Sci.* **1976**, *55*, 469.
5. Yates, D. E.; Levine, S.; Healy, T. *J. Chem. Soc. Faraday Trans. 1* **1974**, *70*, 1807.
6. Davis, J. A.; James, R. O.; Leckie, J. O. *J. Colloid Interface Sci.* **1978**, *63*, 480.
7. Gibbs, J. W. "The Collected Works of J. W. Gibbs"; Yale Univ. Press: New Haven, 1957.
8. Hohl, H.; Stumm, W. *J. Colloid Interface Sci.* **1976**, *55*, 281.
9. Hohl, H.; Sigg, L.; Stumm, W. Chapter 1 in this book.
10. Davis, J. A.; Leckie, J. O. *J. Colloid Interface Sci.* **1978**, *67*, 90.
11. Westall, J.; Hohl, H. *Adv. Colloid Interface Sci.* **1980**, *12*, 265.

RECEIVED October 30, 1978.

Redox Coprecipitation Mechanisms of Manganese Oxides

JOHN D. HEM

U.S. Geological Survey, Menlo Park, CA 94025

A nonequilibrium thermodynamic model for precipitation and oxidation of naturally occurring manganese oxyhydroxides in aqueous systems postulates the maintenance of reaction affinities simultaneously favorable for both the precipitation of Mn_3O_4 and its disproportionation to MnO_2 and $Mn^{2+}(aq)$. Experimental data show changes with pH in the MnO_2-H^+ stoichiometry that are explainable by the model. A solid–solution model can be used to predict the state of oxidation of manganese in mixed-valence oxide, based on pH and $Mn^{2+}(aq)$ activity in associated solutions. Both lead and cobalt raise the oxidation state of manganese oxides when coprecipitated at pH 8 and 8.5. Coupling of lead and cobalt redox processes to Mn_3O_4 disproportionation provides a model that can be used to predict the range of activity of these metals likely to occur in systems where manganese oxides are precipitating.

The aqueous geochemistry of manganese, especially solution and deposition of manganese oxides, was studied extensively during the 1960's and 1970's. Examples are studies by Bricker (*1*), Morgan (*2*), Loganathan and Burau (*3*), and Burns (*4*). Giovanoli et al. (*5, 6, 7, 8*) at Bern, Switzerland, have made important contributions to the knowledge of the crystal structure of these often poorly organized oxides. Many others in Europe, Canada, and throughout the world have also done research on manganese chemistry. However, it is not appropriate here to attempt to review past work in detail.

The chemical reactions of manganese oxidation and precipitation may not attain equilibrium because of unfavorable kinetics. The three possible oxidation states (Mn^{2+}, Mn^{3+}, and Mn^{4+}) which can occur in the solid oxides further complicate the chemical mechanisms. A particularly important property of the Mn^{3+} state is its tendency to disproportionate, so that most oxides in which this form occurs are thermodynamically unstable.

Because of these properties the manganese redox processes do not always attain a state that can be predicted simply as a condition where chemical potentials for the system under study are at or near zero. Under some conditions, favorable chemical potentials can be maintained for both the precipitation and disproportionation reactions.

In natural aqueous systems open to the atmosphere certain non-equilibrium, chemical thermodynamic approaches are applicable. Chemical potentials or reaction affinities can be computed for the reactions that are possible under a given set of conditions. The relative magnitudes of these affinities are useful for showing limiting conditions, understanding catalytic effects, and predicting concentration ranges that could occur. They also are useful in evaluating the possible coupling of manganese oxidation and disproportionation to redox reactions of other metal ions.

Form and Evolution of Manganese Oxides

As noted by other investigators (1) as well as in the laboratory experiments in this work, when a dilute solution of Mn^{2+} ions is held at a pH of 8.0 or 8.5 in the presence of CO_2-free air or oxygen, a precipitate having the crystal structure of hausmannite and a composition closely approximating Mn_3O_4 is formed. The initial step may be the formation of $Mn(OH)_2(c)$, and some $\beta MnOOH$ may be present in the final precipitate. According to Bricker (1) both Mn_3O_4 and $\beta MnOOH$ in aqueous suspensions can change to $\gamma MnOOH$, on long-standing (several months) contact with oxygen. On acidification, hausmannite was shown to disproportionate, giving a species Bricker (1) identified as δMnO_2. This material generally has poorly defined crystallinity and can come close to the 1:2 manganese–oxygen ratio, but does not reach it completely.

There appears to be no generally agreed-upon definition of δMnO_2. Some investigators have equated δMnO_2 with birnessite, which contains sodium and occurs naturally in various locations on land and in some deep-sea ferromanganese nodules. Giovanoli et al. (5) described double-layer structures having the formula $Na_4Mn_{14}O_{27} \cdot 9H_2O$. These investigators also produced sodium-free manganese(III) manganate ($Mn_7O_{13} \cdot 5H_2O$) and manganese(II) manganate ($Mn_7O_{12} \cdot 6H_2O$), both having a structure similar to that of the sodium form (6). The average oxidation state of manganese in these compounds only ranges from +3.43 to +3.71.

Giovanoli et al. (6) believed materials that others called δMnO_2 are fine-grained disordered species like these, but with more Mn^{4+} and less Mn^{3+}, and recommended that the designation δMnO_2 be abandoned.

The oxidation states of manganese in two preparations of δMnO_2 reported by Bricker (1) were $+3.52$ and $+3.62$. The four manganese oxide–hydroxide species that are considered in this chapter represent a simplified sequence of evolution from relatively reduced to relatively oxidized forms. The thermodynamic stability data for the four species are taken from Bricker's (1) determinations and from compatible values from the NBS compilation (9). A material more highly oxidized (γMnO_2), for which Bricker reported an oxidation state for manganese between 3.82 and 3.98, has a standard free energy of formation about 3.3 kJ more negative than that of his δMnO_2. However, he also reported that δMnO_2 could range in oxidation state from 3.48 to 3.98.

Characteristically, the more highly oxidized species prepared in this study were nearly amorphous to x-rays. They contained essentially no sodium. Although such material may be a little less stable than Bricker's (1) δMnO_2, its thermodynamic stability is assumed here to be the same as he determined. It will become evident later that, in general, this assumption produces a reasonable model applicable to laboratory systems. However, natural products may have a considerable range of stability and composition, and because they are generally impure and disordered, their thermodynamic properties can only be approximated.

Figure 1 is a pH–potential diagram for the four solids mentioned previously, showing theoretical manganese solubilities for each between 10^{-3} and 10^{-9} mol/L of dissolved Mn^{2+}. Thermodynamic data used are given in Table I. The diagram omits from consideration the $MnCO_3(c)$ species rhodochrosite, which can be an important control of manganese solubility in carbonate-enriched and relatively reducing environments, and the metastable species $\beta MnOOH$ and γMn_2O_3.

The only aqueous species of significance in Figure 1 below pH 10 is the Mn^{2+} ion. Hydroxide complexes are considered at higher pH's. Organic complexes of Mn^{2+} may be significant in some natural systems, but are not considered in this diagram. Bicarbonate and sulfate complex stabilities are given elsewhere (14).

Figure 1 indicates that manganese oxides are subject to dissolution by chemical reduction. Where organic material is abundant and oxygen is depleted, as in the anoxic sediment in stratified bodies of water, manganese is brought into solution in the pore water and in the oxygen-depleted water in contact with the bottom. When the dissolved manganese comes in contact with oxygen, precipitation of oxide can occur again. The separation between conditions favorable for solution and deposition can be marked sharply. For example, cobbles in streambeds may develop

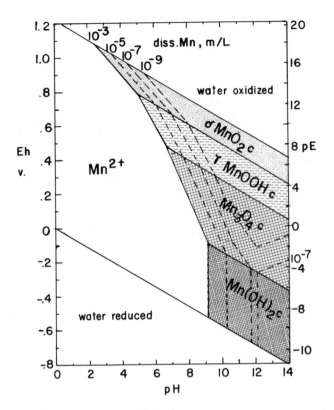

Figure 1. Stability fields and dissolved manganese activity for four oxides or hydroxides in the system $Mn + H_2O + O_2$

black encrustations of manganese oxide on the upper surface exposed to the running water. This crust may terminate abruptly when the rock comes in contact with the bed sediment surface (*15, 16*).

Thus manganese oxides can be subject to repeated cycles of precipitation and dissolution, and even where they are relatively stable, processes of accretion are generally important. In this respect manganese oxide surfaces in aqueous systems are highly active chemically, yet are physically unstable. Manganese oxide nodules and crusts should not be thought of simply as stable substrates where conditions favor only adsorption or ion-exchange reactions like those occurring at surfaces of the silicate minerals.

Because of uncertainties in the interpretation of measured redox potentials in oxygenated heterogeneous systems, such measurements were not made in the systems studied in this work. However, the general behavior of the components, and the results of the changes in solid

Table I. Standard Gibbs Free Energy of Formation of Species Considered

Species	$\Delta G°$, kJ/mol[a]	Source of Data
	Cobalt	
$Co^{2+}(aq)$	−56.11	Naumov et al. (10)
$CoOH^+(aq)$	−237.07	Calc. from Bolzan et al. (11)
$Co(OH)_2(aq)$	−423.13	Naumov et al. (10)
$Co(OH)_3^-(aq)$	−587.9	Naumov et al. (10)
$Co^{3+}(aq)$	−78.20	Naumov et al. (10)
$Co(OH)_2(c)$ (pink)	−455.2	Naumov et al. (10)
$Co_3O_4(c)$	−774.0	Naumov et al. (10)
	Lead	
$Pb^{2+}(aq)$	−24.39	Wagman et al. (12)
$PbOH^+(aq)$	−220.3	Lind (13)
$Pb(OH)_2(aq)$	−402.2	Lind (13)
$Pb(OH)_3^-(aq)$	−575.55	Lind (13)
$Pb(OH)_2(c)$	−452.3	Wagman et al. (12)
$Pb_3O_4(c)$	−601.2	Wagman et al. (12)
$PbO_2(c)$	−217.36	Wagman et al. (12)
	Manganese	
$Mn^{2+}(aq)$	−228.0	Wagman et al. (9)
$MnOH^+(aq)$	−405.0	Wagman et al. (9)
$HMnO_2^-(aq)$	−505.8	Wagman et al. (9)
$Mn(OH)_2(c)$ amorph.	−615.0	Wagman et al. (9)
$Mn_3O_4(c)$ (hausmannite)	−1283.2	Wagman et al. (9)
$MnOOH(c)$ (γ)	−557.7	Bricker (1)
$MnO_2(c)$ (δ)	−453.1	Bricker (1)
	Other Species	
$H_2O(l)$	−237.19	Wagman et al. (12)
$OH^-(aq)$	−157.29	Wagman et al. (12)
$O_2(aq)$	16.3	Wagman et al. (12)
$H^+(aq)$	0.0	Wagman et al. (12)
$H_2O_2(aq)$	−134.10	Wagman et al. (12)
$HO_2^-(aq)$	−67.4	Wagman et al. (12)

[a] 1 kcal = 4.184 kJ.

species from metastable initial precipitates to thermodynamically stable forms, can be predicted from Eh–pH relationships based on theoretical data.

Because Mn_3O_4 and $\beta MnOOH$ are the first products of the air oxidation of Mn^{2+} at pH near 8.5, the effective Eh reached is probably near the stability boundary between these two phases. This boundary is not shown in Figure 1. The free energy of βMnOOH is not given in Table I, but studies currently being conducted in our laboratory as well as earlier work by Bricker (1) show it is less negative than that of γMnOOH. The Mn_3O_4–βMnOOH boundary thus would lie within the γMnOOH stability field in Figure 1, and the short-term solubility of manganese oxide precipitated at a pH near 8 should be predicted by the intercept of the pH with this boundary. During aging of the precipitate, the more stable γMnOOH will form, and the redox equilibrium will shift toward lower values along the boundary between γMnOOH and Mn_3O_4 shown in Figure 1.

Solubility controls of manganese in contact with various oxide species are summarized in Figure 2. From the equation

$$Mn_3O_4 + 2H^+ = 2\gamma MnOOH + Mn^{2+}$$

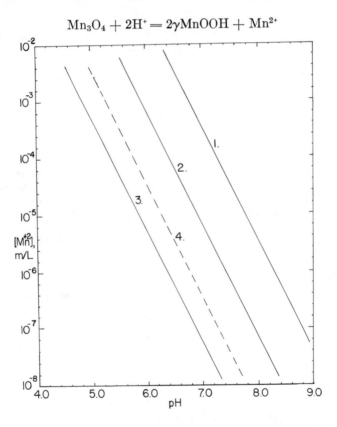

Figure 2. Manganese activity in solution as a function of pH for: (1) equilibrium between $Mn_3O_4(c)$ and $\gamma MnOOH(c)$; (2) disproportionation of Mn_3O_4 where Mn_3O_4 and δMnO_2 are present at unit activity; (3) surface-catalyzed direct oxidation of Mn^{2+} to Mn_3O_4 by aqueous O_2 in aerated water; (4) condition of equal affinity for Reactions 2 and 3 to proceed simultaneously

and data in Table I, the expression

$$[Mn^{2+}] = 10^{10.56}[H^+]^2 \tag{1}$$

may be derived. (Quantities in square brackets are thermodynamic activities of ionic species.) This equation represents the uppermost solubility line 1 in Figure 2. A precipitate of $Mn_3O_4(c)$ may also change directly as follows:

$$Mn_3O_4(c) + 4H^+ = MnO_2(c) + 2Mn^{2+} + 2H_2O$$

giving the relationship (based on data in Table I)

$$[Mn^{2+}] = 10^{8.79}[H^+]^2 \tag{2}$$

This gives a lower solubility line designated as 2 on Figure 2.

In an open natural system, if the manganese oxide particle or coating is to grow in size or thickness, a renewal of the Mn_3O_4 surface is required after Reaction 2 has altered the initial precipitate to MnO_2. Where a flux of oxygenated water containing dissolved Mn^{2+} is available at the surface, these reactants plus the Mn^{2+} produced in Equation 2 can accomplish this by the process

$$3Mn^{2+} + \tfrac{1}{2}O_2(aq) + 3H_2O = Mn_3O_4(c) + 6H^+$$

This leads to the expression

$$[Mn^{2+}] = 10^{6.09}[H^+]^2[O_2]^{-1/6}$$

which, for a dissolved oxygen activity of $10^{-3.50}$ mol/L (near saturation with the atmosphere at ordinary earth-surface temperature), becomes

$$[Mn^{2+}] = 10^{6.67}[H^+]^2 \tag{3}$$

This defines Line 3 in Figure 2.

Reaction 3 represents a lower limiting value for manganese solubility. The commonly observed failure to reach this concentration in the initial precipitation in the laboratory is probably the result of more favorable kinetics for Process 1, in systems where Process 2 has not occurred to a significant extent. The release of Mn^{2+} by Process 2 in natural open systems permits a local increase in dissolved Mn^{2+} activity near the oxide surfaces, and the surplus H^+ released by Process 3 can be absorbed by buffer systems in the inflowing water. Thus the products of Reaction 2 become reactants in Reaction 3.

Processes 2 and 3 constitute a feedback cycle which could be summarized

$$Mn^{2+} + \frac{1}{2}O_2(aq) + H_2O = MnO_2(c) + 2H^+$$

This relationship represents the final product of the manganese oxidation cycle, and gives the limiting stoichiometry of the participating species. However, it cannot be treated meaningfully by a simple mass-law equilibrium approach to predict the kinetically controlled activities of Mn^{2+} and H^+ that would be brought about by operation of the cycle.

Nonequilibrium Thermodynamic Model

Some concepts of nonequilibrium thermodynamics may be useful to develop further the proposed model. For the manganese oxide deposition process to continue, the oxidation (Equation 3) and disproportionation steps (Equation 2) must each have a favorable chemical potential, or positive reaction affinity, A. The reaction affinity is equivalent to the chemical potential but opposite in sign, and can be defined (17) as

$$A = -RT \, (\ln Q - \ln Q_0)$$

where $\ln Q$ is the logarithm of the observed activity quotient, $\ln Q_0$ is the logarithm of the thermodynamic equilibrium constant, R is the gas constant, and T the temperature in Kelvins. At 25°C and 1 atm

$$A = -5.757 \, (\log Q - \log Q_0) \text{ kJ/mol}$$

Reactions 2 and 3 both cannot have zero reaction affinities in the same chemical system unless oxygen is essentially absent ($\sim 10^{-16.20}$ mol/L). The oxides probably would not be stable in such a system. If Reactions 2 and 3 have similar rates and activation energies, as a first approximation, about equal values of A would need to be maintained to assure the continued operation of both processes. One could therefore equate the expressions for A for these two processes to obtain

$$2 \log [Mn^{2+}] - 4 \log [H^+] - 17.58 = 6 \log [H^+] - $$
$$3 \log [Mn^{2+}] - 0.5 \log [O_2] + 18.27$$

or

$$[Mn^{2+}]^5 = 10^{35.86}[H^+]^{10}[O_2]^{-0.5}$$

If a dissolved O_2 activity of $10^{-3.50}$ mol/L is maintained, a nonequilibrium steady-state ($A_2 = A_3 > 0$) condition occurs, where

$$[Mn^{2+}] = 10^{7.52}[H^+]^2 \qquad (4)$$

Because the rates of Reactions 2 and 3 depend on pH, in different ways, one might expect that the above relationship would not be useful to predict manganese activity exactly over a wide pH range; obviously it is affected by other simplifying assumptions, especially with respect to activities of solids. The line defined by Equation 4 is shown in Figure 2 between Lines 2 and 3.

Evidently a reaction affinity favorable for MnO_2 accumulation can be maintained in an open system where the fluxes of $O_2(aq)$ and Mn^{2+} are near stoichiometric equivalence. The minimum activity of $O_2(aq)$ could be lower than that in equilibrium with the atmosphere, but competing redox processes might interfere at low $O_2(aq)$ levels.

Manganese oxide surfaces in aerated water are sites for related biochemical effects. For example, such surfaces would be favorable for the growth of certain types of biota because of the available energy. Manganese-oxidizing species of bacteria similar to iron-oxidizing forms are known to exist in some such systems (*18*). At such surfaces there also could be mechanisms for diverting energy to drive other inorganic redox processes. An evaluation of the latter possibilities can be made using thermodynamic data.

Mixed-Valence Oxide Model

A somewhat different approach to devising a model for the calculation of the steady-state solubility of manganese can be made by considering the activities of components of the mixed-valence oxides.

In developing Equation 2, the standard thermodynamic assumption of unit activity was made for both solids involved in the reaction. Where appreciable quantities of both are available at the solid surfaces in the system, such a simplification may be adequate. However, a solid enriched in either oxidation state of manganese probably will not have equal activities of both. The following is a simplified approach developed to deal with this effect.

When solid species surface activities are considered, Equation 2 would be written

$$\frac{[Mn^{2+}]^2 \alpha MnO_2(c)}{[H^+]^4 \alpha Mn_3O_4(c)} = 10^{17.58} = Q_0$$

$$[Mn^{2+}]^2 = [H^+]^4 \frac{\alpha Mn_3O_4}{\alpha MnO_2} \times Q_0$$

(5)

where the α terms represent effective activities of the two forms of oxide surfaces. If a solid–solution model is applied (*19*) to evaluate the activity of the oxide surface in terms of the manganese oxidation state, the activities can be expressed as mole fractions of manganese at oxidation levels characteristic of the two solids. For any value of the ratio of the mole fractions, a conditional value for the mass-law constant can be calculated.

$$Q_0' = Q_0 \frac{N_{Mn^{2.67+}}}{N_{Mn^{4+}}} \tag{6}$$

The mole fraction ratio may also be expressed in terms of the average oxidation state of manganese in the solid, if it is assumed the bulk composition is similar to that of the surface.

Figure 3 is a graph developed from Equations 5 and 6 showing an expected dissolved manganese activity between pH 4.5 and 9.0 when the

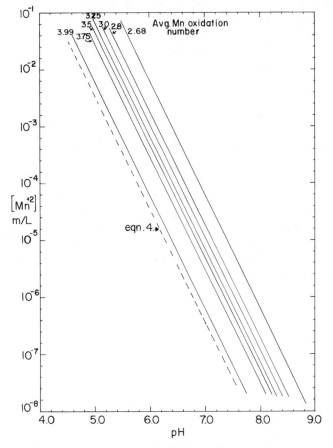

Figure 3. *Manganese activity and pH for oxidation states of manganese from +2.68 to +3.99 in systems dominated by disproportionation processes*

manganese content of solid oxide present ranges in oxidation state from 2.68 (mole ratio 100/1) to 3.99 (mole ratio 1/100).

It is of interest to note that the line for $Mn^{3.99+}$ is not far from the one in Figure 2 that represents equal reaction affinities for the forward and reverse reactions in the postulated manganese precipitation and oxidation cycle.

Laboratory Studies of Manganese Oxidation

Open natural systems of the type to which the proposed models might apply are difficult to simulate and study in the laboratory because of their slow kinetics and low solute activities. However, some features of such systems can be evaluated by extrapolating results of carefully controlled batch experiments in which reactant activities are higher. In these experiments all accessible parameters should be measured and as many as possible held constant. Although the results give some qualitative indications of comparative reaction rates, they are not adequate for determining specific rate constants.

A series of 100-mL aliquots of a $0.01M$ $Mn(ClO_4)_2$ solution was prepared. While CO_2-free air was stirred and bubbled into the solution, the manganese was oxidized and precipitated (at pH 8.0 or 8.5 at 25°C with an automated pH-stat) by adding standard $0.1M$ NaOH. After base addition had stopped and the reaction was essentially completed (usually about 2 hr), the containers were removed from the pH-stat and allowed to stand for a few hours to a few days in the laboratory with occasional stirring. The beakers were covered to minimize evaporation losses, but air was not excluded. At this stage the precipitate was fine grained and had a cinnamon-brown color.

After the brief aging period the solutions in the beakers were adjusted to preselected pH values ranging from 4.0 to 6.0. The pH values were maintained by adding $0.10M$ $HClO_4$ from an automated microburet while bubbling CO_2-free air for a period of 2 hr or more. During this acid titration the precipitates became much darker. Those titrated at the lower pH's were black, and the ones titrated at higher pH were dark brown. The solutions were then removed from the pH-stat and allowed to stand as before, but for several months.

From time to time throughout the experiments the pH was recorded, and a sample of precipitate and supernate was withdrawn from the stirred mixture by means of a syringe, and filtered through a 0.10-μm pore-diameter membrane filter. The filtrate was analyzed for manganese by atomic absorption. The filter membrane with the precipitate was transferred to a small beaker and the oxide was dissolved in a measured excess of standardized oxalic acid and 0.2 mL of concentrated H_2SO_4. This solution was diluted to 100 mL; the manganese was determined on

Table II. Properties of Synthesized Manganese Oxides After Various

Solution	pH	$[Mn^{2+}]$ (mol/L)	Age (Days)	Determined Mn Oxidation No.
K1	7.64	$10^{-5.14}$	1	2.68
K2	7.03	$10^{-4.70}$	4	2.71
K3	6.98	$10^{-4.74}$	6	2.70
K4	7.32	$10^{-5.00}$	6	2.70
K8	7.69	$10^{-5.60}$	6	2.69
5B	6.90	$10^{-4.70}$	175	3.00

one aliquot by atomic absorption, and the unreacted oxalate was determined on another aliquot by titration with permanganate. These data were used to calculate the oxidation state of the precipitated manganese. The experimental uncertainty in this procedure is probably about ±0.03 of an oxidation unit.

The crystal structure of the precipitates was studied by means of x-ray diffraction and electron microscopy. All the initial manganese oxide precipitates displayed patterns characteristic of hausmannite, while some of the precipitates also contained βMnOOH or γMnOOH. After intensive acid treatment the precipitates generally were amorphous.

The experiments were designed to provide data on the composition of both the solution and solid phases and to permit the calculation of stoichiometric yields of the reaction products in relation to amounts of reactants consumed. The results, briefly summarized in Tables II–IV for several typical solutions, support the following conclusions:

1. Fresh precipitates generally approximate the hausmannite oxidation state $Mn^{2.67+}$. The oxidation number increases and pH decreases during aging.
2. Reaction with acid is much more rapid and extensive at the lower pH's.

Table III. Solubility and Oxidation State of Manganese Oxide Followed by

Solution	Holding pH	Final Measured pH	[Mn]	Age (Days)	Determined Mn Oxidation No.
1C	4.01	5.65	$10^{-2.64}$	10	3.44
1C	4.01	5.75	$10^{-2.64}$	46	3.55
1F	4.54	5.81	$10^{-2.86}$	46	2.86
1I	5.11	6.14	$10^{-2.96}$	170	2.86
1L	5.53	6.30	$10^{-3.17}$	42	2.79

and Associated Solutions Precipitated at pH 8.5
Aging Periods

Calculated Mn Oxidation No.	Mineral Species	Reaction Affinities, kJ		
		A_1	A_2	A_3
2.67	Hausmannite	2.40	−15.4	59.4
2.76	Hausmannite	6.85	−6.51	46.0
2.83	Hausmannite	7.65	−4.91	43.6
2.70	Hausmannite	5.25	−9.70	50.8
2.68	Hausmannite	4.45	−11.3	53.2
2.93	Manganite	8.33	−3.54	41.5

3. Oxidation states of precipitated manganese can be approximated (where aging time is sufficient) by means of the solution pH and dissolved manganese activity, using Equations 5 and 6.

4. Stoichiometric data show that the total number of moles of Mn^{2+} liberated is consistently near half of the total H^+ added to maintain a constant pH in all solutions.

5. Stoichiometric data also show that the amount of Mn^{4+} produced in the precipitate varies, but is generally greater than one fourth of the amount of H^+ consumed, and that the amount of Mn^{4+} per mole of H^+ is larger when the solids are reacted at a pH greater than 4.0.

Conclusions 4 and 5 are in accord with the postulated recycling mechanism operating in these test solutions. At pH 4.0 the disproportionation process is much faster than the Mn_3O_4 regeneration; therefore the stoichiometry of the whole process is nearly that of the disproportionation (Reaction 2). At pH 5.5 the rates of Reactions 2 and 3 are more nearly equal, and the stoichiometry is influenced by the H^+ produced when Mn_3O_4 is formed. This H^+ is recycled and decreases the amount required to maintain constant pH. Thus, the excess Mn^{4+} pro-

Precipitates After Maintaining Indicated pH for 2 Hr,
Aging at 25°C

Calculated Mn Oxidation No.	Reaction Affinities (kJ) for Disproportionation					
	At Time 0		2 Hr		Aged	
	A_2	A_3	A_2	A_3	A_2	A_3
3.53	67.8	−65.5	39.0	−22.3	1.48	34.0
3.23					−0.80	37.4
2.92	52.5	−42.5	29.6	−8.10	0.34	35.7
2.78	40.3	−24.2	18.9	7.82	−6.05	45.3
2.74	31.8	−11.5	10.7	20.1	−7.31	47.2

Table IV.   Stoichiometry of Manganese Oxide Disproportionations

| Solution | Reaction pH | Reaction Time (hr) | mmol | | H^+/Mn^{4+} | $H^+/$ $\Delta Mn^{2+}(aq)$ |
			H^+ Used	$Mn^{4+}(c)$ Formed		
IC^a	4.0	2	.769	.188	4.1	1.9
$1C^b$			.769	.208	3.7	2.0
1E	4.5	2	.326	.103	3.2	1.9
K6	5.0	6	.461	.160	2.9	2.0
K1	5.5	9	.386	.203	1.9	2.0
K8	6.0	6	.191	.058	3.3	2.1
Theoretical					4.0	2.0

ᵃ Without aging.
ᵇ Aged 46 days.

duced can be explained readily by the postulated cyclic process. The increasing oxidation state of manganese in the precipitates cannot be ascribed to simple dissolution or ion-exchange processes.

In Tables II and III the determined oxidation state of manganese is compared with the value computed by the relationship used in developing Figure 3. Although there are some differences, the agreement is reasonably close for most of the solutions. Values of the reaction affinity A, for Reactions 1, 2, and 3, are given in Table II for the solutions in contact with relatively fresh Mn_3O_4, and for a solution aged for a longer period in which the solid was identified by x-ray diffraction as mostly $\gamma MnOOH$. A positive affinity for Reaction 1 (to form $\gamma MnOOH$) and for Reaction 3 (the direct precipitation process) existed in each solution. However, these solutions did not attain a favorable affinity for Reaction 2 (disproportionation to form MnO_2).

Affinities for Reactions 2 and 3 in the acidified solutions are shown in Table III for three stages: immediately after acidification, at the end of 2 hr at the pH indicated, and finally, after aging. It is of interest to note the high values for A_2 brought about by acidification and their decrease as the reaction proceeds at low pH. In the two systems held at the higher pH levels (pH 5.1 and 5.5), the values of A_2 and A_3 are both positive after 2 hr, indicating that recycling of released Mn^{2+} could have been taking place. Table IV shows that at these pH's the H^+/Mn^{4+} ratio is lower than would be predicted from disproportionation alone.

Field Observations

Although specific values for dissolved manganese activity in river water are not given here, the general range of concentration observed is from about 10 to 100 $\mu g/L$ (15). Concentrations in unfiltered samples, which sometimes are given in data compilations as "total manganese", include unknown proportions of particulate forms of manganese oxide, and have no value for geochemical studies. High concentrations of manganese in solution in river water (>100 $\mu g/L$) may result from organic complexing, waste discharges, or locally imposed reducing conditions. The range just cited lies between Lines 2 and 3 in Figure 2 in the near-neutral pH range common to river water. Free manganese oxide particles of colloidal size, which may occur in some waters, may pass through the 0.45-μm pore-diameter filters commonly used to separate dissolved from suspended material in such samples. Data presented by Kennedy et al. (20) indicate this effect is less important for manganese than for iron or aluminum, and probably is rarely significant.

Certain streams are affected by inflows that are high in manganese. A paper by Wentz (21) describes certain small streams in the mountainous regions of Colorado that are affected by drainage from abandoned metal mines and tailings or debris dumps. Many of these inflows are high in manganese concentration. Deposition of manganese oxide crusts in streambeds and on exposed rocks occurs in such streams, and the concentrations of manganese in the stream lie in the predicted range, or perhaps below it if there is extensive dilution. Further reference to the chemical data from these streams will be made later.

Coprecipitation Mechanisms

Manganese oxides deposited in natural aqueous systems contain many other metal ions as impurities. Indeed, some of these oxide deposits contain more iron than manganese. The accompanying metal ions may be retained at the oxide surface by sorption processes, and the surface of manganese oxide can exhibit a considerable capacity for cation exchange (22). However, these surfaces are chemically dynamic in that they are being renewed, added to, and changed in oxidation state through the processes discussed here. Such processes seem to be obligatory in a natural system capable of producing and sustaining manganese oxide encrustations. Therefore, the mechanisms by which many of the other metals become incorporated into the manganese oxides are more properly considered as coprecipitation, where the manganese oxide crust is increasing in thickness, or as coupled oxidation, where the manganese precipitate is undergoing redox changes.

Transition metals and other metal ions that have several oxidation states that could be stable in water may participate in at least two different ways in the initial precipitation of Mn_3O_4 in aerated water. Metals that form hydroxides of low solubility (for example, $Pb(OH)_2$) can provide a substrate or nucleation site for the Mn_3O_4 and thus be incorporated into the final manganese oxide as an impurity. Metals in the transition series whose ionic radii are similar to those of Mn^{2+} or Mn^{3+} may substitute for manganese in the hausmannite crystal structure. This may give a solid solution of widely varying composition, or for certain metals, a well-defined manganite compound analogous to the ferrites.

The various components of the mixed precipitates that result from these interactions may subsequently or concurrently participate in the redox alterations. Often these effects raise the oxidation state of the manganese in the precipitate above what would be expected in a pure manganese oxide precipitated in the absence of other metal ions. Chemical thermodynamic calculations can be used to predict or explain these processes. Some effects of lead and cobalt on manganese behavior have been evaluated theoretically and experimentally.

Manganese and Lead

Divalent lead has a very low solubility in the pH range where manganese oxide precipitates might form. Lead hydroxide or basic carbonate can form rapidly when the pH of a solution of Pb^{2+} is increased.

$$Pb^{2+} + 2H_2O = Pb(OH)_2(c) + 2H^+$$
$$[Pb^{2+}] = 10^{8.15}[H^+]^2 \tag{7}$$

(The basic carbonate has a similar solubility in aerated water not enriched with CO_2 species, and probably would be expected to be formed in natural systems rather than the hydroxide.)

Submicron particles of lead hydroxide can act as substrates or nuclei for manganese precipitation. However, the two solids are capable of interacting through electron tranfers:

$$Pb(OH)_2 + 2Mn_3O_4 + 10H^+ = MnO_2(c) + PbO_2(c) + 5Mn^{2+} + 6H_2O$$
$$[Mn^{2+}] = 10^{7.54}[H^+]^2 \tag{8}$$

(assuming unit activity for all solids).

A back reaction to reconstitute the Mn_3O_4 could be

$$PbO_2 + 3Mn^{2+} + 2H_2O = Mn_3O_4 + Pb^{2+} + 4H^+$$
$$[Mn^{2+}]^3 = 10^{11.95}[Pb^{2+}][H^+]^4 \tag{9}$$

which also returns the lead to the dissolved form. The maintenance of the pH at 8.0 by base addition and the supply of $O_2(aq)$ also will continue to promote the direct precipitation of Mn_3O_4.

Reactions 7–9 form a thermodynamic cycle somewhat like Reactions 2 and 3 for the pure manganese system, and can be viewed in a similar way. As long as any of the lead remains in the $Pb(OH)_2$ form, one may postulate that $[Pb^{2+}]$ would be controlled by Reaction 7, and the reaction affinity for this process, A_7, would be zero. At pH 7.00, then, the value of A_8 will be greater than zero when $[Mn^{2+}] < 10^{-6.46}$ and A_9 will be greater than zero when $[Mn^{2+}] > 10^{-7.30}$. Therefore, there is a range of conditions within which Reactions 8 and 9 will catalyze manganese oxidation. At relatively high pH's, where the direct disproportionation (Reaction 3) is slow, this mechanism may be important. The reaction affinity for direct oxidation of Mn^{2+} to Mn_3O_4 (A_3) also will be positive under these conditions.

When $Pb(OH)_2$ solid is not present, control over Pb^{2+} activity may be exerted by redox mechanisms. For example, if the system contains Mn_3O_4 undergoing disproportionation, there could be a diversion of electron transfers from the manganese ions to the lead ions, which may be represented by the equation

$$2Mn_3O_4(c) + Pb^{2+} + 8H^+ = MnO_2(c) + PbO_2(c) + 5Mn^{2+} + 4H_2O$$

From data in Table I this gives

$$[Pb^{2+}] = [Mn^{2+}]^5[H^+]^{-8} \times 10^{-29.53} \tag{10}$$

Equation 10 may be applicable to systems where Reactions 2 and 3 both have positive reaction affinities. At a pH of 6.0 in such a system, the permissible range of manganese activity, for example, is from $10^{-5.33}$ to $10^{-3.21}$. In this system there will be a positive reaction affinity for precipitation of lead, and the affinity will be greatest in the lower part of the manganese activity range. Very low lead concentrations are attainable under optimum conditions.

Figure 4 is a graph developed from Equation 10 showing the minimum lead concentrations that could be reached ($A_{10} = 0$) for various levels of pH and dissolved Mn^{2+} activity. The lines terminate on the left at the minimum manganese activity for Reaction 3 and are dashed on the right in the region where $Pb(OH)_2(c)$ is stable. It should be emphasized that the solubilities indicated are minima which will not be reached in all systems.

Reaction 10 produces Mn^{2+} and could substitute for Reaction 2 in a closed cycle where reaction affinities A_3 and A_{10} are both positive. Because some of the electron transfers that would have gone toward completing the manganese oxide disproportionation are diverted in Reaction 10, lead can probably be expected to slow the manganese disproportionation process.

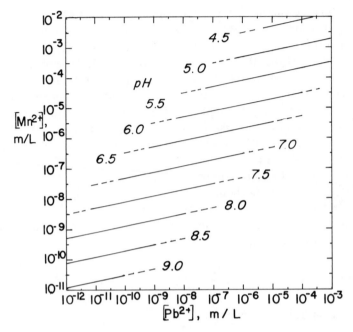

Figure 4.   Activity of lead in solutions from which manganese oxides are being precipitated

Laboratory Tests of the Lead–Manganese Coprecipitation Model

The effect of lead on manganese oxidation and precipitation was explored by preparing a series of solutions like those described for the study of the oxidation mechanisms of manganese alone, the only difference being that lead perchlorate was added to the manganese perchlorate solution in the beginning to give molar ratios of lead to manganese ranging from 1:1 to 1:1000. Subsequent treatment of the solutions was

Table V.   Properties of Coprecipitated Manganese

Solution	$mPb/m(Mn + Pb)$	pH	$[Mn^{2+}]$ (mol/L)
1C	0.001	7.30	$10^{-5.17}$
1B	0.011	7.40	$10^{-5.28}$
1A	0.10	6.90	$10^{-4.99}$
1P	0.15	7.00	$10^{-4.81}$
1N	0.25	7.43	$10^{-5.68}$
1M	0.50	7.28	$10^{-5.36}$

[a] Precipitation at pH 8.0.

identical to that described previously. The samples of solution and precipitate that were withdrawn were analyzed for both metals. Results are given in Tables V–VII.

Previous workers have not agreed on the oxidation state of lead present in manganese oxide precipitates. Cronan (*23*) stated that lead in manganese nodules is in the Pb^{4+} state. However, Van der Weijden and Kruissink (*24*) have concluded from laboratory work and theoretical studies that the Pb^{2+} form is more likely.

The experiments described here provide some indirect evidence that the lead in the precipitates was at least partly in the Pb^{4+} state. The lead-bearing precipitates were amorphous to x-rays. However, there was a substantial increase in the overall oxidation state of lead-bearing precipitates compared to those that did not contain lead. This is in accord with the proposed chemical model that postulates Pb^{4+}.

During the titration with base, the mixed oxide precipitate that was formed was black, and as the data in Table V show, the manganese had an oxidation state above 3.00 in some of the precipitates. The oxidation state determinations in Tables V and VI were made by assuming that all the precipitated lead was present as Pb^{4+}. The calculated oxidation state, based on Equations 5 and 6, agrees well enough with the determined values where the mole percent of lead did not exceed 25; this lends support to the assumption that the lead in these precipitates was in the 4+ oxidation state. However, in the precipitates where the mole percent of lead was higher, the assumption that only Pb^{4+} was present gave manganese oxidation states that did not agree with calculated values.

Lead activity after precipitation (Table V) was generally less than would be predicted for the solubility of $Pb(OH)_2$ at the reported pH.

Acid treatment of the precipitates increased the oxidation state of the manganese. A check between the determined and calculated oxidation numbers was obtained for three of the solutions after aging.

Data in Table VII show that Solutions 1B and 1A (1.1 and 10 mol % lead, respectively) yielded more Mn^{4+} per mole of H^+ used during dis-

and Lead Oxides and Associated Solutions[a]

$[Pb^{2+}]$ (mol/L)	Age	Manganese Oxidation State	
		Determined	*Calculated*
$<10^{-6.5}$	4 hr	2.74	2.74
$<10^{-6.5}$	16 hr	2.75	2.71
$<10^{-6.5}$	16 hr	3.35	3.30
$<10^{-6.5}$	2 days	3.16	2.85
$<10^{-6.5}$	4 hr	2.95	3.12
$<10^{-6.5}$	16 hr	—	3.21

Table VI.   Solubility and Oxidation State of Manganese

Solution	Holding pH	Final pH	$[Mn^{2+}]$
1C	4.01	5.65	$10^{-2.64}$
1B	4.05	5.81	$10^{-2.79}$
1A	4.02	5.71	$10^{-3.04}$
1P	3.00	3.12	$10^{-3.08}$
1N	4.00	4.05	$10^{-3.27}$
	3.50	3.62	$10^{-3.25}$
	3.00	3.22	$10^{-3.16}$
1M	4.00	4.75	$10^{-3.68}$

proportionation at pH 4.0 than Solution 1C. The latter contained 0.1 mol % lead, an amount too small to have a significant influence on the manganese behavior. These results indicate that recycling mechanisms involving H^+ and Pb can affect the manganese disproportionation even at pH 4.

Laboratory Tests of Cobalt–Manganese Coprecipitation Models

Cobalt can occur in divalent or trivalent forms in oxide and hydroxide structures, and the ionic radii of cobalt species are similar to those of corresponding manganese species. Divalent cobalt hydroxide is much more soluble than divalent lead hydroxide. From data in Table I

$$[Co^{2+}] = 10^{13.19}[H^+]^2 \qquad (11)$$

Probably this material is too soluble to play a role in manganese oxide nucleation.

A cobalt analog of hausmannite could be formed by the reaction

$$3Co^{2+} + \tfrac{1}{2}O_2(aq) + 3H_2O = Co_3O_4(c) + 6H^+$$

From data in Table I

$$[Co^{2+}] = 10^{5.71}[H^+]^2[O_2]^{-1/6} \qquad (12)$$

Table VII.   Stoichiometry of Disproportionation

Solution	Mn^{4+} at Start (mmol)	Holding pH	Reaction Time (Hr)	Aging Time (Days)
1C	0.047	4.0	2	10
1B	0.044	4.0	2	49
1A	0.404	4.0	2	49

and Lead Oxide Precipitates after Acid Treatment

[Pb²⁺]	Age	Determined Oxidation No.	Calculated Oxidation No.
$<10^{-6.5}$	10 days	3.44	3.53
$<10^{-6.5}$	49 days	3.36	3.27
$<10^{-6.5}$	49 days	3.69	3.83
$10^{-4.06}$	28 days	—	—
$10^{-3.80}$	2 hr	—	—
$10^{-3.62}$	1 day	—	—
$10^{-3.42}$	30 days	—	—
$10^{-2.86}$	107 days	—	—

Where both cobalt and manganese are present in solution, the coprecipitation of hausmannite and Co_3O_4 might be expected. Sinha et al. (*25*) found that random substitution of cobalt for Mn^{2+} or Mn^{3+} could occur in such material; occurrence of Co^{3+} in manganese oxide crystal-lattice positions was noted by Burns (*4*). Apparently there are no thermodynamic data for mixed cobalt + manganese oxides, but the behavior of the ions can probably be represented over a considerable range of solid composition by a solid–solution model based on the equilibrium between the pure end members. Thus

$$Co_3O_4(c) + 3Mn^{2+} = Mn_3O_4(c) + 3Co^{2+}$$

and where equal activities of solids are assumed,

$$[Co^{2+}]^3 = 10^{-1.15}[Mn^{2+}]^3 \tag{13}$$

Where activities of solid species are unequal, a conditional equilibrium constant could be applied, as noted in the derivation of Equations 5 and 6. Aqueous trivalent cobalt is not thermodynamically stable in the pH range of interest here.

In oxygenated water at a pH high enough to permit precipitation of hausmannite, the interaction of cobalt with manganese at the mixed oxide surface may be represented by the sum of Equations 2, 3, and 12:

$$Mn^{2+} + O_2(aq) + 3Co^{2+} + 4H_2O = MnO_2(c) + Co_3O_4(c) + 8H^+$$

of Manganese in the Presence of Lead

H⁺ Used (mmol)	Mn⁴⁺ Formed (mmol)	H⁺/Mn⁴⁺	Total Mn⁴⁺
0.769	0.188	4.1	0.235
0.520	0.198	2.6	0.242
0.291	0.095	3.1	0.499

From data in Table I,

$$[H^+]^8 = 10^{-17.81}[Mn^{2+}][O_2][Co^{2+}]^3 \tag{14}$$

Where manganese is present in concentrations greater than that of cobalt the simultaneous production of hausmannite would be expected, giving a precipitate containing manganese that is not fully oxidized to the 4+ state.

Divalent cobalt ions could also interact with a hausmannite surface by diverting some of the electron transfers that occur in the disproportionation of the manganese. This process would be favored at lower pH and can be represented as

$$2Mn_3O_4(c) + 3Co^{2+} + 4H^+ = MnO_2(c) + Co_3O_4(c) + 5Mn^{2+} + 2H_2O$$

For this reaction, data in Table I lead to

$$[Mn^{2+}]^5 = 10^{18.73}[Co^{2+}]^3[H^+]^4 \tag{15}$$

From the thermodynamic data it also may be shown that a precipitate containing both Mn_3O_4 and Co_3O_4 would be unstable at low pH. From

$$Mn_3O_4(c) + Co_3O_4(c) + 8H^+ = 2MnO_2(c) + 3Co^{2+} + Mn^{2+}$$

and data in Table I

$$[Mn^{2+}][Co^{2+}]^3 = 10^{34.01}[H^+]^8 \tag{16}$$

This combination of solids can be produced in the laboratory but is less likely to occur in natural systems where the most stable combination would be the fully oxidized MnO_2 and Co_3O_4.

Both Reaction 14 and Reaction 15 are simplified statements of complicated systems where several reaction paths are likely to be available. However, they have some value as predictors of limiting conditions and possible ranges of concentrations to be expected. Generally, when manganese is present in excess, limits will be imposed on the cobalt behavior that are related to the manganese precipitation–disproportionation cycle described by Equations 2 and 3. As noted earlier, these equations predict the range of manganese activities expected at any pH.

Figure 5 is based on Equation 15 and shows the range of activity of cobalt that is compatible with the indicated range of activity of manganese from pH 5.5 to 9.0.

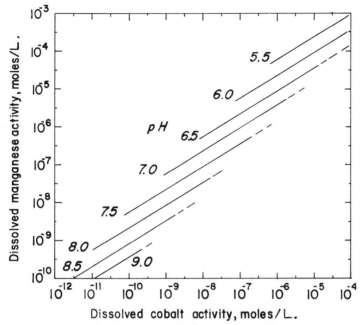

*Figure 5.   Activity of cobalt in solutions from which manganese oxides
are being precipitated*

Results of Manganese and Cobalt Coprecipitation Experiments

To test the applicability of the theoretical chemical relationships involving the two metals, a series of solutions containing manganese and cobalt perchlorate (total metal ion concentration near $10^{-2}M$) was prepared, with Mn:Co molar ratios from about 3:1 to about 50:1. These were precipitated and oxidized at pH 8.5 and later held at lower pH's, just as in the experiments described earlier. Some experiments also were made in which the cobalt was added after the manganese oxide had been precipitated. Experimental data from these solutions are given in Tables VIII–XI.

If Equation 14 applies, the average oxidation state of manganese should be increased with increasing concentrations of cobalt during the initial precipitation. Data in Table VIII show this effect. The oxidation state of manganese computed from the pH and dissolved manganese activity reached approximate agreement with the chemical analysis value, after rather long aging. The activity of dissolved cobalt in all the solutions is much greater than that predicted by Equation 14, but the cobalt activities predicted by Equation 16 are in approximate agreement with the measured values for the first four solutions in the table. Evidently this equation is not applicable for the other solutions.

Table VIII. Properties of Coprecipitated Manganese

			Determined	
Solution	mCo/ m(Mn + Co)	pH	$[Mn^{2+}]$ (mol/L)	$[Co^{2+}]$ (mol/L)
3L	0.02	7.69	$10^{-5.95}$	$<10^{-6.5}$
3K	0.05	6.91	$10^{-4.62}$	$10^{-5.59}$
3D	0.10	7.21	$10^{-5.18}$	$10^{-5.99}$
3E	0.20	6.75	$10^{-4.42}$	$10^{-4.91}$
3F	0.50	7.19	$10^{-5.47}$	$10^{-4.11}$
3G	0.70	6.95	$10^{-5.41}$	$10^{-3.91}$
3H	0.85	6.70	$10^{-4.87}$	$10^{-3.22}$

[a] Precipitation at pH 8.5.

Table IX. Stoichiometry of Disproportionation

Solution	mCo/ m(Mn + Co)	Mn^{4+} at Start (mmol)	Holding pH	Time (Hr)
3C	0.02	0	4.03	7.5
3B	0.098	0.104	4.03	3.0
3A	0.25	0.207	4.02	5.6
4B	0	0	4.05	12

Table X. Solubility and Oxidation State of Manganese

Solution	Holding pH	Final pH	$[Mn^{2+}]$ (mol/L)
3C	4.03	5.68	$10^{-2.61}$
3B	4.03	5.30	$10^{-2.74}$
3A	4.02	4.60	$10^{-2.86}$
4B	4.05	5.20	$10^{-2.61}$

Table XI. Comparison of Observed and Calculated

Solution	Age (Days)	pH
3C	76	5.68
3B	79	5.30
3A	80	4.60
4B	74	5.20

and Cobalt Oxides and Associated Solutions[a]

$[Co^{2+}]$ Calculated[b]	Age (Days)	Manganese Oxidation State	
		Determined	Calculated
$10^{-7.19}$	38	2.88	2.74
$10^{-5.55}$	41	3.06	2.84
$10^{-6.16}$	74	2.91	2.82
$10^{-5.19}$	75	3.13	2.95
$10^{-6.01}$	69	3.48	3.16
$10^{-5.39}$	68	3.47	3.73
$10^{-4.91}$	66	3.74	3.69

[b] Equation 16.

of Manganese in the Presence of Cobalt

H^+ Used (mmol)	Mn^{4+} Formed (mmol)	H^+/Mn^{4+}	Mn^{4+} Final
0.808	0.234	3.5	0.234
0.636	0.201	3.2	0.305
0.624	0.177	3.5	0.384
0.921	0.240	3.8	0.240

and Cobalt Oxide Precipitates After Acid Treatment

$[Co^{2+}]$ (mol/L)	Age (Days)	Manganese Oxidation State	
		Determined	Calculated
$10^{-4.67}$	76	3.58	3.47
$10^{-3.78}$	79	3.59	3.98
$10^{-3.13}$	80	3.81	4.00
$10^{-3.43}$	74	3.54	3.99

Cobalt Activities in Experimental Solutions

Determined		$[Co^{2+}]$ Calculated (mol/L)	
$[Mn^{2+}]$ (mol/L)	$[Co^{2+}]$ (mol/L)	Equation 13	Equation 15
$10^{-2.61}$	$10^{-4.67}$	$10^{-2.99}$	$10^{-3.02}$
$10^{-2.74}$	$10^{-3.78}$	$10^{-3.12}$	$10^{-3.74}$
$10^{-2.68}$	$10^{-3.13}$	$10^{-3.24}$	$10^{-4.88}$
$10^{-2.61}$	$10^{-3.43}$	$10^{-2.99}$	$10^{-3.66}$

Upon acidification to pH 4.0, a relatively rapid disproportionation reaction took place. The reaction times shown in Table IX represent the period during which the pH was held at the indicated level. Manganese in Solution 4B was oxidized and precipitated, and the pH lowered to 4.0, before the cobalt was added. The amounts of Mn^{4+} that were produced during disproportionation and the subsequent aging period (as indicated in Table X) are given in the seventh column of Table IX. These quantities represent different reaction times and can only be interpreted as a general indication that disproportionation occurs in all samples, even in Sample 3A where the precipitate initially contained 25 mol % of cobalt.

The ratio H^+/Mn^{4+} in all samples was less than 4.0, indicating that a cyclic mechanism involving regeneration of Mn_3O_4 and H^+ probably was occurring.

After disproportionation, the solutions gradually increased in pH during aging. This was accompanied by relatively small changes in manganese and cobalt activities. For Solution 3C, the calculated oxidation number for the precipitated manganese agrees rather closely with the measured value (Table X). Agreement for the others, having larger contents of cobalt, is poorer.

Table XI contains calculated and observed cobalt activities after acid treatment. In the solution that had the highest original cobalt/manganese ratio, the value observed is very near that predicted by the Co_3O_4–Mn_3O_4 equilibrium (Equation 13). For Solution 3B the observed activity is very near that calculated from Equation 15. This is also approximately true for Solution 4B. During the titration of solutions like 4B, where the initial manganese precipitate did not contain cobalt, and during the aging of all the acidified solutions, the activity of dissolved cobalt tended to decrease while manganese increased, as Equation 15 predicts. These observations suggest that the coupled redox mechanism occurred in all the acidified solutions. However, the stoichiometric composition of some of the experimental solutions precluded attaining the limiting concentration of cobalt, where the reaction affinity is zero. For example, the total amount of cobalt present in Solution 3C was not great enough to attain the concentrations predicted by either Equation 13 or Equation 15.

Systems in which the behavior of cobalt would be expected to follow Equation 15 most closely would be difficult to reproduce in the laboratory, in a form that permitted good control of conditions and sampling access. However, some field evidence has been obtained that indicates that Equation 15 is useful as a predictor of cobalt activity where manganese oxides are known to be forming in the system, and where reliable measurements of dissolved manganese and pH are available. Analyses for ten small mountain streams in Colorado that were affected by drainage from metal mines gave calculated values for Co^{2+} that agreed well (for the most part) with determined values (26). A similar agreement was observed for three samples of ground-water seepage collected from a gravel deposit in southern Finland, where manganese oxide was being formed at the top of the zone of saturation (27).

Summary and Conclusions

The precipitation of manganese oxides from aerated aqueous systems may be viewed as a two-step process involving oxidation of Mn^{2+} to the Mn^{3+} state, and disproportionation of Mn^{3+} to form Mn^{4+}. Thermodynamic data show that the reaction affinities for both processes will be positive when the fluxes of dissolved oxygen and Mn^{2+} toward the reaction site are at levels commonly attained in river water and some other natural systems.

Mechanisms for coprecipitation of lead and cobalt with manganese oxide can be derived based on thermodynamic calculations. They can explain the increased oxidation state of manganese reached in the mixed oxide precipitates, and they provide a potential control of the solubility of the accessory metals. The effectiveness of the control has been evaluated in a preliminary way by laboratory experiments described here, and by some field observations. Cobalt activity seems to be controlled by manganese coprecipitation in many natural systems. Although more testing by both laboratory experiments and field studies is needed, the proposed mechanisms appear to be applicable to many coupled oxidation–reduction processes.

Literature Cited

1. Bricker, O. P. "Some Stability Relationships in the System Mn-O_2-H_2O at 25°C and 1 Atmosphere Total Pressure," *Am. Mineral.* 1965, *50*, 1296–1354.
2. Morgan, J. J. In "Principles and Applications of Water Chemistry"; Faust, S. D., Hunter, J. V., Eds.; Wiley: New York, 1967; p. 561–624.
3. Loganathan, P.; Burau, R. G. "Sorption of Heavy Metals by a Hydrous Manganese Oxide," *Geochim. Cosmochim. Acta* 1973, *37*, 1277–1293.
4. Burns, R. G. "The Uptake of Cobalt into Ferromanganese Nodules, Soils, and Synthetic Manganese(IV) Oxides," *Geochim. Cosmochim. Acta* 1976, *40*, 95–102.
5. Giovanoli, R.; Stähli, E.; Feitknecht, W. "Über Oxidhydroxide des vierwertigen Mangans mitt Schichtengitter, 1. Natriummangan(II III) Manganat(IV)," *Helv. Chim. Acta* 1970, *53*, 209–220.
6. Giovanoli, R.; Stähli, E.; Feitknecht, W. "Über Oxidhydroxide des vierwertigen Mangans mitt Schichtengitter, 2. Mangan(III) Manganat(IV)," *Helv. Chim. Acta* 1970, *53*, 453–456.
7. Giovanoli, R.; Stähli, E. "Oxide und Oxihydroxide des drei- und vierwertigen Mangans," *Chimia* 1970, *24*, 49–61.

8. Giovanoli, R.; Bürki, P.; Guiffredi, M.; Stumm, W. "Layer Structured Manganese Oxide Hydroxides, IV. The Buserite Group; Structure Stabilized by Transition Elements," *Chimia* **1975**, *29*, 517–520.
9. Wagman, D. D.; Evans, W. H.; Parker, V. B.; Halow, I.; Bailey, S. M.; Schumm, R. H. "Selected Values of Chemical Thermodynamic Properties," *Nat. Bur. Stand. (U.S.) Tech. Note* **1969**, *270-4.*
10. Naumov, G. B.; Ryzhenko, B. N.; Khodakovsky, I. L. "Handbook of Thermodynamic Data," 1971, (translated by G. J. Soleimani) *U.S. Dept. Commer., Off. Techn. Ser., PB Rep.* **1974**, *PB 226 722.*
11. Bolzan, J. A.; Podesta, J. J.; Arvia, A. J. "Hydrolytic Equilibrium among Metal Ions. I. The Hydrolysis of Co(II) Ion in Aqueous Solutions of NaClO₄," *An. Asoc. Quim. Argent.* **1963**, *51*, 43–58.
12. Wagman, D. D.; Evans, W. H.; Parker, V. B.; Halow, I.; Bailey, S. M.; Schumm, R. H. "Selected Values of Chemical Thermodynamic Properties," *Nat. Bur. Stand. (U.S.) Tech. Note* **1968**, *270-3.*
13. Lind, C. J. "Polarographic Determination of Lead Hydroxide Formation Constants at Low Ionic Strength," *Environ. Sci. Technol.* **1978**, *12*, 1406–1410.
14. Hem, J. D. "Chemical Equilibria and Rates of Manganese Oxidation," *U.S. Geol. Surv., Water-Supply Pap.* **1963**, *1667-A.*
15. Hem, J. D. "Deposition and Solution of Manganese Oxides," *U.S. Geol. Surv., Water-Supply Pap.* **1964**, *1667-B*, 5–11.
16. Manheim, F. T. "Manganese-Iron Accumulations in the Shallow Marine Environment," *Mar. Geochem., Proc. Symp.* **1965**, *13*, 217–276.
17. Prigogine, I.; Defay, R. "Chemical Thermodynamics"; Wiley: New York, 1954.
18. Schweisfurth, R. "Manganoxydierend Bakterien," *Z. Allg. Mikrobiol.* **1973**, *13*, 341–347.
19. Garrels, R. M.; Christ, C. L. "Solutions, Minerals, and Equilibria"; Harper and Row: New York, 1965; p. 42–49.
20. Kennedy, V. C.; Zellweger, G. W.; Jones, B. F. "Pore Size Effects on the Analysis of Al, Fe, Mn, and Ti in Water," *Water Resour. Res.* **1974**, *10*, 785–790.
21. Wentz, D. A. "Effect of Mine Drainage on the Quality of Streams in Colorado, 1971–72"; Colorado Water Resources: 1974; Circ. 21.
22. Murray, J. W. "The Interaction of Metal Ions at the Manganese Dioxide-Solution Interface," *Geochim. Cosmochim. Acta* **1975**, *39*, 505–519.
23. Cronan, D. S. In "The Sea: Vol. 5, Marine Chemistry"; Goldberg, E. D., Ed.; Wiley: New York, 1974; p. 512–515.
24. Van der Weijden, C. H.; Kruissink, E. C. "Some Geochemical Controls on Lead and Barium Concentrations in Ferromanganese Deposits," *Mar. Chem.* **1977**, *5*, 93–112.
25. Sinha, A. P. B.; Sinjana, N. R.; Biswas, A. B. "On the Structure of Some Manganites," *Acta Crystallogr.* **1957**, *10*, 439–440.
26. Hem, J. D. "Redox Processes at Surfaces of Manganese Oxide and Their Effects on Aqueous Metal Ions," *Chem. Geol.* **1978**, *21*, 199–218.
27. Koljonen, T.; Lahermo, P.; Carlson, L. "Origin, Mineralogy and Chemistry of Manganiferous and Ferruginous Precipitates Found in Sand and Gravel Deposits in Finland," *Bull. Geol. Soc. Finland* **1976**, *48*, 111–135.

RECEIVED October 24, 1978.

Adsorption Reactions of Nickel Species at Oxide Surfaces

THOMAS L. THEIS and RICHARD O. RICHTER[1]

Department of Civil Engineering, University of Notre Dame,
Notre Dame, IN 46556

Experimental adsorption isotherms of Ni^{+2} on α-quartz and α-FeOOH (goethite) are described adequately by the solvent-ion model of adsorption. The addition of complexing ligands (in this study, sulfate, citrate, nitrilotriacetate, glycine, and cyanide) to the nickel–oxide systems alters adsorption behavior. For sulfate, citrate, and nitrilotriacetate, a competition model for free Ni^{+2} between surface and ligand can account for the observed data. Strong evidence is presented for specific adsorption of glycine– and cyanide–nickel complexes onto goethite. All adsorption reactions are discussed through analysis of theoretical nickel speciation in combination with experimental data, the nature of the oxides, and the structure of the nickel–ligand complexes.

Reactions of metal ions in aqueous media have been shown to be strongly influenced by surface sorption reactions. The adequate description of metal ion behavior in systems where particulates have been included is an important step in the application of laboratory data to natural systems of wide environmental interest. In this study, data on the adsorption of aqueous nickel, Ni^{+2}, onto oxides of silica and iron is presented. Of special interest are the effects which various ligands have on the adsorption reactions. Data are analyzed through the use of the chemical equilibrium computer model REDEQL2 making use of the solvent–ion model of adsorption.

[1] Current address: Department of Civil and Environmental Engineering, Washington State University, Pullman, WA 99164.

Models of Adsorption

There are many models that describe the interaction between solute ions and surfaces. These have been reviewed by several authors (1, 2, 3) and include general ion exchange (4, 5, 6), surface complex formation (7, 8) (the "Swiss" model), and various electrostatic models (Gouy–Chapman–Stern (9), Grahame (10, 11), and James and Healy (12)). For hydrolyzable species sorbing onto hydrous oxide surfaces, the surface complex formation model and the solvent–ion interaction model of James and Healy have been shown to be in good agreement with observations. In this chapter, data are analyzed via the James and Healy model.

In this model, a distinction is drawn between electrostatic forces that arise from coulombic considerations, and those that arise from the energy required to replace (or electrically saturate) the secondary hydration sphere around the ion with the hydration layer attached to the sorbing surface. Since most hydrophobic sols have dielectrics considerably less than bulk solution, the latter energy term, known as solvation energy, is positive and opposes the adsorption reaction. In addition, the model recognizes the existence of specific interaction forces between ion and surface resulting from dipole or covalent bonding. Accordingly, the total energy of adsorption of a species i is given by:

$$\Delta G°_{ADS,i} = \Delta G°_{COUL,i} + \Delta G°_{SOLV,i} + \Delta G°_{CHEM,i} \qquad (1)$$

A mathematical summary of the model is given in the Appendix. The use of the Nernst equation (VIII) is valid for small pH variations around the zero point of charge (ZPC). From Equation XI it can be seen that the magnitude of the solvation energy is proportional to the square of the charge, thereby decreasing fourfold upon hydrolysis of the metal ion. This agrees well with the often-observed phenomenon of increased adsorption of many metal-surface combinations at the pH corresponding to the pK of hydrolysis of the metal ion. Therefore, it is implicit that hydrolysis occurs in conjunction with adsorption.

Although it has a chemical meaning in the James and Healy model, the specific chemical free energy term is in practice an experimental fitting parameter. It can be determined by measuring the extent of adsorption at the ZPC of the oxide surface; however, this method assumes that the magnitude does not change with solution pH. More commonly, theoretical adsorption isotherms of the amount of metal adsorbed vs. pH are fitted to actual data by adjusting the $\Delta G°_{CHEM}$ term.

Aqueous Chemistry of Nickel

The element nickel is a Group VIII metal which is part of the first transition series. Elemental nickel has the outer sphere electron configuration $3d^8 4s^2$ and readily yields the 4s electrons to give the divalent ion

Ni^{+2}, which is the only oxidation state of importance in natural systems. Like the other elements in the first transition series (V, Cr, Mn, Fe, Co), Ni^{+2} is octahedrally coordinated in aqueous systems as $Ni(H_2O)_6^{+2}$ (*13*). This free aquo ion dominates the aqueous chemistry of nickel at neutral pH values; however, complexes of naturally occurring ligands are formed to a small degree ($OH^- > SO_4^{-2} \approx Cl^- > NH_3$).

Among the mononuclear hydrolytic species of nickel, only the stability of $NiOH^+$ is well documented. Baes and Mesmer (*14*) have compiled the following values for the hydrolysis reactions:

$$log \, \beta_n$$

$$Ni^{+2} + H_2O = NiOH^+ + H^+ \qquad\qquad -9.86 \qquad\qquad (2)$$

$$Ni^{+2} + 2H_2O = Ni(OH)_2^0 + 2H^+ \qquad -19 \qquad\qquad (3)$$

$$Ni^{+2} + 3H_2O = Ni(OH)_3^- + 3H^+ \qquad -30 \pm 0.5 \qquad (4)$$

$$Ni^{+2} + 4H_2O = Ni(OH)_4^{-2} + 4H^+ \quad < -44 \qquad\qquad (5)$$

The literature indicates that the solubility of $Ni(OH)_{2(s)}$ depends on the degree of aging of the precipitate. The solubility product is reported to range from $10^{-14.7}$ for a fresh precipitate to $10^{-17.2}$ for an aged one (*15, 16*). The equilibrium relationship of the free nickel ion and hydroxide precipitate can be used in combination with Equations 2–5 to construct the log solubility vs. pH diagrams for Ni(II) as given in Figure 1.

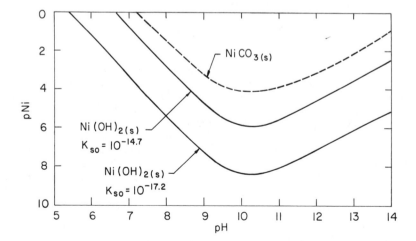

Figure 1. Comparison of solubilities of nickel carbonate ($pK_{so} = 6.87$) and nickel hydroxide for active ($pK_{so} = 14.7$) and aged ($pK_{so} = 17.2$) precipitates ($P_{CO_2} = 10^{-3.5}$ atm)

Carbonate precipitates are important for some metal ions; however, the solubility of nickel carbonate (log $K_{so} = -6.87$[15]) is sufficiently large so that it is of little consequence in natural waters. The dashed line in Figure 1 represents nickel carbonate precipitation for a system in equilibrium with $10^{-3.5}$ atm of CO_2. Under reducing conditions the solubility of nickel would be expected to be controlled by the sulfide, NiS (log $K_{so} = -22.9$[15]).

As indicated, Ni^{+2} forms generally weak complexes with common inorganic ligands. However, complexes of sizeable stability are formed with many organic ligands and a considerable portion of this study has been devoted to their effects on nickel-sorption reactions. Those ligands specifically studied were sulfate (SO_4^{-2}), citrate, nitrilotriacetate (NTA), glycine, and cyanide. Appropriate data are summarized in Table I. These data were obtained from Baes and Mesmer (14), Sillen and Martell (15, 16), and Smith and Martell (17).

Experimental Procedures

The oxides used in this study were α-SiO_2 (α-quartz), obtained commercially, and α-FeOOH (goethite), which was prepared in a manner similar to that of Forbes et al. (18). The silica was washed initially in 0.1N nitric acid. Both oxides were washed with double distilled water, dried at 100°C for 24 hr, powdered with a mortar and pestle, and passed through a 200 mesh (75 μm) sieve. Powdered x-ray diffraction verified the existence of α-quartz and goethite. BET-N_2 adsorption indicated specific surface areas of 1.7 m^2/g for silica and 85 m^2/g for goethite. Corresponding ZPC values, determined by electrophoresis and turbidity measurements, were 1.7 and 5.5. Dielectrics were taken to be 4.3 for silica and 14.2 for goethite (19).

Adsorption studies were carried out in 150-mL borosilicate glass reaction vessels at room temperatures (22°–25°C). The oxide was weighed to give the desired surface area per liter and then placed into the vessel. A 0.11M sodium perchlorate stock solution was added in the proper dilution to give the desired ionic strength. To this was added the appropriate amount of the nickel–ligand mixture. The pH was adjusted with strong acid or base to an initial value and then the vessel was sealed. After shaking for 3 hr, the bottle was removed and the final pH was measured. The contents were filtered through a 0.45-μm membrane filter, and the filtrate was acidified and stored at 4°C prior to analysis. Samples were analyzed for total nickel by flameless atomic absorption.

Data are reported as percent nickel removed vs. pH isotherms. The James and Healy adsorption model as contained in the equilibrium computer model REDEQL2 was used to facilitate data analysis (20). In this way, an assessment of the combined effects of adsorption, complexation, and precipitation could be attempted.

Results

Evaluation of Hydrolysis Data. The application of thermodynamic models such as REDEQL2 to the type of work reported on in this chapter requires a careful assessment of the data base used. Of critical importance in this study are the stability constants given in Table I, especially those defining hydrolysis reactions of nickel. To successfully use REDEQL2 to model laboratory reactions, it was necessary to evaluate the stability constants for $NiOH^+$ and $Ni(OH)_2{}^0$, and the solubility product of $Ni(OH)_{2(s)}$ for the existing conditions. Accordingly, pH values of $10^{-4.77}M$ solutions of Ni^{+2} (at various ionic strengths (I)) were adjusted over a wide range, allowed to react for the 3-hr period, and analyzed for filterable nickel. Figure 2 shows nickel removal vs. pH for three ionic strengths tested. The curves shown are the best fit of the data and were generated using $\log \beta_1 = -9.9$ and $\log K_{so} = -15.2$ ($I = 0$). Since residual nickel at the higher pH values was consistently below the detection limit (2.5 $\mu g/L$), β_2, β_3, and β_4 were assumed to be insignificant at the nickel concentrations used throughout the experimentation and multiple hydroxyl ligand complexes of nickel were ignored. Figure 3 shows the theoretical distribution of nickel(II) species in water with $Ni_T = 10^{-4.77}M$, $I = 0.01$.

Nickel Adsorption onto Silica and Goethite. Figure 4 contains adsorption isotherms for nickel onto α-SiO_2 and α-$FeOOH$ as compared with precipitation which occurs in the absence of oxide surfaces. The lines in Figure 4, as with all figures in this chapter showing sorptive behavior, are predicted using the James and Healy model in conjunction

Table I. **Acidity and Nickel Stability Constants
for Complexing Ligands**[a]

Ligand	Equilibrium	Log K
OH^-	$K_1 = NiL/Ni \cdot L$	4.1
	$K_{so} = Ni \cdot L^2$	-15.2
$SO_4{}^{-2}$	$HL/H \cdot L$	2.2
	$NiL/Ni \cdot L$	2.3
CN^-	$HL/H \cdot L$	9.3
	$NiL_4/Ni \cdot L^4$	31.8
GLY^{-1}	$HL/H \cdot L$	9.9
	$H_2L/H \cdot HL$	2.3
	$NiL/Ni \cdot L$	6.3
	$NiL_2/Ni \cdot L^2$	11.4
CIT^{-4}	$HL/H \cdot L$	16.8
	$H_2L/H \cdot HL$	6.4
	$H_3L/H \cdot H_2L$	4.5
	$H_4L/H \cdot H_3L$	3.0
	$NiHL/Ni \cdot HL$	6.0
	$NiH_2L/Ni \cdot H_2L$	3.8
NTA^{-3}	$HL/H \cdot L$	10.5
	$H_2L/H \cdot HL$	3.2
	$H_3L/H \cdot H_2L$	2.2
	$NiL/Ni \cdot L$	13.1

[a] $T = 25°C, I = 0$.

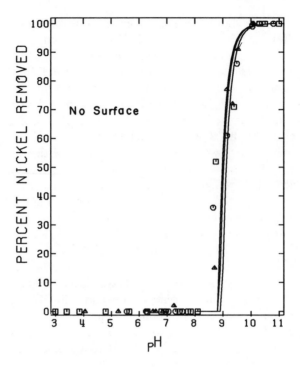

Figure 2.   *Nickel removal as a function of pH for varying ionic strengths (lines are predicted by REDEQL2 using $pK_{so} = 15.2$, $pNi_T = 4.77$; ($\square$) $I = 10^{-3}$; ($\bigcirc$) $I = 10^{-2}$; ($\triangle$) $I = 10^{-1}$)*

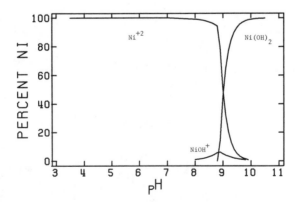

Figure 3.   *Theoretical distribution of nickel as a function of pH ($\{Ni\}_T = 10^{-4.77}M$, $I = 0.01$)*

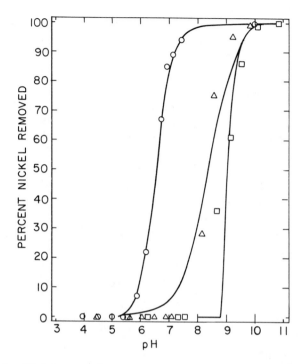

Figure 4. Nickel removal as a function of pH in the presence of oxide surfaces ($\{Ni\}_T = 10^{-4.77}$M, I = 0.01, $\{FeOOH\} = 0.59$ g/L (50 m^2/L), $\{SiO_2\} = 29.4$ g/L (50 m^2/L); (O) Fe; ($\triangle$) Si; ($\square$) no solid)

with the equilibrium subroutines in REDEQL2. The isotherms generated by the model are for ΔG_{CHEM} values of -6.55 kcal/mol and -5.35 kcal/mol for silica and goethite, respectively. It is apparent from Figure 4 that goethite adsorbs greater amounts of nickel at lower pH values than silica in spite of its higher ZPC. This is explained in the James and Healy model by the greater solvation-energy term for sorption onto silica attributable to its low dielectric. By comparison with Figure 3, it can be seen that most adsorption occurs in the pH range where Ni^{+2} is the dominant dissolved form for goethite and where $NiOH^+$ is the dominant dissolved species for sorption onto silica. In the latter case, the reduced charge brings about a reduction in the oppositional solvation-energy term, shifting the overall free energy of adsorption to more negative values.

Effects of Complexing Ligands on Adsorption. There are basically two effects that a complexing ligand can have on the adsorption reactions being studied. If the resulting complex is not specifically adsorbed (that is, $\Delta G_{CHEM} = 0$), then the effect of the ligand is to retard overall adsorption. However, in some instances the ligand has been shown to enhance the adsorption reaction either by directly attaching the complex to the oxide surface or by altering the surface properties and sorption sites such that adsorption is favored more (21). The version of REDEQL2 that was used is based on the former conceptual model, that is, ligands compete with the surface for the metal ion. Substantial disagreement between the model and experimental data suggests that specific adsorption of the complex is occurring. Examples of both retardation and enhancement of nickel adsorption will be given.

SULFATE.   Large amounts of sulfate ion are required to bring about significant complexation of Ni^{+2}. For $Ni_T = 10^{-4.77}M$, $10^{-1}M$ sulfate complexes approximately 50% of the nickel present as the uncharged ion pair, $NiSO_{4(aq)}^0$. The resulting adsorption edges for goethite and silica are shown in Figure 5. Comparison with Figure 4 shows a slight shift of the curves to the right as would be expected. A portion of this is attributable to the higher ionic strength of the sulfate medium ($\sim 0.1M$). The data can be explained adequately by the competition model used in REDEQL2. ΔG_{CHEM} for the complex is taken to be zero.

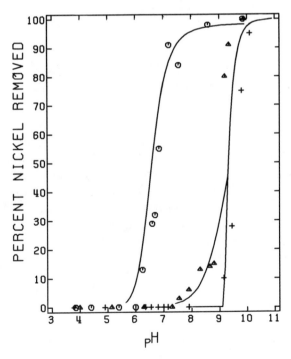

Figure 5.   Nickel removal as a function of pH in the presence of sulfate and oxide surfaces ({Ni}$_T = 10^{-4.77}$M, {SO$_4$}$_T = 10^{-1}$M, I = 0.3, {FeOOH} = 0.59 g/L (50 m²/L), {SiO₂} = 29.41 g/L (50 m²/L); (○) Fe; (△) Si; (+) no solid)

CITRATE.   Citrate can be considered to have the generalized formula H_3L, since a fourth proton, associated with the hydroxyl group, does not ionize in the pH range of water. With Ni^{+2} it forms a stable tridentate chelate ($Ni–CIT^{-1}$) which, at low pH, may be protonated. Figures 6 and 7 show the effects of adding this ligand on adsorption onto silica and goethite, respectively. For both systems, increasing the amount of citrate shifts the curves to the right. For silica, once the pH becomes high enough to produce $NiOH^+$, removal takes place. According to the model, at higher pH values precipitation of $Ni(OH)_{2(s)}$ plays a greater role in the removal as the ligand concentration increases. However, parallel experimental systems containing only the nickel–citrate chelate and no silica showed little precipitation. This anomaly suggests either that the oxide surface is acting as a nucleation site for $Ni(OH)_{2(s)}$ formation, or that specific adsorption of the complex is occurring.

For the nickel–citrate–goethite system (Figure 7), the oxide becomes more competitive with increasing pH because of the greater coulombic attraction predicted by the Nernst equation. Unlike silica, Ni^{+2} is the

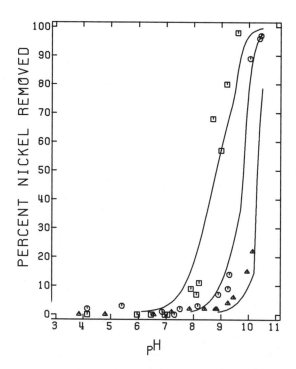

Figure 6.   Nickel removal as a function of pH in the presence of silicon dioxide and citrate ($\{Ni\}_T = 10^{-4.77}M$, $I = 0.01$, $\{SiO_2\} = 29.41$ g/L (50 m^2/L); ($\square$) CIT $= 10^{-5}$; ($\bigcirc$) CIT $= 10^{-4}$; ($\triangle$) CIT $= 10^{-3}$)

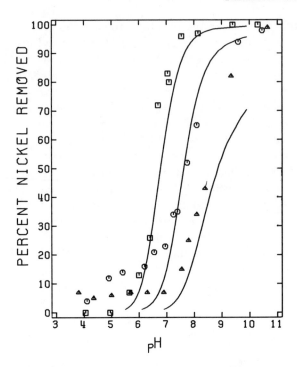

Figure 7. Nickel removal as a function of pH in the presence of iron oxide and citrate ({Ni}$_T$ = 10$^{-4.77}$M, I = 0.01, {FeOOH} = 0.59 g/L (50 m^2/L); (□) CIT = 10^{-5}; (○) CIT = 10^{-4}; (△) CIT = 10^{-3})

main species in solution. At the highest citrate concentration studied (10^{-3}M), the data show slightly greater nickel removal than predicted by the model. As with silica, it appears possible that there is a small amount of adsorption of the nickel–citrate complex onto the oxide surface. In general, however, the competition model used is sufficient to account for pertinent data trends.

NITRILOTRIACETATE. Nitrilotriacetic acid (NTA) is a triprotic weak acid which is a well known sequestering agent. It forms a quadradentate chelate with Ni^{+2} of the form Ni–NTA^{-1} which is very stable (log K = 13.1). If sufficient NTA is added to solution, virtually complete (~100%) chelation of Ni^{+2} takes place in the neutral and alkaline pH range. Results of the nickel–NTA–oxide systems are shown in Figures 8 and 9 for silica and goethite, respectively.

For silica, it can be seen that the effect of NTA on the adsorption of nickel is qualitatively similar to that of citrate, although, as would be expected, the extent of the adsorption shift to the right is considerably more pronounced. A limited amount of NiOH$^+$ adsorption is deferred

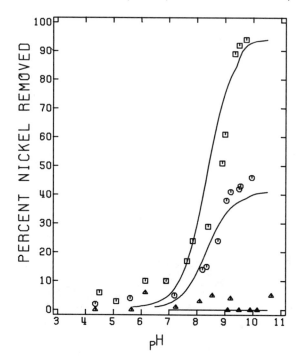

Figure 8. Nickel removal as a function of pH in the presence of silicon dioxide and NTA ($\{Ni\}_T = 10^{-4.77}M$, $I = 0.01$, $\{SiO_2\} = 29.41$ g/L (50 m²/L); (□) NTA = 10^{-6}; (○) NTA = 10^{-5}; (△) NTA = 10^{-4})

to high pH values since the nickel concentration available for adsorption is lower in the presence of the complexing ligand. At the high NTA concentration ($10^{-4}M$), neither adsorption nor precipitation occurs.

As with silica, the model is able to account well for the observations made in the goethite system. The adsorption intensity between the oxide and Ni^{+2} will never increase to a point sufficient to remove the nickel from the NTA. Even though K_{ADS} increases with pH (as the oxide surface becomes more negative), the strength of the Ni–NTA^{-1} complex will increase since the amount of deprotonated NTA (L^{-3}) will also become greater. Nickel removal in Figure 9 is slightly greater than predicted at the higher NTA concentrations, again, as with Ni–CIT^{-1}, suggesting the possibility that some adsorption of the complex takes place. However, the small discrepancies between theory and experiment in Figures 7 and 9 could just as well be explained through slight adjustments in the thermodynamic data base used.

The interactions that take place in the systems described can be viewed conveniently in a graphical manner as shown in Figure 10. Here

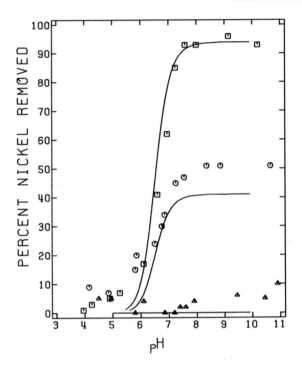

Figure 9.   Nickel removal as a function of pH in the presence of iron oxide and NTA ($\{Ni\}_T = 10^{-4.77}$M, I = 0.01, $\{FeOOH\} = 0.59$ g/L (50 m^2/L); ($\square$) NTA = 10^{-6}; ($\bigcirc$) NTA = 10^{-5}; ($\triangle$) NTA = 10^{-4})

the speciation of nickel in both the NTA–silica and NTA–goethite systems is given. Qualitatively similar diagrams for citrate and sulfate effects could also be made.

GLYCINE. Glycine, like all the amino acids, displays amphoteric properties over the pH range of interest. It can be represented as H_2L^+ having two available protons. The deprotonated ligand forms both bidentate (Ni–GLY$^+$) and quadradentate (Ni–(GLY)$_2^0$) chelates of moderate stability with Ni^{+2}, although at the glycine–nickel ratios used in this study Ni–GLY^{+1} is the dominant complex.

The results of the addition of 10^{-5} and 10^{-4} mol/L of glycine to $10^{-4.77}$M nickel are shown in Figure 11. Both theory (solid lines) and data confirm that a tenfold increase in total glycine results in an increase of one pH unit for hydroxide precipitation. At the lower concentration of glycine, insufficient complexation takes place to defer the precipitation reaction.

When α-SiO$_2$ is added to the system, the results shown in Figure 12 are obtained. Adsorption and precipitation of nickel are affected only slightly by 10^{-5}M glycine. At 10^{-4}M both the model and the data suggest

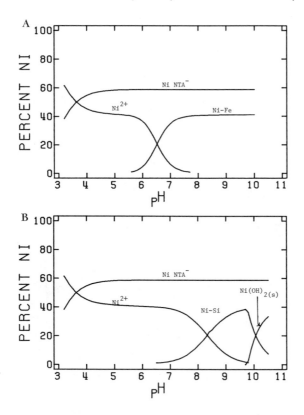

Figure 10.   Theoretical distribution of nickel in the presence of (A) iron oxide and NTA ($\{Ni\}_T = 10^{-4.77}$M, $\{NTA\}_T = 10^{-5}$ M, I = 0.01, $\{FeOOH\}$ = 0.59 g/L (50 m^2/L) and (B) silicon dioxide and NTA ($\{Ni\}_T = 10^{-4.77}$M, $\{NTA\} = 10^{-5}$M, I = 0.01, $\{SiO_2\} = 29.41$ g/L (50 m^2/L))

that only $NiOH^+$ is adsorbed up to a maximum of 15% of the total nickel at pH 8.5, with precipitation controlling above pH 10. These interactions are shown more clearly in the speciation diagram of Figure 13. The model and data are in good agreement, indicating $Ni–GLY^+$ does not adsorb specifically onto silica.

Figure 14, when compared with Figure 4, shows that the model predicts that $10^{-5}M$ glycine will not affect the removal of nickel by goethite. This is confirmed by the experimental data; however, for higher glycine concentrations the model and the data are in poor agreement. The theory used predicts that as the pH approaches the dissociation pH of the amino group of the glycine ($\sim$9), the extent of complexation is great enough to complex the nickel effectively and prevent adsorption of either Ni^{+2} or $NiOH^+$ by the goethite. The model permits adsorption of these species only. However, it is likely that there is interaction between the oxide surface and the $Ni–GLY^+$ chelate. Electrostatic attrac-

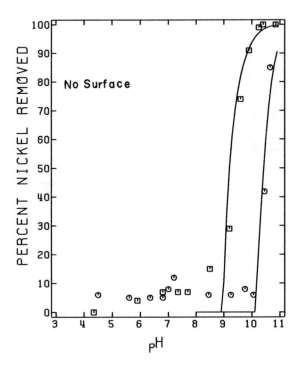

Figure 11. Nickel removal as a function of pH in presence of glycine
({Ni}$_T$ = 10$^{-4.77}$M, I = 0.01; ($\square$) GLY = 10^{-5}; ($\bigcirc$) GLY = 10^{-4})

tion would occur between the positively charged complex and the negative surface, but it would not be as great for Ni^{+2} since it has a lower charge. Using the equations in the Appendix it can be shown that coulombic forces are insufficient to account for the removals observed, and that a specific adsorption energy (ΔG_{CHEM}) of approximately -3 kcal/mol for Ni–GLY$^+$ and goethite is required.

CYANIDE. Hydrogen cyanide is weakly acidic ($pK_a = 9.3$). Cyanide ion reacts to form complexes with many transition metals. The quaternary nickel–cyanide complex, $Ni(CN)_4^{-2}$, is extraordinarily stable (log $\beta_4 = 31.8$) and dominates the simple aqueous chemistry of nickel as shown in Figure 15. At total cyanide concentrations of $10^{-4}M$ and above, hydrolysis reactions of Ni^{+2} are deferred indefinitely.

When silica is added to the nickel–cyanide system, the results are quite predictable as shown in Figure 16. At the lower cyanide concentration ($10^{-5}M$) there is enough cyanide to complex about 10% of the nickel present, so adsorption of NiOH$^+$ onto silica and precipitation of $Ni(OH)_{2(s)}$ are affected only slightly. As would be expected, at higher cyanide concentrations the nickel is made completely soluble at all pH values studied.

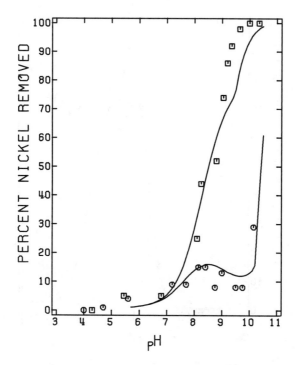

Figure 12. Nickel removal as a function of pH in the presence of silicon dioxide and glycine ($\{Ni\}_T = 10^{-4.77}M$, $I = 0.01$, $\{SiO_2\} = 29.41$ g/L (50 m²/L); ($\square$) GLY = 10^{-5}; ($\bigcirc$) GLY = 10^{-4})

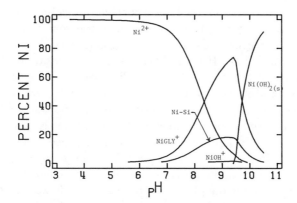

Figure 13. Theoretical distribution of nickel in the presence of silicon dioxide and glycine ($\{Ni\}_T = 10^{-4.77}M$, $\{GLY\}_T = 10^{-5}M$, $I = 0.01$, $\{SiO_2\} = 29.41$ g/L (50 m²/L))

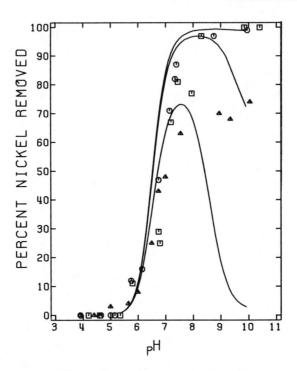

Figure 14.   Nickel removal as a function of pH in the presence of iron oxide and glycine ($\{Ni\}_T = 10^{-4.77}$M, I = 0.01, $\{FeOOH\} = 0.59$ g/L (50 m^2/L); ($\square$) GLY = 10^{-5}; ($\bigcirc$) GLY = 10^{-4}; ($\triangle$) GLY = 10^{-3})

Behavior in the goethite–nickel–cyanide system (Figure 17) is much less obvious. Agreement between model and theory exists only at the low cyanide concentration. At higher concentrations the model is able to predict only the neutral pH range in which the soluble nickel–cyanide complex does not adsorb. Of considerably greater interest is the enhanced adsorption of nickel at pH 6 and below, where no adsorption takes place in the absence of cyanide (Figure 4). The negatively charged complex will adsorb onto the now positively charged goethite surface. However, somewhat analogous to the glycine system, the favorable electrostatic gradient is not sufficient to account for the removals observed. It is apparent that $Ni(CN)_4^{-2}$ is capable of adsorbing specifically to goethite. From the equations in the Appendix, a ΔG_{CHEM} term of -7 kcal/mol can be calculated. The slight downturn in the experimental data for the $10^{-3}M$ cyanide case below pH 4 may be real, since it would be expected that the $Ni(CN)_4^{-2}$ complex will break down as HCN is formed at low pH (Figure 15). The small but significant removal that occurs above pH 9 may also be attributable to specific adsorption of the complex;

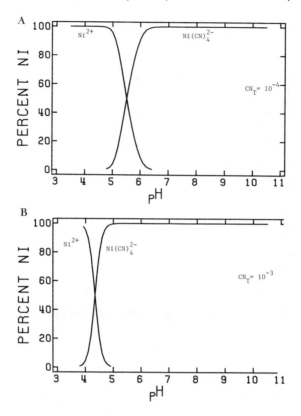

Figure 15. Theoretical distribution of nickel in the presence of (A) 10⁻⁴M cyanide (${Ni}_T = 10^{-4.77}M$, I = 0.01) and (B) 10⁻³M cyanide (${Ni}_T = 10^{-4.77}M$, I = 0.01)

however, in this case the removal is much less because it is against an unfavorable electrostatic gradient. In view of this, the fact that no adsorption occurs in the neutral pH range cannot be readily explained, in spite of agreement with the model, and therefore points to the possibility that ΔG_{CHEM} may vary with pH. Interactions among cyanide, nickel, goethite, and pH are summarized in Figure 18.

Discussion

The results of the specific adsorption of nickel onto silica and goethite are in basic agreement with similar work reported for other metals (2, 18). The James and Healy model is capable of describing the interactions well. Similarly, when complexing ligands are added to nickel–silica systems the results can be adequately explained by a competition model in

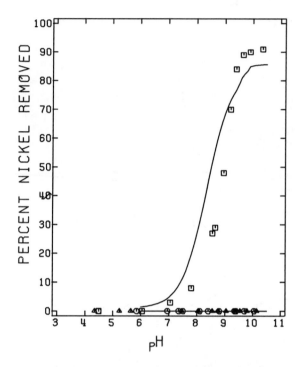

Figure 16. Nickel removal as a function of pH in the presence of silicon dioxide and cyanide ({Ni}$_T$ = 10$^{-4.77}$M, I = 0.01, {SiO$_2$} = 29.41 g/L (50 m²/L); (□) CN = 10^{-5}; (○) CN = 10^{-4}; (△) CN = 10^{-3})

which the nickel–ligand complex does not adsorb specifically to the silica surface.

For the case of nickel–ligand–goethite systems, the results for sulfate, citrate, and NTA could be accounted for by the model used, although some small degree of specific adsorption appeared to take place for citrate and NTA complexes. The evidence for this is not conclusive. However, for both nickel glycine and nickel cyanide, specific energies of adsorption of −3 and −7 kcal/mol, respectively, were calculated.

Reasons for the adsorption of a complex in one case, but not in another, are not always straightforward. In the case of silica it was shown that energetic considerations favor adsorption when NiOH$^+$ is the dominant species in solution. Any complex that defers or eliminates the hydrolysis of Ni^{+2} would be expected to affect adsorption of nickel accordingly. Specific bonding of a complex to a surface oxygen could be via a hydrogen or a dipole effect; however, the bond energies evidently are not large enough to overcome the repulsive solvation and electrostatic forces. Direct bonding of a functional group on the complex to the silicon atom is unlikely because of the very large Si–O bond energies of silica which have a large ionic character.

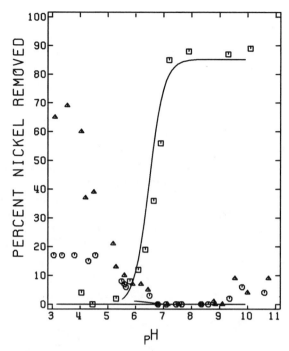

Figure 17.   Nickel removal as a function of pH in the presence of iron oxide and cyanide ($\{Ni\}_T = 10^{-4.77}$M, I = 0.01, $\{FeOOH\} = 0.59$ g/L (50 m^2/L); ($\square$) CN = 10^{-5}; ($\bigcirc$) CN = 10^{-4}; ($\triangle$) CN = 10^{-3})

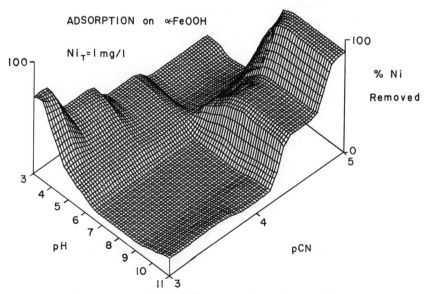

Figure 18.   Three-dimensional depiction of the nickel–cyanide–goethite system (conditions as given in Figure 17)

The bonding of complexes to the surface of goethite is far more likely than to the surface of silica because of the more neutral ZPC and higher dielectric, which act to moderate electrostatic repulsion when it exists, and because of the available d orbitals of iron which extend further from the nucleus and which are capable of accepting electrons from appropriate donor atoms of the complex. The structures of the citrate, NTA, glycine, and cyanide complexes of nickel are shown in Figure 19. Nickel–citrate and nickel–NTA both have a -1 charge, so at pH values above the ZPC (5.5) electrostatic repulsion takes place. In the case of Ni–NTA^{-1}, the only possible bonding site is via one of the water molecules attached to the nickel. This would require deprotonation of a water molecule, bringing about a greater negative charge and more electrostatic repulsion. In contrast, the Ni–CIT^{-1} chelate has an available binding site at the terminal carboxyl group. The fact that little, if any, specific adsorption of this complex takes place on goethite implies that electrostatic forces are too great to overcome.

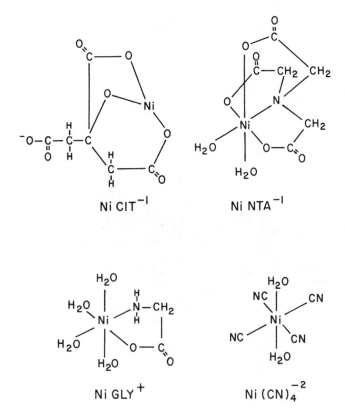

Figure 19. *Conformational structures of complexes used in this study. Data from Ref. 13 (cyanide), Ref. 22 (citrate and NTA), and Ref. 23 (glycine).*

For Ni–GLY$^+$ the electrostatic gradient is more favorable, and it appears that specific bonding to the surface through the amine group is possible either by replacement of one of the protons and formation of a covalent bond with iron or by the direct formation of a hydrogen bond with a surface oxo group. The former mechanism is more favorable for Ni–NTA^{-1} since bonding can take place without disruption of the chelate ring. However, the magnitude of ΔG_{CHEM} (−3 kcal/mol) would favor the latter mechanism.

The structure of the nickel–cyanide complex is square planar—the four Ni–C bonds being of equal length at 90° to one another. The polar waters are slightly closer to the central nickel atom (*13*). The excellent electron donor properties of the nitrogen atoms that face toward the solution imply very strongly that covalent surface bonding to the iron takes place. The relatively high ΔG_{CHEM} (−7 kcal/mol) supports this. The specific adsorption which was observed at high pH, that is, against electrostatic forces, would require an even greater value. Cyanide ion is known to form metal–cyanide chains in which both the carbon and nitrogen atoms bond covalently to neighboring metal ions. It is simplest to postulate specific adsorption of the complex to two surface sites according to the reaction

$$\equiv Fe_2(OH)_2{}^{+2} + Ni(CN)_4{}^{-2} = \equiv Fe_2-(CN)_2-Ni-(CN)_2 + 2H_2O$$

An alternate mechanism which cannot be discounted based on the evidence at hand involves the replacement of surface oxo with cyano groups, thereby altering surface properties, followed by adsorption of nickel according to

$$\equiv Fe_2(CN)_2 + Ni(CN)_4{}^{-2} + 2H^+ = \equiv Fe_2-(CN)_2-Ni-(CN)_2 + 2HCN$$

The effect on nickel sorption is the same in either case. These possibilities are undergoing further study.

Concluding Remarks

The description of natural systems is made decidedly more difficult when specific adsorption of metal complexes onto oxide surfaces occurs. It may often be possible, in a qualitative way, to predict those systems in which complex adsorption will take place; however, the quantitative extent of adsorption can be established only through direct experimentation. The number of combinations of even common metals, ligands, and surfaces is quite large, making the task a formidable one. It may be worthwhile to devise methods of estimating complex-adsorption energies similar to those used now to estimate stability constants.

Appendix. Equations of the James and Healy Model

A. Adsorption Isotherm

$$\Gamma_{M,i} = \frac{\Gamma^{MAX} K_{ADS,i} [M_i]}{1 + \sum_i K_{ADS,i} [M_i]} \quad mol/m^2 \tag{I}$$

where $K_{ADS,i} = \exp(-\Delta G^{\circ}{}_{ADS,i}/RT)$ L/mol $\tag{II}$

$$\Gamma^{MAX} = \frac{1}{\pi(x)^2 N} \quad mol/m^2 \tag{III}$$

and $\Delta G^{\circ}{}_{ADS,i} = \Delta G^{\circ}{}_{COUL,i} + \Delta G^{\circ}{}_{SOLV,i} + \Delta G^{\circ}{}_{CHEM,i}$ $\tag{IV}$

B. Coulombic Energy

$$\Delta G^{\circ}{}_{COUL,i} = z_i F \Delta \psi_x \quad J/mol \tag{V}$$

where $\Delta \psi_x = 2 \dfrac{RT}{zF} \ln \left[\dfrac{(\exp(y) + 1) + (\exp(y) - 1) \exp(-\kappa x)}{(\exp(y) + 1) - (\exp(y) - 1) \exp(-\kappa x)} \right]$

$$\tag{VI}$$

and $\qquad y = zF\psi_0/2\,RT$ $\tag{VII}$

$$\psi_0 = 2.3\,RT/F\,(pH_{ZPC} - pH) \quad V \tag{VIII}$$

$$\kappa = 0.328 \times 10^{10}\,(I)^{0.5} \quad m^{-1} \tag{IX}$$

$$x = r_i + 2r_{water} \quad m \tag{X}$$

$z_i = $ sign and charge of adsorbing ion

$z = 1$ (for 1:1 inert electrolyte)

$I = $ ionic strength

C. Solvation Energy

$$\Delta G^{\circ}{}_{SOLV,i} = \frac{(z_i e)^2 N}{16\,\pi\,\epsilon_0} \left[\left(\frac{1}{x} - \frac{r_i}{2x^2} \right) \left(\frac{1}{\epsilon_I} - \frac{1}{\epsilon_b} \right) + \frac{1}{2} \cdot \frac{1}{x} \left(\frac{1}{\epsilon_S} - \frac{1}{\epsilon_I} \right) \right]$$

$$J/mol \tag{XI}$$

where $\epsilon_I = \dfrac{\epsilon_b - 6}{1 + B\,(d\psi/dx)_x{}^2} + 6$ $\tag{XII}$

$$\text{and } \frac{d\psi}{dx_x} = \frac{-2\kappa RT}{zF} \sinh\left(\frac{zF\Delta\psi_x}{2RT}\right) \text{V/m} \qquad \text{(XIII)}$$

$$B = 1.2 \times 10^{-17}$$

D. Chemical Energy

$\Delta G°_{CHEM,i}$ = fitting term which is constant for the free and hydrolyzed species of a component. It represents the energy change resulting from chemical interaction forces such as van der Waals and H-bonding.

Acknowledgment

This research was supported in part by Grant EY-76-02-S-2727 of the Department of Energy.

Literature Cited

1. James, R. O.; Stiglich, P. J.; Healy, T. W. "Analysis of Models of Adsorption of Metal Ions at Oxide/Water Interfaces," *Faraday Discuss. Chem. Soc.* 1975, *59*, 142.
2. James, R. O.; MacNaughton, M. G. "The Adsorption of Aqueous Heavy Metals on Inorganic Minerals," *Geochim. Cosmochim. Acta* 1977, *41*, 154.
3. Hohl, H.; Stumm, W. "Interaction of Pb^{+2} with Hydrous γ-Al_2O_3," *J. Colloid Interface Sci.* 1976, *55*, 281.
4. Kurbatov, M. H.; Wood, G. B.; Kurbatov, J. D. "Application of the Mass Law to Adsorption of Divalent Ions on Hydrous Ferric Oxide," *J. Chem. Phys.* 1951, *19*, 258.
5. Greenberg, S. A. "The Chemisorption of Calcium Oxide by Silica," *J. Phys. Chem.* 1956, *60*, 325.
6. Dugger, D. L. "The Exchange of Twenty Metal Ions with the Weakly Acidic Silanol Group of Silica Gel," *J. Phys. Chem.* 1964, *68*, 757.
7. Schindler, P. W.; Fürst, B.; Dick, R.; Wolf, P. U. "Ligand Properties of Surface Silanol Groups—I. Surface Complex Formation with Fe^{+3}, Cu^{+2}, Cd^{+2}, Pb^{+2}," *J. Colloid Interface Sci.* 1976, *55*, 469.
8. Huang, C. P.; Stumm, W. "Specific Adsorption of Cations on Hydrous γ-Al_2O_3," *J. Colloid Interface Sci.* 1973, *43*, 409.
9. Shaw, D. J. "Introduction to Colloid and Surface Chemistry," 2nd ed.; Butterworths: London, 1978.
10. Grahame, D. C. "The Electrical Double Layer and the Theory of Electrocapilarity," *Chem. Rev.* 1947, *41*, 441.
11. Grahame, D. C. "On the Specific Adsorption of Ions in the Electrical Double Layer," *J. Chem. Phys.* 1955, *23*, 1166.
12. James, R. O.; Healy, T. W. "Adsorption of Hydrolyzable Metal Ions at the Oxide–Water Interface—III: A Thermodynamic Model of Adsorption," *J. Colloid Interface Sci.* 1972, *40*, 65.
13. Cotton, . A.; Wilkinson, G. "Advanced Inorganic Chemistry," 2nd ed.; Wiley–Interscience: New York, 1977.
14. Baes, C. F.; Mesmer, R. E. "The Hydrolysis of Cations"; Wiley–Interscience: New York, 1976.
15. Sillen, L. G.; Martell, A. E. "Stability Constants of Metal Ion Complexes"; The Chemical Society: London, 1964.

16. Sillen, L. G.; Martell, A. E. "Stability Constants of Metal Ion Complexes, Supplement No. 1"; The Chemical Society: London, 1971.
17. Smith, R. M.; Martell, A. E. "Critical Stability Constants, IV: Inorganic Complexes"; Plenum: New York, 1976.
18. Forbes, E. A.; Posner, A. M.; Quirk, J. P. "The Specific Adsorption of Inorganic Hg(II) Species and Co(III) Complex Ions on Goethite," *J. Colloid Interface Sci.* **1974,** *49,* 403.
19. Vuceta, J. "Adsorption of Pb(II) and Cu(II) on α-Quartz from Aqueous Solutions: Influence of pH, Ionic Strength, and Complexing Ligands," Ph.D. Thesis, California Institute of Technology, Pasadena, CA, 1976.
20. McDuff, R. E.; Morel, F. M. M. "Description and Use of the Chemical Equilibrium Program REDEQL2," Technical Report EQ-73-02, W. M. Keck Laboratory, California Institute of Technology, 1973.
21. Davis, J. A.; Leckie, J. O. "Effect of Adsorbed Complexing Ligands on Trace Metal Uptake by Hydrous Oxides," *Environ. Sci. Technol.* **1978,** *12,* 1309.
22. Chaberek, S.; Martell, A. E. "Organic Sequestering Agents"; Wiley: New York, 1959.
23. Martell, A. E.; Calvin, M. "Chemistry of the Metal Chelate Compounds"; Prentice-Hall: New York, 1952.

RECEIVED December 7, 1978.

Poliovirus Adsorption on Oxide Surfaces

Correspondence with the DLVO–Lifshitz Theory of Colloid Stability

JAMES P. MURRAY[1] and G. A. PARKS

Departments of Geology, Medical Microbiology, and Applied Earth Sciences,
Stanford University, Stanford, CA 94305

*Adsorption of enteric viruses on mineral surfaces in soil and
aquatic environments is well recognized as an important
mechanism controlling virus dissemination in natural systems. The adsorption of poliovirus type 1, strain LSc2ab,
on oxide surfaces was studied from the standpoint of
equilibrium thermodynamics. Mass-action free energies are
found to agree with potentials evaluated from the DLVO–
Lifshitz theory of colloid stability, the sum of electrodynamic
van der Waals potentials and electrostatic double-layer
interactions. The effects of pH and ionic strength as well as
electrokinetic and dielectric properties of system components
are developed from the model in the context of virus adsorption in extra-host systems.*

The transmission of viral disease in contaminated drinking water is a
well-recognized phenomenon, recently reviewed by Berg et al. (1). Virus
dissemination in soil and aquatic environments is controlled, to a large
extent, by adsorption to mineral surfaces (2). The objective of this
investigation is to determine the principal mechanisms that control
adsorption, and to build predictive models that characterize the relevant
interactions. We find good agreement between mass-action free energies
and potentials evaluated from the DLVO–Lifshitz theory of colloid

[1] Current address: Division of Applied Sciences, Harvard University, Cambridge,
MA 02138.

0-8412-0499-3/80/33-189-097$09.25/0
© 1980 American Chemical Society

stability $(3, 4, 5)$ which is built upon reasonable geometric models. Predictions can now be made regarding the kinds of environments that are conducive to transport of enteric viruses and those that are not.

Materials and Methods

Virus and Assay. Poliovirus type 1, strain LSc2ab, was obtained from F. L. Schaffer, Naval Biomedical Laboratories, Oakland, California. This virus is characteristic of enteroviruses in terms of size (27 nm in diameter), architecture (6), adsorption, the electrokinetic properties of many types (discussed later), and in terms of stability in aquatic environments (7). A diagram of this virus adsorbing to a flat surface is shown in Figure 1.

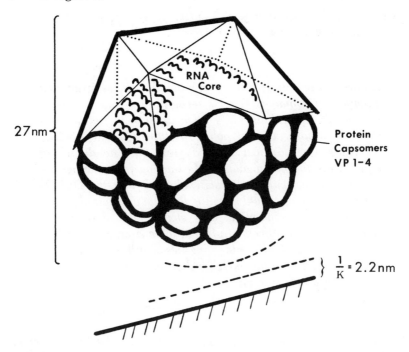

Figure 1. Cutaway model of poliovirus adsorbing to a flat surface. The virus is a highly ordered structure containing single stranded RNA packaged into an icosahedral spheroid composed of protein subunits (6). Double-layer thickness $(1/\kappa)$ at 0.02 I buffer conditions is shown to scale.

The virus was propagated and assayed in rhinovirus-adapted HeLa cells obtained from V. V. Hamparian, Medical Microbiology, Ohio State University, Columbus, Ohio. It was labeled with ^{3}H-uridine (RNA) and ^{14}C-amino acid mixture (protein), both from New England Nuclear, in the presence of 1 μg mL^{-1} actinomycin D, a gift from Merck, Sharp, and Dohme Research Laboratories. The virus was purified by differential ultracentrifugation, extraction with trichlorotrifluoroethane ($C_2Cl_3F_3$), rate zonal sedimentation in a sucrose gradient, isopycnic banding in a CsCl gradient, and exhaustive dialysis against 0.02 ionic strength (*1*), pH 7, NaCl plus NaHCO$_3$ buffer. Details of the purification sequences have been described earlier (*8*). Cramer (*9*) also presents an excellent review of classical methods in virus purification. Clean peaks in preparatory sucrose gradients demonstrate that, initially, purified virus preparations were relatively free of unincorporated label. In addition, they were not clumped but were present as discrete virus particles (virions).

Assays were done by the monolayer plaque method of Dulbecco and Vogt adapted by McClain and Schwerdt (*10*), which measures virus infectivity in terms of plaque forming units (PFU). Cells were maintained in Eagle's minimal essential medium with 10% fetal bovine serum, obtained from Grand Island Biol. Co., and Microbiol. Assoc. Maintenance medium was augmented with 400 units mL^{-1} potassium penicillin G and 100 μg mL^{-1} streptomycin sulfate to inhibit possible bacterial contamination. Serial dilutions for assays were prepared in Hanks' balanced salt solution obtained from Grand Island Biol. Co. plus 4% cell maintenance medium. According to virus-container adsorption studies (*11*) and expected albumin concentrations present in the mixture (*12*), adsorption of virus to the sides of serial dilution tubes should be blocked completely. Plaques were negatively stained with neutral red and were counted three days after infection.

To quantitate virus particle concentrations on a molal basis, which is appropriate for thermodynamic analysis of adsorption data, we measured virus particle concentrations spectroscopically (*13*) with concentrated, highly purified preparations, using an absorbance constant of 1.32×10^{13} virions per absorbance unit-cm at 260 nm (*8*). Particle to PFU ratios varied from 355 to 18621.

Radioactivity was measured by liquid scintillation techniques with a Tri-Carb model 3003, using Insta-Gel as the cocktail, both of which were manufactured by Packard Instruments.

Adsorbents. Adsorbents chosen for this study were selected to cover a wide range of dielectric and electrokinetic properties as well as chemical reactivities so that they could assist in resolving the major mechanisms operating in adsorption of virus to the various materials. Because many of them are also abundant in nature, the information should be relevant in an environmental context. α-SiO$_2$, α-Fe$_2$O$_3$, β-MnO$_2$, CuO, α-Al$_2$O$_3$, silicon metal, and aluminum metal were selected. The SiO$_2$ was Min-U-Sil 5 powder (a gift of Pennsylvania Glass Sand Corporation) which was cleaned by multiple overnight refluxing in aqua regia and

16M HNO_3, with a thorough rinse with double-distilled deionized water (ddd H_2O) between and after refluxing steps. Fe_2O_3, MnO_2, and CuO powders (<50 μm), containing rare aggregates in the case of Fe_2O_3, were ACS reagent-grade materials. They were aged for 14 days at 90°–95°C in ddd H_2O and were washed extensively to remove fine particulates. In addition, the Fe_2O_3 was roasted overnight at 800°C prior to aging in ddd H_2O. Al_2O_3, Linde A alumina powder (<20 μm), manufactured by Union Carbide, was obtained from R. L. Bassett, U.S. Geological Survey, Menlo Park, California. It was aged overnight in ddd H_2O at 25°C and rinsed once in ddd H_2O prior to use. Silicon metal semiconductor wafers, 30 ohm cm N, and 0.3 ohm cm P types, 5 cm in diameter, were manufactured by Monsanto Chemical Co. They were sterilized by rinsing with 70% aqueous ethanol and exposing them overnight to a Westinghouse Sterilamp (254 nm, $\sim$10 J m^{-2} sec^{-1}), and cleaned by rinsing with sterile 0.2M NaOH and ddd H_2O. With the exception of silicon metal, all solids were sterilized by dry heat, and were stored dry. Oxide structural forms, as designated by α,β prefixes, were confirmed by x-ray diffraction. Quantities of the various materials used in adsorption experiments fell within the following ranges: SiO_2 (0.3–3 g); Fe_2O_3 (0.1–1.4 g); MnO_2 (0.05–0.2 g); CuO (0.05–1.6 g); Al_2O_3 (0.02–0.04 g); aluminum powder (0.1–0.2 g); and aluminum metal foil (0.18–0.29 g). Exact quantities of each adsorbent in every experiment were determined by weighing in the reaction vessels.

Solution Phase. Adsorption experiments were done in aqueous solutions containing 1.093 $\times$ 10^{-2}M $NaHCO_3$ plus 9.156 $\times$ 10^{-3}M NaCl (0.02 I buffer) (some buffers were prepared with 1% less NaCl) and 1.516 $\times$ 10^{-2}M $NaHCO_3$ plus 2.941 $\times$ 10^{-1}M NaCl (0.305 I buffer) (twice physiological ionic strength) equilibrated with 5% CO_2 in air at 15° $\pm$ 2°C. Buffer compositions were determined by chemical titration and calculations that included both activity coefficients and complexation. Buffer solutions were prepared with ddd H_2O, sterilized by 0.22 μm Millipore filtration, and stored in borosilicate glass vessels. For transition metal oxide experiments, buffers were pre-equilibrated with solids for at least 4 hr prior to beginning the experiments.

In addition to the solution components just listed, $H_4SiO_4^0$ (aq), an uncharged dissolved species, should be present in buffer solutions. Equilibrium considerations of amorphous silica suggest a solubility of 10$^{-2.7}$ mol kg^{-1} for glass, an order of magnitude greater than α-SiO_2, which has an equilibrium solubility of 10$^{-3.7}$ mol kg^{-1} (14). Where Al_2O_3 was used as an adsorbent, the work of Wiese and Healy (15) on γ-Al_2O_3 solubility suggests that the total dissolved aluminum concentration in the system would be approximately 8 $\times$ 10^{-7} mol kg^{-1}. Equilibrium solubilities at pH 7 in their system are approached in approximately 1 min. Since aluminum metal has a well-defined oxide coating (16), dissolved aluminum species should be comparable in magnitude in this system as well.

In the case of transition metal oxides, solubility considerations of Fe_2O_3 and MnO_2 suggest that the total dissolved transition metal concentrations in these systems should be far too low, less than 10^{-10} mol kg^{-1} (17, 18, 19, 20), to be measured by feasible analytical procedures. Buffers equilibrated with CuO had total dissolved copper concentrations of 1.25×10^{-5} mol kg^{-1} for 0.02 I buffers and 1.47×10^{-5} mol kg^{-1} for 0.305 I buffers, as determined by atomic absorption spectrophotometry. This agrees well with values predicted from equilibrium considerations for CuO of 1.05×10^{-5} mol kg^{-1} and 1.65×10^{-5} mol kg^{-1}, respectively. The calculated values included correction for activity coefficients and carbonato, chloro, and hydroxo complexes, and have been described in more detail elsewhere (8).

Controls demonstrated that buffer constituents had no significant effect on the virus titers for time periods equal to or greater than those used for adsorption experiments.

Surface Area Determination. Specific surface areas were required to determine the fraction of adsorbent surface covered by the virus at equilibrium. This in turn is required for adsorption free-energy evaluations. BET-N_2 (21) methods were used where applicable and Kozeny (22) permeametry methods were used for confirmation. Values are listed in Table I. Although the values measured by the two methods are not directly comparable, trends shown by the two sets of values are similar. The BET method measures all surface accessible to adsorbing N_2 molecules and generally is considered to be the most reliable

Table I. Surface Areas of Adsorbents

Solid	*BET*/m^2g^{-1}	*Kozeny*/m^2g^{-1} *(duplicates)*
α-SiO_2	3.95 ± 0.79[a]	1.43, 1.44
α-Fe_2O_3	2.33 ± 0.47	0.61, 0.79
β-MnO_2	0.38 ± 0.08	0.11, 0.10
CuO	0.29 ± 0.06	0.16, 0.16
α-Al_2O_3[b]	14.6 ± 2.9	2.38, 2.34
Al metal powder[c]	(0.132)	0.037, 0.038
Al metal foil[d]	0.0012 m^2g^{-1}	
Si metal[d]	0.0039 m^2g^{-1}	

[a] The $\pm$ values are standard deviation estimates.
[b] BET data supplied by R. L. Bassett, U.S. Geological Survey, Menlo Park, California.
[c] BET value is estimated from Kozeny determination and average BET/Kozeny ratio obtained on other materials.
[d] Estimated by direct measurement times a roughness factor of 2.

measurement of surface area for adsorption studies (23); the Kozeny method measures only the surface adjacent to interconnecting flow paths in a packed column of material. In our adsorption calculations BET measurements were used except where noted.

BET measurements were not feasible for aluminum and silicon metal samples because their surface areas were below detection limits. The specific surface area of aluminum powder was obtained by correcting the Kozeny value with the average BET/Kozeny ratio of 3.5 ± 1.6. Surface areas of aluminum metal foil and silicon metal semiconductors were obtained by direct measurement, multiplied by a roughness factor of two, which was estimated on the basis of geometric considerations of etched metallic surfaces.

Theory and Results

The overall free energy of an adsorption process (ΔG_{ads}) can be broken down into various components, for example,

$$\Delta G_{ads} = \Delta G_{dl} + \Delta G_{VdW} + \Delta G_{hyd} + \Delta G_{cov\text{-}ion} \ldots - T\Delta S_0 \quad (1)$$

each of which is contributed by a different type of interaction. In this study, the term "free energy" is used to represent the Gibbs function rather than the Helmholtz function. In Equation 1, the subscripts dl, VdW, hyd, and cov-ion refer to double-layer, van der Waals, hydration, and covalent–ionic interactions, respectively. $T\Delta S_0$ refers to configurational and other entropy changes not found in other components, and will be discussed in greater detail later.

We apply this expression to the problem of virus adsorption by first determining ΔG_{ads} from mass-action arguments, and then by evaluating various components independently to determine to what degree each controls adsorption. Where quantitative estimation of the magnitudes of various terms is not possible, we examine the degree to which they contribute by critical hypothesis-and-test experiments or analogy with similar systems.

The free energy of adsorption is obtained with the site-fraction equation of state

$$\frac{\theta}{x(1-\theta)^x} = \frac{C_{eq}}{C_{H_2O}} \exp\left(\frac{-\Delta G_{ads}}{RT}\right) \quad (2)$$

derived by Dhar, Conway, and Joshi (24) from mass-action arguments. θ, x, C_{eq}, and C_{H_2O} refer to fractional surface coverage by virions, the number of interfacial water molecules displaced per virion adsorbed, the equilibrium concentrations (mol kg^{-1}) of virions in the solution phase, and the equilibrium concentration of water, respectively. C_{eq} and θ are determined from adsorption experiments, plaque assays, particle to PFU ratios, and specific surface areas (BET where available).

For evaluation of θ we have assumed that the virus would be hexagonally closest-packed in a two-dimensional array on the adsorbent surface where θ equals unity. The quantity x was estimated to be 50 from space filling models where approximately 3% of a 27-nm diameter icosahedral virion face is estimated, on the basis of globular protein structure (*12*), to approach the surface of the solid close enough to displace interfacial water molecules. If our estimate is changed to 10 or 250, instead of a most probable value of about 50, Equation 2 predicts that the ΔG_{ads} will be shifted by only $\pm$ 4 kJ mol^{-1}, which is comparable to the uncertainty given by experimental data scatter.

Adsorption–Desorption Experiments. Adsorption–desorption isotherms were measured at 15° $\pm$ 2°C in 0.02 I and 0.305 I, pH 7 buffers. Solids were added to 15-mL borosilicate screw-capped tubes, weighed, and rinsed twice with pre-equilibrated buffer. Virus (10^5 to 10^8 PFU) plus 10 mL of pre-equilibrated buffer were then added and the tubes covered with flamed aluminum foil to prevent leakage. Adsorption was allowed to proceed as samples were rotated at 0.5 or 1 rpm for 2 hr and then centrifuged at 1000 rpm for 10 min in an IEC PR-2 refrigerated centrifuge. Supernatants were sampled for residual infectivity and radioactivity and then aspirated. The volumes of buffer remaining about the solids were determined and then fresh buffer not containing virus was added to the reaction vessels. The reactions were allowed to proceed in the reverse direction for 2 hr, and then were centrifuged and sampled again as detailed above.

With aluminum metal foil, experiments were done in T-25 polystyrene tissue-culture flasks obtained from Corning Glass Works; the reaction times were extended to 10 hr for both the adsorption and desorption steps. With silicon metal semiconductors, experiments were done in 60 $\times$ 15 mm tissue-culture petri dishes in which the chips were attached with polystyrene cement. These dishes were placed in a sealed container to prevent loss of CO_2, and the reaction times were changed to 4 hr for each step. In both the aluminum and silicon metal experiments, the reaction vessels were rocked gently (mechanically) at 1.5 Hz to insure good contact between solid and virus. In all experiments, volumes were determined by weighing or pipetting, and density corrections were made where appropriate. Aseptic (sterile) techniques were used throughout.

The 2-hr time periods selected for most adsorption experiments were based on the observation that, in general, 1 hr is required to adsorb isoclonal poliovirus from isotonic solutions to cell monolayers for assay.

We extended this 1 hr period by a factor of two to guarantee that equilibrium would be approached closely. On solids where adsorption took place on sheets of material the times were extended to account for the increase in mean free path of the adsorbing virus.

Adsorption–desorption experiments such as those just described are far more appropriate than simply following the progress of an adsorption reaction to determine whether the reaction has reached equilibrium. Simply following the virus concentration in solution as a function of time until the titer of virus free in suspension is no longer changing is insensitive to possible hysteresis. Hysteresis would be expressed by marked differences in the titer in solution, hence also that adsorbed, depending on whether adsorption or desorption is followed. This phenomenon occurs quite frequently in problems concerning gas adsorption (25). We might have missed the remarkable difference in adsorption and desorption characteristics of virus on aluminum metal had the time-course approach been taken.

However, if the same value is obtained when an adsorption and desorption experiment is performed, then we know that the time period selected for the experiments is longer than that required for the reactions to approach equilibrium closely, and that the value obtained is an accurate representation of the free energy difference between virus adsorbed and virus in the solution phase.

Residual radioactivity data corresponded to residual plaque assay data for adsorption–desorption experiments involving SiO_2 and silicon metal, the weakest adsorbents. On other solids, virus was adsorbed to such a large extent that if infectivity data were to correspond exactly to radioactivity data, the amount of ^3H-RNA and ^{14}C protein left in solution would be far below detectable levels. However, a barely detectable but statistically significant amount of these materials remained in the supernatants. Because viruses are thermodynamically unstable and spontaneously degrade at finite, temperature-dependent rates (25), we suspect this residual probably represents degradation products of a formerly infectious virus which adsorb considerably less strongly than the virus itself. This observation states the importance of using plaque assay data, as we have used here, to evaluate virus adsorption.

For all experiments, controls demonstrated that the level of agitation used in adsorption experiments as well as the buffers themselves (containing various dissolution products) had no significant effect on virus infectivity.

Dissolution products of the various adsorbents could alter the adsorption characteristics of solids or virus during adsorption reactions. Effects on the solids are completely accounted for because zeta potentials of the solids as well as adsorption–desorption reactions were carried out with pre-equilibrated buffers. The principal solids for which free energies were used in resolving the principal mechanisms of adsorption were SiO_2, Fe_2O_3, and MnO_2. Dissolution products of SiO_2 are predominantly uncharged, and those of Fe_2O_3 and MnO_2 are so low in concentration, as suggested earlier, that possible effects on the zeta potential of the virus, which could influence adsorption, do not appear to be reasonable. It is possible that trace dissolution products of aluminum, Al_2O_3, and CuO could influence virus adsorption to some extent, although no effects of this sort were seen in regard to enhancing virus adsorption to container walls by pre-equilibrated buffers. Further developments concerning this possibility have been left for future investigation or are discussed in later sections where appropriate.

Isotherms and Adsorption Free Energies

Adsorption and desorption data points are presented in Figures 2–4 and are well matched with linear, least-squares fit isotherms, namely, θ vs. C_{eq} plots. Linear isotherms are expected for situations having low surface coverage such as ours (seen in very low values of θ exhibited by the data), where the probability of adsorbate–adsorbate interactions (aggregation) is small. It is conceivable that actual variation of θ with C_{eq} may be more complex than a linear function; however, data scatter is such that more detailed analysis of adsorption–desorption data is not justified. In any case, linearity, as characterized by 1:1 slope on the log–log plots, fits the data quite well. These plots are required for presentation of data that span six orders of magnitude in each direction.

In these diagrams, log C_{eq} values of -17 correspond to titers of approximately 5 PFU mL^{-1}, and values of -11 correspond to titers of about 5×10^6 PFU mL^{-1}. Correspondence is only approximate here because of the variation of particle to PFU ratios from preparation to preparation. To represent the difference in the relative strengths of the materials used in this study, we normalized the adsorption data by evaluating expected equilibrium titers at a surface coverage of 10^{-5}, and one particle to PFU ratio of 10^3. These data are presented in Table II. We see titer differences of at least five orders of magnitude between the weakest and strongest adsorbents.

Within the normal range of data scatter, adsorption and desorption data points are coincident. This suggests that the reactions are reversible and approach equilibrium in all cases, except with CuO at high ionic strength and with aluminum metal. Free energies of adsorption (ΔG_{ads}) for the oxides are evaluated by applying Equation 2 to plotted isotherms at infinitesimally low ($\leq 10^{-5}$) surface coverage (*see* Table III). Standard deviations are calculated from data scatter about regressions.

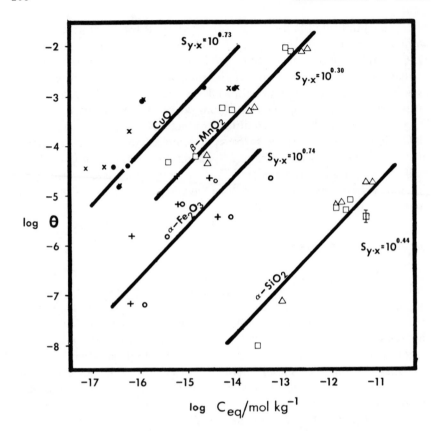

Figure 2.   Adsorption ($\triangle,\bigcirc,\bullet$)–desorption ($\square,+,\times$) isotherms of polio-virus in pH 7, 0.02 I NaCl + NaHCO₃ buffer.

θ represents the fraction of adsorbent surface covered by virus particles and C_{eq} is the molal concentration of virus particles in supernatants of adsorption experiments. Log C_{eq} of -17 corresponds to approximately 5 PFU mL^{-1} and log C_{eq} of -11 corresponds to approximately 5×10^6 PFU mL^{-1}. Approximate nature of the correspondence of concentrations to PFU is a result of variation of particle to PFU ratios of different preparations. $S_{y \cdot x}$ is the standard deviation of the scatter of log C_{eq} values about regression lines, calculated with $n - 2$ degrees of statistical freedom.

On SiO_2, Fe_2O_3, and Al_2O_3, kinetic analysis of multiple extraction data shows no statistically significant effect on the titer of a virus preparation adsorbed to the solids for 2 hr (26). However, on MnO_2 and CuO we see that approximately 85–95% appear to be degraded during adsorption, but this is not reflected in the reversibility and equilibrium conditions applied to our isotherms (26). Apparently this is attributable to the fact that degradation reactions appear to progress quite rapidly at first, and only slowly after 2 hr. Therefore the amount degraded between adsorption and desorption steps may be small. An extended discussion

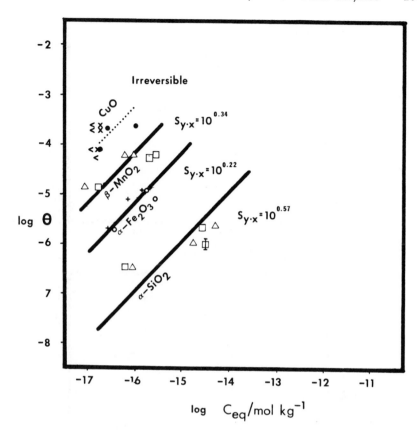

Figure 3.   Adsorption ($\triangle$,$\bigcirc$,$\bullet$)–desorption ($\square$,$+$,$\times$) isotherms of poliovirus in pH 7, 0.305 I buffer. Additional information is included in the caption of Figure 2. ($<$) indicates points below detection limit.

Table II.   Predicted Log Concentrations of Poliovirus Infectivity in Equilibrium with Various Solids at pH 7.0

Solid	Log Equilibrium Virus Titer[a]	
	0.02 I[b] (PFU mL⁻¹)	0.305 I (PFU mL⁻¹)
α-SiO$_2$	7.15 ± 0.43	4.07 ± 0.55
α-Fe$_2$O$_3$	3.56 ± 0.71	2.34 ± 0.21
β-MnO$_2$[c]	2.50 ± 0.29	1.39 ± 0.36
CuO[c]	1.30 ± 0.71	$\rightarrow - \infty$
α-Al$_2$O$_3$	3.66 ± 0.26	ND
Al metal[c]	$\rightarrow - \infty$	ND
Si metal	≥ 5.90	ND

[a] Assuming $\theta = 10^{-5}$, virus particle to PFU ratio $= 10^3$, and standard temperature and pressure conditions. $\pm$ values are standard deviations.
[b] This ionic strength is characteristic of that found in many secondary wastewater effluents.
[c] Degradation that took place during adsorption is included here.

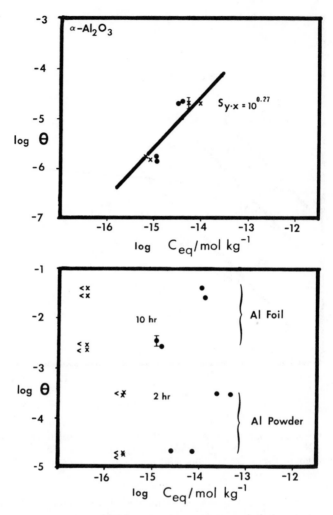

Figure 4. Adsorption (●)–desorption (×) isotherms on Al_2O_3 and Al metal in pH 7, 0.02 I NaCl + $NaHCO_3$ buffer. Reaction with the metal is strong and irreversible. Additional information is included in the caption of Figure 2. (<) indicates points below detection limit.

of virus degradation by inorganic surfaces is presented elsewhere (26). Accounting for this degree of degradation on MnO_2 and CuO, we find that the adsorption free energies for these materials listed in Table III may be about 5 kJ mol^{-1} too negative. Although this is approximately twice the normal range of data scatter found about isotherms used to evaluate free energies, it is not materially large enough to affect the conclusions of this study, as seen by examining the measured free energy values.

Table III. Free Energies of Adsorption of Poliovirus on Oxide Surfaces as Determined by Mass-Action Equilibria

Solid	$\Delta G_{ads}/kJ\ mol^{-1}$ (0.02 I)	$\Delta G_{ads}/kJ\ mol^{-1}$ (0.305 I)[a]
α-SiO$_2$	-32.4 ± 2.4[a]	-50.0 ± 3.1
α-Fe$_2$O$_3$	-52.9 ± 4.0	-59.9 ± 1.2
β-MnO$_2$[b]	-58.9 ± 1.7	-65.3 ± 2.1
CuO[b]	-65.8 ± 4.0	strong[c]
α-Al$_2$O$_3$	-52.3 ± 1.5	ND[d]
Al metal	very strong[c]	ND
Si metal	≥ -39.5	ND

[a] $\pm$ values are standard deviations.

[b] Virus degradation on these solids (*26*) indicates that free energy values tabulated here may be ~5 kJ mol^{-1} too negative.

[c] Reactions are clearly irreversible. Both radioactivity and infectivity data show adsorption to be quite strong.

[d] Not determined.

For silicon metal we estimated a limiting value for the ΔG_{ads} because the quantity of virus adsorbed in this system was too low to be measured accurately. On aluminum metal, the amount of infectious virus, ^{3}H-RNA, or ^{14}C protein was quite small compared to the amount added initially. Only in one case during desorption experiments was any detectable virus found, and here the supernatant contained only 4 PFU mL^{-1}. This is clearly indicative of strong adsorption and irreversibility of the reaction. The significance of these findings is presented in the following sections of this chapter.

Application of DLVO Theory. Our approach to determine the contribution of double-layer interaction and van der Waals potentials to ΔG_{ads} involves comparing differences in the magnitudes of ΔG_{ads} found on the same solid but with different solution conditions, to potentials (U), or theoretical free energy components, evaluated from the DLVO–Lifshitz theory of colloid stability.

To evaluate DLVO–Lifshitz potentials we approximate the highly complex interaction of a spherical-icosahedral, deformable virus adsorbing to a real surface (Figure 1) with sphere-plate models (Figure 5). The complex real interactions, even if they were well defined, cannot be quantified by present ab initio quantum mechanical procedures.

To adjust for errors introduced by modeling these highly complex interactions with simple geometries, we consider the characteristic interaction-separation distances in expressions derived for the models to be adjustable parameters. The DLVO–Lifshitz potentials are evaluated for characteristic distances found to fit data for SiO$_2$, where ΔG_{ads} and U determinations are most accurate. These distances are then used for all other solids. Since the assumptions used in the double-layer and van der Waals models (described in following sections) are quite

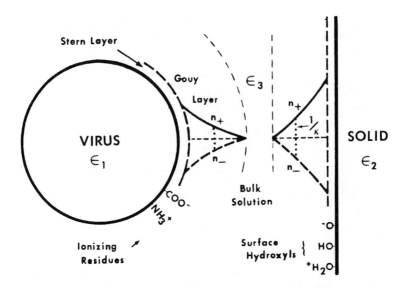

Figure 5.   Sphere-plate model of virus adsorbing to a flat surface, show-ing inner (Stern) layer and outer (Gouy) double layers.

The Stern layer consists of bound electrolyte and oriented water molecules at both surfaces. Gouy layers are localized regions containing an excess of counter-ions (n_+) and a deficit of coions (n_-) compared to concentrations in the bulk solution. Overlap of the double layers at close approach of the virus and solid is responsible here for a repulsive component of the adsorption free energy according to the DLVO theory. The functions $\epsilon_{1,2,3}$ represent complex dielectric susceptibility functions of the virus, solid phase, and immersion medium, and are characteristic of their respective electronic and molecular structures. These functions, according to the Lifshitz theory, determine the magnitudes of van der Waals components of the free energy. These components vary substantially with the nature of the substrate, and appear to play the most significant role in virus adsorption to materials examined in this study.

different, we have selected different values for the characteristic distance for each model. This will be discussed in greater detail in later sections.

The test of the models is then twofold: (1) Do the differences in DLVO potentials account for the observed differences in free energy for all solids? (2) Are the values selected for characteristic distances realistic, considering what is understood about the atomic structure of solid–liquid interfaces, for example, 0.1 to 1 nm? If these conditions are met we have evidence that the DLVO theory explains virus adsorption to various inorganic surfaces, provided that equally convincing evidence is not obtained when considering other types of interaction.

Electrostatic Component of ΔG_{ads}.   A major contribution to ΔG_{ads} in colloidal systems arises from the interaction of electrostatic double layers that form about charged particles immersed in electrolyte solutions. Oxide and protein surfaces principally develop charges by ionizing prototropic groups on their surfaces (27).

Double-layer interaction potentials (U_{dl}) are evaluated with Gregory's (28) constant-charge, linear-superposition assumption (LSA) expression

$$U_{dl} = N_a \frac{128 \ (a_v + d_s) \pi n k_B T}{\kappa^2} \ \gamma_1\gamma_2 \exp \ (-\kappa d_{es}) \tag{3}$$

for sphere-plate boundary conditions. N_a is Avogadro's number, a_v is the virion radius (13.5 nm), here augmented by 0.5 nm to account for the thickness of the Stern layer d_s (29), n is the cation or anion number per unit volume, k_B is Boltzmann's constant, T is the absolute temperature, and d_{es} is the characteristic electrostatic separation distance, here taken to be zero at coincidence of the Stern surfaces of the interacting particles. The functions $\gamma_{1,2}$ are equal to tanh $(y_{1,2}/4)$, where $y_{1,2}$ are the reduced potentials $ze\psi_{1,2}/k_B T$, z being the charge number per ion, e the magnitude of electron charge, and $\psi_{1,2}$ the Gouy–Chapman surface potentials. Here we define $\psi_{1,2}$ as equal to the Stern plane potentials and assume them to be approximated closely by zeta potentials (ζ) based on Smith's work (29) on oxide electrokinetics. κ is the well-known Debye screening inverse length $(2e^2nz^2/\epsilon_0\epsilon_v k_B T)^{1/2}$, where ϵ_0 is the relative (to vacuum) dielectric constant measured at zero frequency and ϵ_v is the dielectric permitivity of free space.

We selected Gregory's expression because it predicts potentials intermediate between those predicted by expressions based on exact constant-charge and exact constant-potential assumptions. This corresponds with observations of Visser (30), who demonstrated that the experimental adhesion force variation for polystyrene latex spheres on cellophane as a function of ionic strength is intermediate between force variation calculated with the DLVO approach and these two limiting assumptions.

A recent development in double-layer interaction theory, known as the surface-regulation approach (31), also predicts interaction potentials that are intermediate in value. Data required for application of the regulation approach are not available.

Zeta potentials for evaluating double-layer interaction potentials were determined for conditions paralleling those used in 0.02 *I* adsorption–desorption experiments. Readings for the zeta potentials were taken immediately after rinsing the solids in pre-equilibrated 0.02 *I* buffer, at 2 hr, and at 5 hr, and were then averaged. Significant, progressive variation of ζ with time was not observed. Conventional microelectrophoresis techniques and a Zeta-Meter were used for these measurements. Since zeta potentials were not measurable directly at 0.305 *I*, we estimated their magnitudes at this high ionic strength using Hunter and Wright's correlation (32). This correlation shows that zeta potentials of various oxide surfaces vary linearly with the logarithm of the counterion concentration, the $\zeta = 0$ intercept occurring at a concentration of approximately $1M$ in the absence of specific adsorption.

In our system, we have some indication that carbonate or bicarbonate does adsorb specifically to Fe_2O_3 and CuO (8). However, the activity of these species is essentially the same in both buffer systems. Because Hunter and Wright's correlation corrects for the expected decrease in ζ with increasing electrolyte, we should get reasonable values of ζ at the high ionic strength because we are keeping the bicarbonate and carbonate activities constant and are changing only the concentrations of inert electrolyte.

The zeta potential of the virus at 0.02 I was measured by zonal electrophoresis in a sucrose gradient. In its evaluation we corrected the retarded, relaxed Smoluchowski equation (33) for the sucrose-dependent viscosity gradient (8). Again, the Hunter and Wright correlation was used to estimate the value of the zeta potential at 0.305 I.

Zeta potentials of the oxides and virus are presented in Table IV. Using these values, double-layer interaction potentials were evaluated and are presented in Table V. They are repulsive in all cases.

We now can determine whether this method evaluates electrostatic adsorption free energy components by comparing the differences in mass-action free energies at the two ionic strengths with equivalent differences in double-layer interaction potentials. We assume here that other components do not change significantly as a function of ionic strength. The comparison is shown in Table VI and gives excellent agreement.

The characteristic model distance d_{es} is taken to be zero at coincidence of the Stern surfaces, implying that the actual virus-to-solid separation distance would be approximately 1 nm on the basis of Smith's work (29). This is quite reasonable, and all solid–virus interaction potentials and free energy differences are matched with a single model distance.

Our estimates of the zeta potential at high ionic strength could contain some degree of inaccuracy. However, the magnitudes of these potentials, and therefore the magnitudes of the double-layer interaction potentials calculated at the high ionic strength, are nonetheless quite low. This suggests that possible errors in the differences of the two potentials at the two ionic strengths introduced by Hunter and Wright's correlation are minimal and can be neglected. These differences are the key variables used for the fit of the DLVO theory.

Now that the degree to which electrostatic double layers control the adsorption of virus on our solid surfaces is known, we will investigate the types of interaction responsible for contributing the principal residual free energy. Lipatov and Sergeeva (34) present substantial evidence

Table IV.   Zeta Potentials of Oxides and Virus in pH 7 NaHCO₃–NaCl Solutions

	0.02 I (mV)	0.305 I[b] (mV)
α-SiO$_2$	-44.5 ± 8.5[a]	-13.4 ± 2.6
α-Fe$_2$O$_3$	-28.8 ± 2.5	-8.7 ± 0.8
β-MnO$_2$	-20.7 ± 5.9	-6.2 ± 1.8
CuO	-17.6 ± 6.1	-5.3 ± 1.8
Poliovirus 1 LSc2ab	-5.9 ± 0.9	-1.8 ± 0.3

[a] The $\pm$ values are measured standard deviations for oxides, and estimates for the virus according to the accuracy of determining electrophoretic migration distances in duplicate experiments.
[b] Estimated from Hunter and Wright's (*32*) correlation. Here standard deviations are estimated from variations taken at the lower ionic strength.

Table V.   Double Layer Interaction Potentials Between Virus and Solid Phases According to the LSA Expression

Solid	U$_{dl}$(0.02 I)/kJ mol^{-1}	U$_{dl}$(0.305 I)/kJ mol^{-1}
α-SiO$_2$	18.9 ± 4.8[a]	1.9 ± 0.5
α-Fe$_2$O$_3$	12.7 ± 2.2	1.2 ± 0.2
β-MnO$_2$	9.2 ± 3.0	0.9 ± 0.3
CuO	7.9 ± 3.0	0.7 ± 0.3

[a] The $\pm$ values are standard deviations.

Table VI.   Comparison of Differences in ΔG_{ads} and U$_{dl}$ at 0.02 I and 0.305 I

Solid	ΔG_{ads}(0.02 I − 0.305 I)/kJ mol^{-1}	U$_{dl}$(0.02 I − 0.305 I)/kJ mol^{-1}
α-SiO$_2$	17.6 ± 4.0[a]	17.0 ± 4.8
α-Fe$_2$O$_3$	7.0 ± 4.3	11.5 ± 2.2
β-MnO$_2$	6.3 ± 2.6	8.4 ± 3.0
CuO	—[b]	7.2 ± 3.0

[a] The $\pm$ values are standard deviations.
[b] Cannot be determined because the isotherm at 0.305 I is irreversible.

that an increase in the molecular weight of polymers is strongly correlated with an increase in their tendency to adsorb. They attribute this increase to electrodynamic van der Waals interactions. In the following section, we will determine the extent to which these interactions govern virus adsorption.

Electrodynamic Component of ΔG_{ads}. Since electrostatic components of ΔG_{ads} are repulsive for all experimental cases presented here, yet the virus is quite strongly adsorbed, other strong attractive components must also be involved in adsorption. Viruses are considered to be disperse colloidal systems, and in these systems, colloid stability theory suggests electrodynamic van der Waals interactions to be of major importance (30). A combined Lifshitz–Hamaker approach (30) is used to investigate van der Waals potentials.

Dzyaloshinski, Lifshitz, and Pitaevski (5) have quantified van der Waals interactions using the methods of quantum field electrodynamics, characterizing the potentials by summation over dielectric susceptibility functions over all frequencies. This approach generally is referred to as the Lifshitz theory. The van der Waals force at small separations (less than 5 nm), at zero temperature (K), and between two materials (designated by subscripts 1 and 2) immersed in a third medium (designated by subscript 3) can be written

$$F_{132} = \frac{\hbar}{16\,\pi^2 d_{ed}{}^3} \int_0^\infty \int_0^\infty \chi^2 \left[\frac{(\epsilon_1 + \epsilon_3)\,(\epsilon_2 + \epsilon_3)}{(\epsilon_1 - \epsilon_3)\,(\epsilon_2 - \epsilon_3)} \exp\chi - 1 \right]^{-1} d\xi d\chi \tag{4}$$

for semi-infinite plate–plate boundary conditions.

In this expression, $\hbar$ is Planck's constant $h/2\pi$, d_{ed} is the interaction separation distance, χ is a complex integration variable (5), and $\epsilon_{1,2,3}$ are complex dielectric susceptibility functions $\epsilon = \epsilon'(\omega) + i\epsilon''(\omega)$ as a function of the complex frequencies $\omega = \omega_R + i\xi$.

Evaluating the integral and describing the dielectric susceptibility as a function of the imaginary frequency component (5), we have

$$F_{132} = \frac{\hbar}{8\pi^2 d_{ed}{}^3} \bar{\omega}_{132} \tag{5}$$

where

$$\bar{\omega}_{132} = \int_0^\infty \frac{[\epsilon_1(i\xi) - \epsilon_3(i\xi)]\,[\epsilon_2(i\xi) - \epsilon_3(i\xi)]}{[\epsilon_1(i\xi) + \epsilon_3(i\xi)]\,[\epsilon_2(i\xi) + \epsilon_3(i\xi)]} d\xi \tag{6}$$

Because the integrated form over χ has the same form as the older Hamaker relations, which have been solved for various different geometric boundaries, the Lifshitz theory can be used to obtain Hamaker coefficients which can then be used to calculate potential energy differences.

The Hamaker formulation for plate–plate geometries is

$$F_{132} = \frac{A_{132}}{6\pi d_{ed}} \tag{7}$$

and the three-body Hamaker coefficient (A_{132}) is related to the Lifshitz frequency integral by

$$A_{132} = \frac{3\hbar}{4\pi} \bar{\omega}_{132} \tag{8}$$

for short-distance, unretarded interactions. Because nearly all the van der Waals potentials in our system are developed at very short separation distances compared to those where retardation becomes significant (5), this treatment is appropriate for our system.

In principle, all that is required for a direct Lifshitz solution is the variation of ϵ with ω for all materials involved and their precise spatial arrangements. The variation of ϵ with ω can be closely approximated from spectroscopic and dielectric information.

However, the spectroscopic, dielectric, and geometric information required for exact ab initio calculation of virus–solid interactions using Lifshitz theory is not available. Nevertheless, we can estimate their magnitudes with the sphere-plate Hamaker expression (35)

$$U_{\text{vdw}} = -N_a \frac{A_{132}}{6} \left[\frac{2a_v(a_v + d_{\text{ed}})}{d_{\text{ed}}(d_{\text{ed}} + 2a_v)} - \ln \frac{(d_{\text{ed}} + 2a_v)}{d_{\text{ed}}} \right] \quad (9)$$

where A_{132} is the complex Hamaker coefficient and d_{ed} is the characteristic interaction separation distance (0.1 nm for this model).

The complex Hamaker coefficients are predicted from individual self-interacting Hamaker coefficients (for example, A_{11}) evaluated by Visser (30) from direct Lifshitz solutions or Ninham and Parsegian's (36) macroscopic approximations. We used combining rules derived from thermodynamics and the Lifshitz theory by Bargeman and Van Voorst Vader (30, 37)

$$A_{132} = \frac{3h}{16\pi^2 b_i} \frac{(A_{11}^{1/2} - A_{33}^{1/2})(A_{22}^{1/2} - A_{33}^{1/2})}{A_{33}^{1/2}(A_{11}^{1/2} - A_{22}^{1/2})} \ln \left[\frac{1 - \frac{16\pi^2 b_i}{3h}(A_{22}A_{33})^{1/2}}{1 - \frac{16\pi^2 b_i}{3h}(A_{11}A_{33})^{1/2}} \right]$$

$$(10)$$

and

$$A_{131} = \frac{(A_{11}^{1/2} - A_{33}^{1/2})^2}{1 - \frac{16\pi^2 b_i}{3h}(A_{11}A_{33})^{1/2}} \quad (11)$$

where b_i is a fitting parameter equal to approximately 0.32×10^{-16} sec^{-1} used to predict these coefficients. Visser's values of A_{11} and our coefficients are shown in Table VII. Virus and polystyrene are assumed to have similar dielectric properties on the basis of atomic composition and molecular structure. Unfortunately, available dielectric and spectroscopic information does not allow discrimination among transition metal oxides.

We can compare potentials determined from these values and the Hamaker sphere-plate expression to residual free energy components (ΔG_{res}) obtained by subtracting double-layer contributions (U_{dl}) from

Table VII. Individual Self-Interaction Hamaker Coefficients Taken from Visser (30) and Calculated Mixed Coefficients

Material	$A_{11}/erg \times 10^{-13}$	Theory
Water	4.38	Lifshitz
Polystyrene (virus)	6.15 to 6.60	Lifshitz
Oxides, etc.	10.6 to 15.5	Lifshitz
Metals	16.2 to 45.5	Lifshitz
Quartz	8.0 to 8.8	macroscopic approximation

Materials	$A_{132}/erg \times 10^{-13}$	Theory
Quartz–H_2O–virus	0.33 to 0.49	Bargeman and Van Voorst Vader
Oxides–H_2O–virus	0.53 to 1.06	Bargeman and Van Voorst Vader
Metals–H_2O–virus	0.91 to 2.97	Bargeman and Van Voorst Vader
Polystyrene–H_2O–virus	0.17 to 0.26	Bargeman and Van Voorst Vader

the overall ΔG_{ads}, to determine to what extent van der Waals interactions can account for the remaining potentials controlling adsorption. The comparison is presented in Table VIII, and again, the agreement is excellent. This suggests that most of the free energy of adsorption can be accounted for by addition of van der Waals and double-layer interaction potentials.

On silicon metal we do not observe significantly stronger adsorption potentials than on SiO_2. The Lifshitz theory predicts that we should (30). Silicon metal is known to have an amorphous oxide coating (38), which for material with a history similar to ours was shown by ellipsometry to be 3–6 nm thick.

Table VIII. Comparison of Lifshitz–van der Waals Potentials to Residual Free Energy Components after Correcting for Double Layer Repulsion

Solid	$U_{vdw}(Lifshitz)/kJ\ mol^{-1}$	$\Delta G_{res}/kJ\ mol^{-1}$
α-SiO_2	−43.0 to −63.9	−51.9 ± 3.2[a]
α-Fe_2O_3	−69.1 to −138	−61.0 ± 1.2
β-MnO_2	−69.1 to −138	−66.1 ± 2.1
CuO	−69.1 to −138	−73.7 ± 5.0
Al metal	−118 to −387	very strong[b]
Polystyrene	−22.2 to −33.9	weak[c]

[a] The ± values are standard deviations.
[b] Irreversible reaction, ΔG_{ads} could not be determined.
[c] No adsorption measurable on polystyrene flasks.

On the other hand, aluminum metal also has an oxide coating. Smith (29) found that the layer is approximately 5 nm thick on ethanol–$HClO_4$-electropolished metal and about 20 nm thick on acid–dichromate-etched metal (also determined by ellipsometry). However, adsorption of virus on aluminum is considerably stronger than on Al_2O_3 and is quite characteristic of metals as predicted by the Lifshitz theory.

The inconsistency in the adsorptive properties of these materials may involve local "thin spots" on aluminum metal, allowing significant dipolar contributions from the metal itself to interact with the virus, or may involve kinetic limitations in the approach of the virus to silicon metal surfaces. This question should be clarified in further investigations.

The value obtained for our characteristic model distance, 0.1 nm, is smaller than the distances used by other investigators, for example, 0.2 nm (39) and 0.4–1 nm (30), and is smaller than the distance of approximately 1 nm which matches Gregory's LSA evaluation of our electrostatic double-layer interactions. Because the assumptions used to derive the Hamaker–Lifshitz approach and the LSA expressions are substantially different, we cannot expect the two fitted-model parameters to be the same. In addition, we have good physical reasons to suspect that our distance parameters for the Hamaker sphere-plate model will be smaller than required for the LSA model. Among these is the damping of van der Waals potentials by ions located between the virus and the solid phase (40). Furthermore, because the virus is not a perfect sphere (it is probably deformable) and the surfaces to which it adsorbs are not planar, we should expect greater contact area than that given by the sphere-plate model. This should reduce the magnitude of the apparent fitting distance. Since the van der Waals potentials are probably more sensitive to distance than double-layer potentials, we can expect the reduction in fitted-model distance to be greater for van der Waals interactions.

An increase in ionic strength should reduce the magnitudes of van der Waals potentials to some extent (40). The amount of reduction expected cannot be predicted theoretically with much accuracy because the detailed structure and composition of the region between the virus and surface in the adsorbed configuration are unknown. Future investigations may show this phenomenon, especially at exceptionally high ionic strengths.

Configurational and Other Entropy Changes Not Found in Other Potentials

The rotational and vibrational entropy of an adsorbed virion is probably somewhat less than it is in the free state. Because of this, and because the order of the system will possibly increase when the virus is adsorbed, we expect there will be other entropy contributions to ΔG_{ads} and that the sign of ΔS_0 will be negative, although its magnitude cannot be predicted at this time.

We do not see the effects of this term in matching ΔG_{res} to van der Waals potentials, but experimental error and the uncertainty in dielectric data and interaction geometry could accommodate significant ΔS_0 contributions. It is possible that the temperature dependence of ΔG_{ads} could resolve magnitudes of ΔS_0, but only if its magnitude is large enough so that it could not be accounted for by the uncertainty in predicting double-layer, van der Waals, and hydration terms.

Hydration Energy. In our mass-action expression, we estimated that 50 interfacial water molecules were displaced by adsorption of one virion. Energy required for this transfer should be considered. LeNeveu et al. (*41*) indicate that the chemical potentials of hydration-associated water molecules adjacent to surfaces of phospholipid bilayers vary from 0.2 kJ mol^{-1} at 0.4-nm bilayer separation to 0.01 kJ mol^{-1} at 1-nm bilayer separation relative to the chemical potential of bulk water. If we assume that the hydration energies will be similar for our systems, and the displaced water molecules are located between these distances of separation, we obtain an additional 0.5–10 kJ mol^{-1} repulsive energy resulting from partial dehydration during adsorption. Again, uncertainty in predicting double-layer and van der Waals contributions could easily accommodate significant hydration terms.

Possible Contributions from Other Interactions

Even though the DLVO theory explains the viruses on various surfaces quite well, it might be argued that the fit is possibly coincidental. For this reason we examined other possible interaction mechanisms to determine whether adsorption can be explained in other ways. Other mechanisms considered include electrostatic-induced image interactions, covalent–ionic interactions, hydrophobic interactions, and hydrogen bonding.

Ion-exchange concepts and formulations are inappropriate for virus adsorption because of the larger size of viruses compared to ions. Electrostatic interactions that are responsible for ion exchange in smaller scaled systems are manifested in our system in terms of double-layer interactions.

Electrostatic Induction Potentials. Since many of our materials are semiconductors (Fe_2O_3, MnO_2, CuO, silicon metal), and aluminum metal is a conductor (*42–46*), the possibility that potentials arising from electrostatic induced-image forces could contribute significantly to adsorption free energies should be considered. These forces have been demonstrated to be important in Xerox toner particle adhesion to photoreceptor surfaces (*47*).

These potentials are a result of charge carrier (electron and hole) migration in the solid phase attributable to the electrical fields entering the solid surface from an adjacent charged body. In our case, the virus develops electrical charge resulting from the ionization of prototropic groups in aqueous solutions.

Because the electrical field entering the solid phase is necessarily damped by an increase in ionic strength, we expect that if electrostatic induction controls adsorption, we would have weaker adsorption at high ionic strength than at low ionic strength. Since this is exactly the opposite of what we obtained experimentally, we have good reason to believe that this type of interaction does not control virus adsorption to our oxides.

However, electrostatic induction may contribute significantly in some cases, particularly where we have conductors as adsorbents. On aluminum metal it may be contributing, but the metal's high effectiveness as an adsorbent can also be accounted for by strong van der Waals interaction.

Covalent–Ionic Interaction. Covalent–ionic interactions appear to be important in the adsorption of small molecules on various materials, for example, octyl hydroxamate on Fe_2O_3 (48), where the functional group of the adsorbate is known to form strong aqueous complexes with metal ions (in this case Fe^{3+}) present in surface sites of the respective substrate.

Virus coat proteins have ionizing amino acid residues expressed on the exterior that might be able to participate in bonding to metal sites of an adsorbent. Choppin and Philipson (49) showed that sulfhydryl groups on the virus exterior react with p-chloromercuribenzoate and iodoacetamide.

Eyring and Wadsworth (50) and Little (51) present spectroscopic evidence that thiols bond to ZnO surfaces and xanthates bond to PbS surfaces by way of hydroxyl-exchange reactions in which metal–sulfide bonds are formed, and metal–hydroxyl bonds are broken during adsorption. We expect that trends exhibited in the relative stabilities of amino acid and hydroxo complexes of various dissolved cations should parallel trends in the relative reactivities and adsorption free energies, should this type of bonding control adsorption.

Hydroxyl ions tend to form far more stable complexes or precipitates with metal ions (Fe^{n+}, Al^{3+}, Mn^{n+}, and Si^{4+}) considered present in surface sites of many of our materials than do amino acids or similar functional-group analogues (17, 18). Divalent copper, presumably present in CuO surface sites, reacts with sulfhydryl-containing amino acids (52) to form Cu^+ complexes and ligand dimers; therefore dissolved

Table IX. Extraction of Purified

Sample		Titer/PFU mL^{-1}
A B	extracted with $C_2Cl_3F_3$	3.97 ($\pm$0.98) $\times$ 10^5 4.83 ($\pm$1.12) $\times$ 10^5
C D	control, w/o $C_2Cl_3F_3$	7.28 ($\pm$1.48) $\times$ 10^5 5.52 ($\pm$1.22) $\times$ 10^5

Note: The $\pm$ values are theoretical standard deviations based on random event considerations and measured pipetting variation (8). PFU = plaque forming units; CPM = counts per minute.

complex stability tells us little about the strengths of possible surface bonds. However, Cu^{2+} does form substantially more insoluble sulfides than hydroxides (17, 18), suggesting that sulfhydryl–Cu^{2+} surface bonds should be stronger than hydroxyl–Cu^{2+} surface bonds. Extended discussions of ΔG values for these reactions are presented elsewhere (8).

This suggests that if covalent–ionic interactions control adsorption, we could expect viruses to be strongly adsorbed to CuO, but only weakly adsorbed, if at all, to our other oxide surfaces. This is clearly inconsistent with observed trends in virus adsorption, suggesting that here, covalent–ionic interactions are not involved to any major extent, especially considering the good correspondence obtained with the DLVO–Lifshitz theory.

Hydrophobic Interactions. Hydrophobic interactions, responsible for stabilizing the tertiary structure of many proteins and for stabilizing membranes (53), are shown not to be involved in virus adsorption by critical hypothesis-and-test experimentation. Prior to discussing the experiment it should be recognized that occasionally these interactions have been confused with electrodynamic van der Waals interactions, even in recent adsorption literature. In this study, we refer to the aggregation of nonpolar surfaces or molecules resulting from the minimization of reoriented and thus higher (than bulk) free energy water structure adjacent to the nonpolar surfaces as hydrophobic bonding. Tanford (53) presents an extended discussion of this type of interaction and its importance in micelle formation, etc.

If purified poliovirus is not substantially concentrated at a trichlorotrifluoroethane ($C_2Cl_3F_3$)–water interface, since fluorocarbons are among the most hydrophobic substances known, it is rather improbable that hydrophobic interactions could be involved in poliovirus adsorption to any material. This is particularly true with our oxide surfaces, which in being wet with water, are demonstrated to be hydrophilic in character. This $C_2Cl_3F_3$ extraction procedure is commonly used in enterovirus purification, and is highly effective in removing hydrophobic materials (for example, lipids) from partially purified preparations.

Poliovirus with Trichlorotrifluoroethane

$^3H\text{-}RNA/CPM\ mL^{-1}$	$^{14}C\text{-}protein/CPM\ mL^{-1}$
173.0 ± 12.2	48.9 ± 3.6
168.5 ± 11.9	48.0 ± 3.5
238.7 ± 16.7	74.3 ± 5.3
194.4 ± 13.6	85.1 ± 6.1

Dilute, highly purified poliovirus, which would occupy less than 0.03 cm² if hexagonally closest packed in a two dimensional array, was emulsified with equal volumes (10 mL) of 0.02 I, pH 7 buffer and $C_2Cl_3F_3$ in borosilicate extraction tubes at 25°C. The emulsion was allowed to separate spontaneously, which took about 5 min, and the aqueous phase was sampled for residual virus. In Table IX we see only a small decrease in infectivity and radioactivity in the $C_2Cl_3F_3$-extracted samples, compared to controls run in parallel but not containing $C_2Cl_3F_3$. This experiment was repeated and essentially identical results were obtained.

The surface of the virus is concluded to be hydrophilic in character because of this. Because the oxide surfaces are all wet with water, they are therefore also hydrophilic. We have little evidence to support the conjecture that hydrophobic interactions might be involved to any large extent.

Hydrogen Bonding. The involvement of hydrogen bonding cannot be approached conclusively by critical experiment or calculation, though perhaps investigation of virus adsorption in the presence of a competing solute that participates in hydrogen bonding (for example, urea or guanidine) might be useful in this regard.

However, we can predict the importance of hydrogen bonding from its involvement in analogous systems. Since the prototropic groups on both the solid phase and the virus are necessarily hydrogen-bonded to water, one must break these bonds prior to forming hydrogen bonds between the virus and the solid directly. In a similar system, internal hydrogen bonding in globular proteins, von Hippel (54) states that the bond interchange with water has been shown "by recent calculations and measurement to be small, if not zero."

Where hydrogen bonding is known to be important in aqueous systems, for example, the bonding between base pairs in complementary DNA, we have a highly ordered arrangement which maximizes the occurrence of close bonding contacts. It is unlikely that the geometric array of surface hydroxyls on oxides would correspond to the array of prototropic groups on poliovirus. This suggests that hydrogen bonding is probably unimportant in virus adsorption to our materials, but further investigation is required to prove this with materials in general.

Environmental Implications

We have presented considerable evidence that poliovirus type 1, strain LSc2ab, adsorbs on inorganic surfaces according to the electrodynamic and electrostatic potentials defined by the DLVO–Lifshitz theory of colloid stability. We shall now present a general discussion concerning the predicted implications these findings have in regard to the overall problem of virus transport in the environment.

Our adsorption system at 0.02 I was designed to be comparable in ionic strength to many contaminated water conditions (55), and yet defined enough so that the mechanisms governing adsorption to various surfaces could be well defined. Temperatures for our adsorption experiments were 15° ± 2°C, which is characteristic of average natural water temperatures in temperate climates.

Experimental isotherms are linear with respect to virus concentration over several orders of magnitude. In addition, the fraction of the surfaces covered in these experiments is exceptionally small as seen directly in Figures 2–4. These data suggest that we have no significant level of particle–particle interactions during adsorption, which would produce nonlinear concave-upward isotherms (27). And this suggests that our adsorption experiments are characteristic of isolated virus particles. Since an isolated particle–surface interaction should be completely independent of the virus concentration, our observed adsorption characteristics should be appropriate for describing the adsorption of viruses at the exceptionally low concentrations, generally less than 10 PFU mL^{-1} (56), found in contaminated waters.

Our analysis describes virus adsorption from the standpoint of chemical equilibrium. Since adsorption equilibrium appears to be approached closely in our systems in less than or equal to 2 hr, and since the residence time of viruses in natural water systems is greater than 2 hr for many cases (for example, lakes, groundwaters, rivers, etc.), equilibrium considerations are entirely appropriate. In other situations, where residence times of the virus in the system are small compared to expected times required for adsorption to approach equilibrium (for example, sand filters in water treatment, water distribution systems, etc.), the DLVO–Lifshitz theory may still be applied directly. The work of Fitzpatrick and Spielman (57) concerning filtration and that of Zeichner and Schowalter (58) concerning colloid stability in flow fields demonstrate this clearly. Their developments of hydrodynamic trajectory analysis coupled to DLVO–Lifshitz considerations can be extended readily to problems concerning the transport of viruses in saturated or unsaturated flow in soils, as well as to those concerning the optimization of virus removal in filtration systems. Nonequilibrium aspects should be the subject of future investigations.

Effects of Virus Characteristics on Adsorption. We noted previously that enteroviruses are quite consistent in regard to general shape and size, about 27 nm in diameter (6), and that they (1) are thought to represent the most significant category of viruses in regard to water-borne virus transmission. However, in some cases viruses that could be possible environmental hazards, for example, adenoviruses—70 to 80 nm in diameter (6) and rotaviruses—60 to 70 nm in diameter (59, 60), are obviously larger in size.

The combined potentials given by Equations 3 and 9 predict that at close separation, we expect an approximately linear increase in adsorption potential with particle radius, that is, larger viruses according to the DLVO–Lifshitz theory should adsorb more strongly, given that all other significant parameters of the system are the same. Physically this corresponds to an increase in adsorption potential with an increase in the shared area of close contact between the virus and surface.

However, great care must be used in applying this prediction. The architecture of a relatively large virus could well be more complex than the architecture of a given small virus. There is a chance that the complex virus, because of surface irregularities, has a smaller area of close contact with a given solid surface than a small virus, leading to weaker adsorption potentials for the large virus. In addition, the electrokinetic properties of the two viruses may well be markedly different, which could cause changes in the electrostatic potentials large enough to mask the effects of increased shared contact area. The studies of Wallis and Melnick (61) concerning the variation in effectiveness of aluminum hydroxide and phosphate precipitates as adsorbents of different virus types give a general trend of larger virus—more virus adsorbed, but there is a large degree of scatter in this correlation. This is quite consistent with our predictions and their limitations.

The increase in the strength of adsorption with an increase in shared close contact area suggests that if virus particles are aggregated or clumped, they should adsorb more strongly than discrete particles. Because the status of the degree of aggregation in natural water systems with low virion concentrations has yet to be defined, environmental effects of this prediction are difficult to ascertain. However, some implications should be obvious such as the predicted greater retention of aggregated particles in transport through porous media.

Electrokinetic properties of the virus, which we showed to be involved directly in the electrostatic double-layer contribution to virus adsorption, should play a significant role in virus adsorption in many instances. Enteroviruses studied in regard to electrokinetic properties appear to have two isoelectric points (IEP), each of which represents interconvertible conformational states A and B. This was shown initially by the classic work of Mandel (62). His values of the isoelectric points of poliovirus type 1, strain Brunhilde, are presented in Table X along with subsequent findings of other investigators for other virus types. Estimated values presented for poliovirus type 1, strain LSc2ab, were calculated from our zeta potentials of an infectious virus and non-infectious material by applying Hunter and Wright's correlation to estimate the surface potential (ψ_0) of the virus as detailed in the section concerning predicted effects of pH and ionic strength. The Nernst equation (27) was then used to estimate the isoelectric points. This procedure has been described in greater detail elsewhere (8). Values obtained are quite similar to values reported for other enteroviruses, except for echovirus type 1, strain Farouk.

Table X. Comparison of Isoelectric Points of Various Enteroviruses

Virus	State	IEP	Reference
Poliovirus 1 LSc2ab	A noninfectious[a] material	~ 6.6 ≤ 5.7	(8)
Poliovirus 1[b] Brunhilde	A B	7.0 4.5	(62)
Poliovirus 1 Brunenders	A B	7.4 3.8	(63)
Poliovirus 1 CHAT	A B	7.5 4.5	(64)
Poliovirus 2 Sabin T2	A B	6.5 4.5	(65)
Poliovirus 1 Mahoney	A B	8.5 ± 0.1 ?	(66)
Coxsackievirus A21	A B	6.1 4.8	(65)
Echovirus 1 Farouk	A ? B	5.6 5.1	(67)

[a] All values reported here were obtained by isoelectric focusing except for those of strain LSc2ab, which were estimated from zeta potentials as detailed in the text.

[b] Mandel found essentially identical results as these when he determined IEPs by looking at zonal electrophoresis polarities as a function of pH in different buffers. This suggests that ampholytes used in isoelectric focusing (polyamino–polycarboxylic acids), and his buffers, had little effect on the expression of conformations.

Mandel's conformational bimodality is predicted to be of critical importance in regard to virus adsorption to inorganic surfaces in the environment. Gerba et al. (*68*) recently observed that echovirus type 1 adsorbed on sandy soil to a much lesser extent than poliovirus type 1, strain LSc. This corresponds to the low value of the A state conformation of echovirus type 1 presented in Table X compared to that of poliovirus. Since this bimodality of virus IEP's is intrinsically coupled to solution pH, we will expand the discussion of its importance in the section concerned with the effects of pH on adsorption.

An unanswered question does remain regarding the extension of our findings to predict the adsorption characteristics of human hepatitis type A. This virus is the etiologic agent of infectious hepatitis, which is considered to be the most serious problem in the transmission of water-borne virus disease (*1*). It is similar to other enteroviruses in terms of size (27 nm in diameter), density in CsCl gradients (1.34 g cm^{-3}), stability in the presence of chemical and physical agents, and probable nucleic acid type (*69*). However, electrokinetic properties of this virus have yet to be characterized. This information is required before accurate predictions of electrostatic components of adsorption can be made for this virus.

Effects of Substrate Properties on Adsorption. Preliminary predictions can be made to some extent on the basis of isoelectric points of various solid surfaces regarding electrostatic components of adsorption. In Table XI we present isoelectric points of solids commonly found in soils and sediments. For mineral surfaces, the IEP is defined to be that pH at which a given solid particle has no net electrical potential relative

Table XI. Isoelectric Points of Various Solids Present in Soils and Aquatic Systems

Solid Phase	*IEP*	*Reference*
Quartz α-SiO$_2$	2–3.5	(*70*)
Corundum α-Al$_2$O$_3$	5–9.2	(*70*)
Albite NaAlSi$_3$O$_8$	2.0	(*71*)
Microcline KAlSi$_3$O$_8$	2.4	(*71*)
Kaolinite Al$_4$(Si$_4$O$_{10}$)(OH)$_8$	$<$ 2–4.6	(*27*)
	$\sim$ 5 edges	(*71*)
Montmorillonite (Na,K)$_{x+y}$(Al$_{2-x}$Mg$_x$)$_2$- [(Si$_{1-y}$Al$_y$)$_8$O$_{20}$](OH)$_4 \cdot n$H$_2$O	$\leqslant$ 2.5	(*71*)
Hematite α-Fe$_2$O$_3$	4.2–9.3	(*27*)
Goethite α-FeOOH	5.9–8.3	(*27*)
Pyrolusite α-MnO$_2$	7.3 $\pm$ 0.2	(*72*)
Birnessite δ-MnO$_2$	1.5 $\pm$ 0.5	(*72*)
Tenorite CuO	9.5 $\pm$ 0.4	(*70*)

to bulk solution in essentially pure water, and is usually determined by microelectrophoresis or electroendosmosis experiments. The electrokinetic properties of inorganic surfaces and proteins (important here because they are found on the exterior of enteroviruses) are extremely complicated and beyond the scope of this discussion. Scatchard and Black (73), Parks (27, 70, 71), Hunter and Wright (32), and Yates and Healy (74) present extended discussions on the origins of electrical charge, the relationship between charge, potential, and solution composition, and what is represented by isoelectric points and zeta potentials.

When the electrostatic components of adsorption are considered by themselves, solids that have high values of IEP could well be better adsorbents than those with low values. For example, Al_2O_3 could be considered a better adsorbent than SiO_2 on this basis alone, which corresponds to the experimental findings reported earlier. High values for IEP may correspond with a greater probability of the solids having a net positive zeta potential at the pH of a natural water system. If the virus has a negative ζ potential under these conditions, electrostatic as well as electrodynamic components of ΔG_{ads} could be attractive, leading to greater adsorption compared to a solid that would have a negative zeta potential under these same conditions.

Again, great care must be taken concerning predictions based on isoelectric point considerations alone. For example, in the case of virus adsorption on kaolinite, which is known to have different charge characteristics associated with crystalline edges than with plates normal to them (27), an isoelectric point that represents contributions from both edges and plates may not at all characterize electrostatic components of virus adsorption free energies. In addition, on kaolinite and other clay minerals, differences in the origin of the charge compared to oxides (27), and the inaccessibility of interlayer surface to molecules as large as viruses (75) may lead to additional complications.

Differences in the abilities of various materials to generate large van der Waals potentials are probably the most important factors, in many cases, controlling the relative effectiveness of these materials as virus adsorbents. The Lifshitz theory, which has been shown earlier in this chapter to predict adsorption characteristics of poliovirus on a wide range of materials quite well, (Table VIII), predicts the following general series of adsorbent effectiveness (30)

$$\text{Metals (strong)} > \text{Sulfides} \geq \text{Transition Metal Oxides} >$$
$$SiO_2 > \text{Organics (weak)}$$

on the basis of the dielectric properties of the materials. Since the magnitudes of the van der Waals potentials are larger than double-layer interactions (Tables V and VIII) at 0.02 *I*, the Lifshitz predictions may well be more important than electrokinetic considerations in adsorption of viruses to different materials for most natural water conditions.

Predictions concerning the low effectiveness of organics as adsorbents are well supported by experimental evidence in addition to their dielectric properties. Under conditions detailed earlier for our adsorption experiments, we saw no significant adsorption to polystyrene (8) or polyethylene reaction vessels (26). (*See* also Table IX and discussions concerning weak virus adsorption at the water–$C_2Cl_3F_3$ interface and hydrophobic interactions.)

In addition to the dielectric properties and experimental evidence described previously, most organic substances found in natural waters are negatively charged under natural water conditions (27). This also predicts that most organics should be weak adsorbents under these conditions. However, under special solution conditions (for example, at physiological ionic strength, especially at reduced pH), even materials such as polystyrene and cellulose acetate (Millipore filters) can remove substantial fractions of virus from aqueous suspension. Materials such as aluminum metal should be far more effective as adsorbents than these materials, even under conditions that assist adsorption to the organic polymers. It should be mentioned here that these discussions concerning the predicted weak effectiveness of organics are appropriate for characterizing general, nonspecific interactions, and should not be applied to specialized cases of materials designed for chemical coupling.

Effects of Ionic Strength, pH, and Solution Composition. The tendency of virus to adsorb strongly to various materials at high ionic strength is well known. Carlson et al. (76) showed that silicate minerals strongly adsorbed T2 phage and poliovirus type 1, Sabin strain, at NaCl concentrations of greater than 0.02*M* and at CaCl concentrations of greater than 0.002*M* at pH 7. The results we presented earlier also demonstrate this with poliovirus and other inorganic surfaces. Both results correspond well with the DLVO–Lifshitz theory. Information presented in Tables V and VII shows that the electrostatic double-layer contribution to the free energy in brackish water conditions (*I* = 0.305, seawater *I* = 0.7 (14)) becomes unimportant compared to van der Waals contributions. However, at exceptionally high ionic strength increased amounts of electrolyte between the adsorbed virion and the surface may dampen dipole–dipole (van der Waals) interactions to such an extent that adsorption is decreased (40).

The effects of pH changes are also predicted by the DLVO theory. In general, high pH favors free virus and low pH favors adsorbed virus (2). However, there have been several reports in the literature that this is not a general phenomenon. For example, Berg (77) reported a maximum in adsorption at about pH 7 with poliovirus on Millipore filters, with less adsorption occurring at higher or lower pH levels. Buras (78) found maxima near pH 7 and pH 4.5 with magnetite, and a decreasing tendency for virus to adsorb at intermediate pH's.

The DLVO theory explains these results quite readily. To demonstrate this, we will discuss predictions of the adsorption characteristics of poliovirus on SiO_2 as a function of pH. To make these predictions we have adopted free energy values and electrodynamic and electrostatic adsorption potential equations found in the earlier sections of this study to characterize virus adsorption. These equations are then used to predict the effect of altering pH on the basis of electrochemical theory. We selected the measured van der Waals potential of -52 kJ mol^{-1} and added the double-layer potential as evaluated by Gregory's LSA expression to calculate an estimate of the fraction of virus adsorbed (F_{ads}) with Equation 2. We selected a system containing 1 m^2 of SiO_2 and 1 L of $0.02M$ NaCl for the demonstration.

To evaluate the double-layer potential as a function of pH, zeta potential estimates of solid and virus are required. The zeta potentials of SiO_2 were obtained by direct microelectrophoresis as a function of pH (8). For the virus, this information is not available directly from experimental measurements. Here we applied the Nernst equation (27) to Mandel's (62) isoelectric points (A state = pH 7.0, B state = pH 4.5) to estimate surface potentials (ψ_o):

$$\psi_o = 2.3 \frac{RT}{F} (\text{IEP} - \text{pH}) \qquad (12)$$

where F is Faraday's constant. We can then obtain preliminary estimates of ζ from ψ_o by applying Hunter and Wright's correlation. For some oxide surfaces, if we extrapolate a linear function to a counterion concentration of $10^{-6}M$ the value obtained is quite close to a surface potential obtained by a best-fit Stern–Graham approach (32). First-order estimates of U_{dl} as a function of pH can then be made by first estimating ψ_o with the Nernst equation, then applying the correlation just mentioned, and by finally applying Gregory's LSA expression.

Predicted results for this system are shown in Figure 6. The effects of Mandel's two conformational states dominate the picture. At high pH, the virus is predominantly freely suspended. Near pH 7, the A state isoelectric point virus is strongly adsorbed. As the pH is decreased to about 6, we observe an A → B state transition, drawn here where Mandel found it in his experiments, that predicts decreased levels of adsorption compared to that seen for pH 7. At pH 5 and below, strong adsorption is again predicted, this with the B state virus as well as the A state virus. This bimodality in conformational state explains the results of both Berg and Buras. However, at this stage of development the calculated values plotted for F_{ads} should be regarded as approximate and tentative.

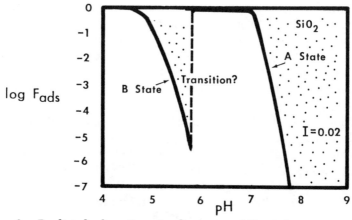

Figure 6.   Predicted adsorption of poliovirus on SiO_2 surfaces as a function of pH according to the DLVO theory.

F_{ads} *represents the fraction of the total virus present that is adsorbed to 1 m^2 of SiO_2 surface from 1 L on 0.02M NaCl. Dotted areas represent regions where adsorption is predicted to be weak. At pH 5 to 6 and 8 to 9, virus and SiO_2 are similar (negative) in regard to Gouy layer charge. The transition in the central region of the diagram represents conformational change from A to B state, originally described by Mandel (62).*

Solution components may alter the surface properties of various solids or viruses in a given system so that predictions based on dielectric properties, isoelectric points, pH, and ionic strength considerations may be inaccurate.

For example, dissolved carbonate species present in our 0.02 I buffer system, predominantly CO_2, $H_2CO_3^0$, HCO_3^{2-}, CO_3^{2-}, and various carbonato complexes, had a marked effect concerning adsorption to transition metal oxides. The zeta potential of CuO in 0.02 I buffer was -17.6 ± 6.1 mV, while in 0.02M NaCl that contained only traces of total dissolved carbonate (approximately $10^{-5}M$), it was $+32.0 \pm 5.8$ mV. This shows marked alteration of the electrical structure of double layers by some carbonate species. The same effects were seen to lesser extents on Fe_2O_3 and MnO_2 (8). Double-layer interaction potentials calculated with zeta potentials measured in 0.02 I reaction buffers matched adsorption free energy differences well, and these potentials included the effects of carbonate species. Where effects of dissolved constituents have not been accounted for intrinsically, predictions made on the basis of the DLVO–Lifshitz considerations alone must be made with care.

Dissolved and suspended organic material may also affect the double-layer or the van der Waals contributions to adsorption, should they form coatings sufficiently thick that close contact of the virus with the adsorbent is blocked. From considerations of the low dielectric nature of organics and the predominantly negative charge of humic substances (79) common in soils and aquatic environments, we can the general prediction that natural organics may well reduce the amount of virus adsorbed to a given substrate compared to an identical system that does not contain the organics. However, in special cases such as addition of anthropogenic cationic polymers such as polyethylenimines (80) to a given system, virus adsorption could well be enhanced. This of course depends on other system variables being conducive to this enhancement.

In any case, we did not attempt to predict adsorption characteristics of all viruses on all solids in any aqueous system in this study. In this discussion of environmental implications, we merely attempted to extract salient features of the importance of the DLVO–Lifshitz theory in regard to problems concerning virus adsorption. Additional implications and specialized cases will be the subjects of future investigations.

Conclusions

Poliovirus adsorption to many oxide surfaces is controlled principally by the combination of electrodynamic van der Waals interactions and electrostatic double-layer interactions, as demonstrated by the excellent correspondence of the DLVO–Lifshitz theory with experimentally determined adsorption free energies.

The Lifshitz theory, which characterizes van der Waals interactions, quantitatively predicts the wide differences in adsorption characteristics of various materials that are results of these interactions. Double-layer interactions appear to be well predicted by Gregory's LSA, constant-charge expression and measured zeta potentials of viruses and oxides. Other contributions to the adsorption free energies such as valence bonding, induced-image forces, hydrophobic interaction, and configurational entropy appear to be of secondary importance in our system.

The DLVO–Lifshitz theory should be regarded as a principal mechanism governing the adsorption of viruses on various inorganic surfaces. This finding has direct application to problems concerning transport of viruses in aquatic systems and soils. It is possible that it could lead to the design and optimization of adsorption–filtration processes for removing viruses and other particulates from contaminated water.

Acknowledgments

R. O. James (currently at C.S.I.R.O. Textile Physics Division, Ryde, N.S.W., Australia) is thanked for his assistance with concepts of the DLVO–Lifshitz theory; Gordon Brown and Milton Kerker are thanked for their helpful comments; and John L. Murray, John R. Murray, and H. H. Huang are thanked for their financial assistance. Part of the materials used in this study were purchased with a Sigma-Xi Grant-in-Aid of Research. The final phase of the project was supported by the ⌐.S. Environmental Protection Agency Grant R-805016.

Literature Cited

1. "Viruses in Water;" Berg, G., Bodily, H. L., Lennette, E. H., Melnick, J. L., Metcalf, T. G., Eds.; A.P.H.A.: Washington, D.C., 1976, 256 p.
2. Gerba, C. P.; Wallis, C.; Melnick, J. L. *J. Irrig. Drain. Div., Am. Soc. Civ. Eng.* 1975, *101*, IR 3, 157.
3. Derjaguin, B.; Landau, L. D. *Acta Physicochim. URSS* 1941, *14*, 633.
4. Verwey, E. J. W.; Overbeek, J. Th. G. "Theory of the Stability of Lyophobic Colloids"; Elsevier: New York, 1948; 205 p.
5. Dzyaloshinskii, I. E.; Lifshitz, E. M.; Pitaevski, L. P. *Adv. Phys.* 1961, *10*, 165.
6. Fenner, F.; McAuslan, B. R.; Mims, C. A.; Sambrook, J.; White, D. O. "The Biology of Animal Viruses," 2nd ed.; Academic: New York, 1968; 834 p.
7. Akin, E. W.; Benton, W. H.; Hill, W. F., Jr. In "Virus and Water Quality: Occurrence and Control," *Univ. Ill. Bull.* 1971, *69*, 59.
8. Murray, J. P. Thermodynamics of Poliovirus Adsorption, Ph.D. Dissertation, Stanford University, Palo Alto, CA, 1978.
9. Cramer, R. In "Techniques in Experimental Virology"; Harris, R. J. C., Ed.; Academic: New York, 1964; p. 145.
10. McClain, M.; Schwerdt, C. E. *Fed. Proc., Am. Soc. Exp. Biol.* 1954, *13*, 505.
11. Charney, J.; Machlowitz, R.; Spicer, D. S. *Virology* 1962, *18*, 495.
12. White, A.; Handler, P. H.; Smith, E. L. "Principles of Biochemistry," 5th ed.; McGraw–Hill: New York, 1973; p. 806.
13. Joklik, W. K.; Darnell, J. E. *Virology* 1961, *13*, 439.
14. Stumm, W.; Morgan, J. J. "Aquatic Chemistry"; Interscience: New York, 1970; 583 p.
15. Wiese, G. R.; Healy, T. W. *J. Colloid Interface Sci.* 1975, *52*, 452.
16. Smith, T. *Surf. Sci.* 1976, *55*, 601.
17. Sillen, L. G.; Martell, A. E. "Stability Constants of Metal–Ion Complexes"; The Chemical Society: 1964, 754 p.; Supplement No. 1, 1971; 865 p.
18. Martell, A. E.; Smith, R. M. "Critical Stability Constants"; Plenum: New York, 1974–1977; Vols. 1–4.
19. Wagman, D. D.; Evans, W. H.; Parker, V. B.; Halow, I.; Bailey, S. M.; Schumm, R. H. *Nat. Bur. Stand. (U.S.), Tech. Note* 1968, *270–273*.
20. Langmuir, D. *U.S. Geol. Surv. Prof. Pap.* 1969, *650-B*, B180.
21. Brunauer, S.; Emmett, P.; Teller, E. *J. Am. Chem. Soc.* 1938, *60*, 309.
22. Amer. Soc. for Testing Mat., C-204-73, *13*, 182 1973,
23. Jura, G. "The Determination of the Area of the Surfaces of Solids," In *Phys. Meth. Chem. Anal.* 1951, *2*, 255.
24. Dhar, H. P.; Conway, B. E.; Joshi, K. M. *Electrochim. Acta* 1973, *18*, 789.
25. Ginoza, W. In "Methods in Virology"; Maramorosch, K., Koprowski, H., Eds.; 1968, *IV*, 139.
26. Murray, J. P. "Physical Chemistry of Virus Adsorption and Degradation on Inorganic Surfaces: Its Relation to Wastewater Treatment"; in press.
27. Parks, G. A. In "Chemical Oceanography"; Riley, G., Skirrow, G., Eds.; Academic: New York, 1976; Vol. 1, p. 241.
28. Gregory, J. *J. Colloid Interface Sci.* 1975, *51*, 44.
29. Smith, A. L. *J. Colloid Interface Sci.* 1976, *55*, 525.
30. Visser, J. In "Surface and Colloid Science"; Matijevic, I. E., Ed.; Interscience: New York, 1976, *8*, 3.
31. Chan, D.; Perram, J. W.; White, L. R.; Healy, T. W. *J. Chem. Soc., Faraday Trans. 1* 1975, *71*, 1046.
32. Hunter, R. J.; Wright, H. J. L. *J. Colloid Interface Sci.* 1971, *37*, 564.
33. Wiersema, P. H.; Loeb, A. L.; Overbeek, J. Th. G. *J. Colloid Interface Sci.* 1966, *22*, 78.
34. Lipatov, Yu. S.; Sergeeva, L. M. "Adsorption of Polymers"; Trans. from Russian, Halstead: New York, 1974; 177 p.

35. Hamaker, H. C. *Physica* **1937**, *IV*, 1058.
36. Ninham, B. W.; Parsegian, V. A. *Biophys. J.* **1970**, *10*, 646.
37. Bargeman, D.; Van Voorst Vader, F. *J. Electroanal. Chem.* **1972**, *37*, 42.
38. Lukes, F. *Surf. Sci.* **1972**, *30*, 91.
39. van Oss, C. J.; Good, R. J.; Neumann, A. W.; Weiser, J. D.; Rosenberg, P. *J. Colloid Interface Sci.* **1975**, *59*, 505.
40. Richmond, P. In "Colloid Sci., II," Everett, D., Ed.; The Chem. Soc.: London, **1975**, 130.
41. LeNeveu, D. M.; Rand, R. P.; Parsegian, V. A.; Gingell, D. *Biophys. J.* **1977**, *18*, 209.
42. Samsonov, G. V. "The Oxide Handbook"; IFI Trans. from Russian, Plenum: New York, 1973; 524 p.
43. Klose, P. H. *J. Electrochem. Soc.* **1970**, *117*, 854.
44. Jarzebski, Z. M. "Oxide Semiconductors"; Pergamon: New York, 1973; 285 p.
45. Kaye, G. W. C.; Laby, T. H. "Tables of Physical Constants"; 13th ed.; John Wiley: New York, **1966**, 249 p.
46. Kittel, C. "Introduction to Solid State Physics," 5th ed.; John Wiley: New York, 1976; 608 p.
47. Goel, N. S.; Spencer, P. R. "Toner Particle-Photoreceptor Adhesion," Invited Paper, ACS Annual Mtg., Philadelphia, **1975**, 67 p.
48. Raghavan, S.; Fuerstenau, D. W. *J. Colloid Interface Sci.* **1975**, *50*, 319.
49. Choppin, P. W.; Philipson, L. *J. Exp. Med.* **1961**, *113*, 713.
50. Eyring, E. M.; Wadsworth, M. *Trans. Am. Inst. Min. Metall. Pet. Eng.* **1956**, *5*, 531.
51. Little, L. H. "The Infrared Spectra of Adsorbed Species"; Academic: New York, 1966; 428 p.
52. Vallon, J. J.; Badinand, A. *Anal. Chim. Acta* **1968**, *42*, 445.
53. Tanford, C. "The Hydrophobic Effect: Formulation of Micelles and Biological Membranes"; John Wiley: New York, 1973; 200 p.
54. von Hippel, P. H. In "Protein-Ligand Interactions"; Sund, H., Blauer, G., Eds.; Walter de Gruyter: Berlin, New York, 1975; p. 452.
55. Todd, D. K. "The Water Encyclopedia: A Compendium of Useful Information on Water Resources"; Water Information Center: Port Washington, New York, **1970**, 559.
56. Berg, G. *Proc. Sixth Internat. Water Poll. Res. Conf.* **1972**, B/14/28, 8.
57. Fitzpatrick, J. A.; Spielman, L. A. *J. Colloid Interface Sci.* **1973**, *43*, 350.
58. Zeichner, G. R.; Schowalter, W. R. *J. Am. Inst. Chem. Eng.* **1977**, *23*, 243.
59. Flewett, T. H.; Bryden, A. S.; Davies, H.; Woode, G. N.; Bridger, J. C.; Derrick, J. M. *Lancet* **1974**, 7872.
60. Wyatt, R. G.; Kapikian, A. Z.; Thornhill, T. S.; Sereno, M. M.; Kim, H. W.; Charnock, R. M. *J. Infect. Dis.* **1974**, *130*, 532
61. Wallis, C.; Melnick, J. L. "Concentration of Viruses on Aluminum Hydroxide and Aluminum Phosphate Precipitates," in "Transmission of Viruses by the Water Route"; Berg, G., Ed.; John Wiley: New York, 1967; p. 129.
62. Mandel, B. *Virology* **1971**, *44*, 554.
63. La Colla, P.; Marcialis, M.; Mereu, G. P.; Loddo, B. *Experientia* **1972**, *28*, 1115.
64. Ward, R. L. *J. Virol.* **1978**, *26*, 299.
65. Korant, B. D.; Lonberg-Holm, K., personal communication, 1977.
66. Floyd, R.; Sharp, D. G. *Appl. Environ. Microbiol.* **1978**, *35*, 1084.
67. Young, D. C.; Sharp, D. G. Pres. at ASM Nat'l. Mtg. Honolulu, 1979.
68. Gerba, C. P.; Goyal, S. M., personal communication, 1977.
69. Provost, P. J.; Wolanski, B. S.; Miller, W. J.; Ittensohn, O. L.; McAleer, W. J.; Hilleman, M. R. *Proc. Soc. Exp. Biol. Med.* **1975**, *148*, 532.
70. Parks, G. A. *Chem. Rev.* **1965**, *67*, 121.

71. Parks, G. A. In "Equilibrium Concepts in Natural Water Systems," *Adv. Chem. Ser.* **1967,** *67,* 121.
72. Healy, T. W.; Herring, A. P.; Fuerstenau, D. W. *J. Colloid Interface Sci.* **1966,** *21,* 435.
73. Scatchard, G.; Black, E. S. *J. Phys. Colloid Chem.* **1949,** *53,* 88.
74. Yates, D. E.; Levine, S.; Healy, T. W. *Trans. Faraday Soc.* 1 **1974,** *70,* 1046.
75. Albert, J. T.; Harder, R. D. *Soil Sci.* **1973,** *115,* 130.
76. Carlson, G. F.; Woodard, F. E.; Wentworth, D. F.; Sproul, O. J. *J. Water Pollut. Control Fed.* **1968,** *40,* R89.
77. Berg, G. *J. Sanit. Eng. Div., Am. Soc. Civ. Eng.* **1971,** 79, SA-6, 867.
78. Buras, N., personal communication, 1977.
79. Mortensen, J. L.; Himes, F. L. In "Chemistry of the Soil," 2nd ed., *ACS Monogr.* **1964,** *160,* 206.
80. Glaser, H. T.; Edzwald, J. K. *Environ. Sci. Technol.* **1979,** *13,* 299.

RECEIVED November 24, 1978.

Metal Transport Phases in the Upper Mississippi River

S. J. EISENREICH[1] M. R. HOFFMANN, D. RASTETTER,
E. YOST, and W. J. MAIER

Environmental Engineering Program, Department of Civil and Mineral
Engineering, University of Minnesota, Minneapolis, MN 55455

Six metals of geochemical and water quality interest in the upper Mississippi River were partitioned into soluble, adsorbed, solid organic, oxide coating, and crystalline phases, applying a chemical fractionation scheme to particulates and ultrafiltration to the dissolved ($< 0.4 \mu m$) phase. Crystalline and oxide fractions were dominant transport phases for aluminum and iron, while solution and solid organic fractions were dominant for manganese, copper, cadmium, and lead. The fraction of metal transported in available phases (noncrystalline) ranged from 10 to 25% for aluminum and from 30 to 70% for iron, while the amount of manganese, copper, cadmium, and lead generally exceeded 90%. Ultrafiltration studies showed that the highest concentrations of copper, cadmium, and lead occurred in the 1–10K mol wt fraction and correlated with organic carbon. The magnitude of residual copper complexation capacities ($\sim 1.0 \mu M$) and conditional stability constants ($\sim 10^{10}$) suggests that natural organic ligands are likely multidentate, containing nitrogen and sulfur functional groups.

The partitioning of major and trace metals between dissolved and particulate phases in aquatic systems is an important factor controlling metal availability and toxicity to biota. Metal concentrations and speciation in natural waters, whether derived from indigenous or anthro-

[1] To whom correspondence should be sent.

0-8412-0499-3/80/33-189-135$10.50/0
© 1980 American Chemical Society

pogenic sources, are regulated by aqueous-phase (complexation, ion-pair formation, redox) and solid-phase (precipitation, adsorption/absorption, ion exchange) reactions. The relative importance of each process will depend on the aqueous chemistry of the metal, the chemical milieu of the natural water, the amount and type of surface available, and hydrologic factors. For example, the solid phase has been implicated as controlling the concentrations of trace metals in freshwater (1–6) and sea water (7). Jenne (1) has proposed that adsorption on or inclusion in the hydrous oxides of manganese and iron is the major controlling mechanism determining trace metal concentrations in freshwater. Gibbs (2) and Shuman et al. (3) concluded similarly that trace metals were concentrated in the crystalline and hydrous oxide phases with solid organic-associated metal contributing to a lesser degree. Recent data (8, 9, 10, 11) suggest that naturally occurring organic matter in freshwater such as fulvic acids can complex or adsorb trace metals, increasing their concentraction, transport, and potential availability. Natural organic trace-metal stability constants have been found to be greater than corresponding inorganic complexes, and these associations can be significant even in the presence of excess concentrations of competing major cations (12). Biological availability/toxicity may be affected by adsorption, complexation, or incorporation of trace metals by/into dissolved or colloidal organic matter. Complexation of micronutrients by soluble organic ligands may increase the physiological availability of trace metals to aquatic organisms when the ratio of organic matter to metal is reasonably high (13, 14, 15), while inverse relationships have been found between complexation capacity and relative toxicity of copper and cadmium to planktonic algae (16, 17, 18).

This chapter presents information on: (1) the partitioning of iron, aluminum, manganese, copper, cadmium, and lead between dissolved and particulate phases; (2) molecular size characteristics and stability relationships of metals and organic matter; and (3) the speciation of metals in the particulate phase. The importance of organic-metal and particulate-metal interactions in river water is discussed also.

Experimental

Sampling. The area under study is an 885 km stretch of the upper Mississippi River extending from near the source of the river (river km 2197) to just below the Twin Cities metropolitan area (river km 1312). Sampling sites were chosen to represent terrestrial influences of varying types on the concentrations and speciation of metals in the river (Figure 1). Site 1 is located at river km 2197, 12.9 km southwest of Bemidji, Minnesota, near the source of the river. The site was chosen as being representative of a pristine area with natural background inputs affecting the total organic carbon (TOC) and metal

Figure 1. Sampling sites in the upper Mississippi River

content of the river derived from relatively undisturbed wetland areas. The watershed consists primarily of coniferous forests (72% by area) with minor cropland (15.6%), pasture (3.5%), and other land uses (6.5%). Site 2 is located at river km 1496 and was chosen to characterize the effect of agricultural land use on river quality; minimally treated sewage discharges occur upstream. The watershed surrounding this segment of the river is devoted primarily to farming (47.6%), with smaller amounts devoted to forest (22.4%), pasture (17.5%), and other land uses (13.1%). Site 3 lies within the urban Twin Cities area but upstream of major industrial and municipal discharges. Site 4 is located downstream of the Twin Cities metropolitan area and 8 km downstream of a major metropolitan sewage treatment plant discharging 0.8–0.9 $\times$ 10^6 m^3 of treated effluent per day. Diffusion models and aerial photographs (*19*) show that treated sewage was mixed 70–100% with Mississippi River water at Site 4. Site 5 on the Minnesota River was added to the sampling network in the spring of 1978 to evaluate the effects of Minnesota River inputs to the Mississippi River at Site 4. Overall, the sampling network encompasses 885 km and 13 dams, which effectively separate the chemical/physical processes occurring at each site, except for Sites 3 and 4. Flows increase by a factor of 100 from Site 1 to Site 4 based on annual flow data.

Water samples were obtained from low-lying bridges near midstream at all sites at a depth of 0.5 m with a cleaned and hand-operated plastic pump fitted with Tygon tubing, or with a linear polyethylene (LPE) bucket. Contamination from the pump or bucket was negligible. An aliquot of the water sample was filtered through preweighed 0.4-μm pore size membrane filters (Nuclepore) either in situ or in the laboratory within 2 hr. Nuclepore membrane filters were used because they exhibit low and constant metal blanks within a batch, and have a relatively uniform pore-size distribution. The dissolved fraction, operationally defined as the quantity passing a 0.4-μm filter, was collected directly into acid-leached LPE bottles in polycarbonate filtration assemblies, and acidified to 1% acid with redistilled HNO_3. Unfiltered water samples for TOC analysis were collected in 125-mL carbon-free glass bottles with aluminum foil-lined caps. Samples for general parameter analysis were stored in 500-mL acid-leached LPE bottles. Bulk water samples of 100–240L were stored, acid-cleaned LPE carboys. Metal, carbon, and general parameter samples were collected in duplicate and stored at 4°C until analyzed.

Analysis. The membrane filters plus nonfilterable residue, operationally defined as particulate matter, were dried to constant weight in a desiccator and digested by sequential treatment with ·HCl, –HNO_3, –HF (750 μL, 250 μL, and 100 μL, respectively) using a Teflon bomb and a dry-air oven (20). The dissolved and digested metals were determined by atomic absorption spectrophotometry (AAS) with a PE Model 360 equipped with a Model 2100 heated graphite furnace and D_2 background correction, or in the flame mode with a Varian Model AA 175. Dissolved calcium, magnesium, sodium, and potassium analyses were performed by inductively coupled plasma emission (Applied Research Labs QA—137) in a 2N HCl matrix. Digested metal concentrations were adjusted for filter blanks. Metal standards were prepared daily to correspond to the treatment of field samples. General parameters were determined according to Standard Methods (21) and TOC measured with a Beckman Model 915A TOC analyzer.

Suspended solids were isolated from 100–240 L of water for particulate metal characterization by continuous-flow centrifugation using a Sorvall SS-3 Centrifuge equipped witha KSB-3 continuous-flow assembly. Centrifuge speed and sample flow rate were adjusted such that particles with a density greater than or equal to 1.5 g/cc and with a Stokes' diameter greater than or equal to 0.4 μm were collected. Centrifuged water, when filtered through 0.4-μm Nuclepore membranes, showed no change in metal content. To prevent metal contamination from the rotor assembly and latex tubing, machined polycarbonate liners were used in the rotor chambers, and Teflon or glass liners were used elsewhere. Solids were transferred to polycarbonate tubes and dried under vacuum prior to chemical fractionation and analysis.

Ultrafiltration. Millipore 90-mm high-flux ultrafiltration cells and membranes listed in Table I were used for all ultrafiltration experiments. A Prince-

Table I.   Characteristics of

Membrane	Pore Size (nm)	Nominal Molecular Weight Cutoff (AMU)	pH Range
UMO5	1.0	500 (0.5K)	2–12
PSAC	1.4	1,000 (1K)	2–10
PTGC	2.8	10,000 (10K)	1–14
PSED	3.4	25,000 (25K)	2–10
PTHK	6.2	100,000 (100K)	1–14

ton Applied Research Polarographic Analyzer Model 174A equipped with a Metrohm E410 HMD electrode and a Scientific Products Magnestir were used for anodic stripping voltammetry (ASV). An HP-7040A X–Y recorder was used to record scans.

All water samples were stored at 4.0°C before analysis. The centrifugate was filtered through a 0.40-μm membrane immediately before ultrafiltration. After pretreatment, a sample of known volume ($0.4 \leq v \leq 1.4$ L) was introduced into a well-cleaned ultrafiltration cell, and the system was pressurized with nitrogen to 25 or 40 psig, depending on the membrane to be used. The ultrafiltrate was collected in an acid-washed polyethylene bottle. When the retentate volume was reduced to approximately 50% of the initial volume, the system was depressurized and the retentate was transferred to three different acid-washed bottles and stored for AAS, ASV, and TOC analysis. The AAS portion was acidified immediately. Volumes were determined with graduated cylinders. The ultrafiltrate from this step was fractionated sequentially using membranes with the next lower pore size until the smallest pore size was used. Volumes of both the retentate and ultrafiltrate were recorded at each stage. Membranes were soaked at pH 2.4 for 24 hr before use. Both membranes and cells were rinsed with 2 L of Milli-Q water (4 cell volumes). The final 60 mL was collected for blank determinations.

Free and labile copper, cadmium, and lead were determined by ASV in the differential pulse mode using a HMDE, a plating potential of -1.3 V vs. SCE, a plating time of 90 sec with stirring, and a 30-sec quiescent period before stripping. Other operating parameters were: scan rate, 5 mV/sec; modulation amplitude, 25 mV; current range, 0.1–0.5 amp; and drop time, 0.5 sec. Sample mixing was achieved by bubbling deoxygenated nitrogen gas and by magnetic stirring with a 10-mm Teflon bar.

Complexation capacity measurements for copper were made in the direct current mode with a plating potential of 0.8 V vs. SCE, a plating time of 150 sec, a 30-sec quiescent period, a scan rate of 20 mV/sec, a modulation amplitude of 25 mV, a current range of 0.2 μ amps, and a drop time of 0.3 sec. For complexation capacity measurements a carbon dioxide buffer/nitrogen buffer was used. Carbon dioxide and nitrogen (Matheson) were mixed using Matheson rotometers #610 (for carbon dioxide) and #602 (for nitrogen). After mixing, the mixture was deoxygenated with a vandadate solution containing 2.3 g V_2O_5, amalgamated zinc, and 12M HCl in 250 mL of ultrapure water obtained from a Millipore Milli-RO/Milli-Q water purification system. Standard copper solutions (10^{-4}–$10^{-5}M$) were made from AR copper wire dissolved in a small amount of concentrated HNO_3. $NaClO_4$ (G. F. Smith) was recrystallized and used as the inert electrolyte at a concentration of 30mM.

Ultrafiltration Membranes

Membrane Composition	Manufacturer	Operating Pressure (psig)
polyelectrolyte	Amicon	40
polyelectrolyte	Millipore	40
aromatic polymer	Millipore	25
polyelectrolyte	Millipore	25
aromatic polymer	Millipore	25

Fractionation of Particulate Matter. Duplicate samples of river particulates collected by continuous-flow centrifugation were leached sequentially according to the following schema:

Treatment	Fraction	Ref.
0.5N MgCl$_2$, 7 hr	Adsorbed/ion exchangeable	(2)
0.4N Na$_4$P$_2$O$_7$, 10 hr pH 7	Particulate organic	(22)
0.3N HCl, 30 min, 90°C	Metal oxide	(23)
Conc. HNO$_3$, –HCl, –HF	Crystalline	(20)

Approximately 0.5–1.0 g of solid matter was added to acid-leached polycarbonate centrifuge tubes followed by 25 mL of leaching reagent. The mixture was then equilibrated under the conditions specified. Solid residues were separated from the leachate by centrifugation, the supernatant removed with a pipette, and the residues resuspended and washed twice more with extracting reagent. The supernatant and wash were combined, acidified to pH 1.0, and analyzed. In some instances, particulate samples consisted of composites from several sampling dates, and therefore were weighted to the amount of solid residue obtained on each sampling date.

Results and Discussion

Metal Partitioning Between Dissolved and Particulate Phases. General water quality and chemical characteristics of the upper Mississippi River and Minnesota River are shown in Figure 2 with statistical summaries for all data given in Table II. Sites 1–4 are typified by high TOC ($\bar{x} = 10.8$ to 13.3 mg/L as C), low suspended solids (SS) ($\bar{x} = 5.5$ to 49 mg/L), high total inorganic carbon (TIC) ($\bar{x} = 31$ to 49 mg/L as C), and moderate chloride ($\bar{x} = 2.3$ to 19 mg/L) and sulfate ($\bar{x} = 2.8$ to 42 mg/L) concentrations. The concentrations of TOC and TIC exhibit little fluctuation on a mean basis from Sites 1–4, whereas sulfate, chloride, specific conductance, and SS generally increase with distance downstream from the headwaters. Calcium ($\bar{x} = 34$ to 54 mg/L) and magnesium ($\bar{x} = 11$ to 19 mg/L) mean concentrations were moderately high, corresponding to the predominance of glacial calcitic and dolomitic rock in most of the watershed, and fluctuate little from site to site, whereas sodium ($\bar{x} = 5$ to 15 mg/L) and potassium (1.4 to 4 mg/L) varied by a factor of three from Sites 1–4. The variability in the mean sodium and potassium concentrations represents changes in clay composition of soil types and the influence of the metropolitan area on river composition. Metal concentrations and speciation should be controlled by these factors. Thus Sites 1–4 have high alkalinities, high TOC and major cation concentrations, and a relatively low SS load. In contrast, Site 5 on the Minnesota River, draining the intensively cultivated agricultural area of

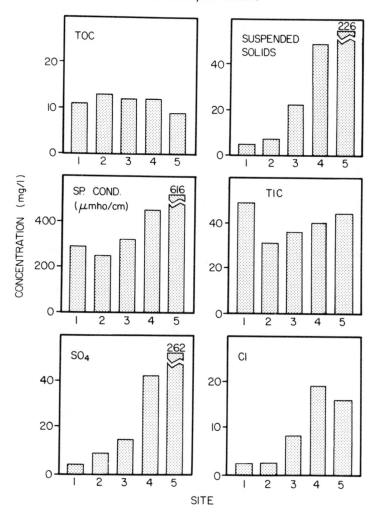

Figure 2. Mean concentrations of major cations and general water quality parameters in the upper Mississippi River

central Minnesota, exhibited high concentrations of most parameters. Site 5 was added to the sampling network in spring 1978, and high SS and sulfate concentrations may partially be attributable to spring runoff. These values agree with data obtained quarterly over the past ten years (24). Site 5 was included to evaluate its influence on metal transport at Site 4, which is attributable to a combination of factors resulting from upstream transport, sewage effluent, and input from the Minnesota River.

The dissolved and particulate concentrations of iron, aluminum, manganese, copper, cadmium and lead obtained for Sites 1–5 for 1977–1978 are shown in Figure 3 with statistical summaries given in Table III.

Table II. Major Elements and General Parameters

	Site 1			Site 2		
	$\overline{X}$	SD	N	$\overline{X}$	SD	N
Ca	49	10.6	4	34	5.5	4
Mg	18	5.4	4	10.7	2.1	4
Na	5.5	1.7	4	4.7	.8	4
K	1.4	.2	4	1.6	.3	4
TOC	10.8	4.4	4	13.3	4.3	4
TIC	49	11.1	4	31	7.0	4
Cl	2.3	1.0	4	2.3	1.6	4
SO_4	2.8	1.6	4	8.2	.5	4
SS	5.5	2.3	4	6.9	3.8	4
SC (μmho/cm)	290	100	4	250	40	4
TEMP (°C)	9.6	7.8	4	9.7	7.6	4
pH	7.5	.3	4	8.0	.5	4
FLOW (cms)	—	—	—	95	—	1

Table III. Dissolved and Particulate Metal in

	Site 1			Site 2		
	$\overline{X}$	SD	N	$\overline{X}$	SD	N
Filterable						
Al	3.4	2.1	4	20	9.3	4
Mn	140	190	4	37	14.8	4
Fe	170	64	4	170	28	4
Cu	.8	.6	4	2.1	.7	4
Cd	.13	.08	4	.26	.16	4
Pb	.7	.4	4	1.0	.6	4
Nonfilterable						
Al	490	510	4	620	602	4
Mn	13.4	11.4	4	49	25	4
Fe	320	88	4	350	230	4
Cu	.4	.5	4	.3	.4	4
Cd	.07	.04	4	.16	.12	4
Pb	1.7	2.3	4	2.5	4.2	4
Total Metal						
Al	490	520	4	640	608	4
Mn	150	190	4	86	14.2	4
Fe	490	99	4	520	250	4
Cu	1.2	.7	4	2.4	1.1	4
Cd	.20	.08	4	.42	.18	4
Pb	2.3	2.0	4	3.5	4.7	4

in the Upper Mississippi River—1977–1978 (mg/L)

Site 3			Site 4			Site 5		
$\overline{X}$	SD	N	$\overline{X}$	SD	N	$\overline{X}$	SD	N
42	5.6	17	54	8.7	10	79	20	3
14.8	2.5	17	19	3.5	10	28	9.8	3
7.4	1.8	17	15.3	5.0	10	12.9	4.1	3
2.1	.7	17	4.0	.8	10	6.7	0	3
12.1	2.6	17	11.5	2.2	10	8.9	2.1	3
36	5.4	17	40	5.8	10	44	8.6	3
7.6	1.5	17	19	4.8	10	16	3.1	3
14.4	3.0	17	42	18	10	260	180	3
22	28	17	49	56	10	230	150	3
320	40	17	450	70	10	620	150	3
7.1	7.0	17	8.8	6.4	10	10.6	2.2	3
7.8	.2	17	7.8	.2	10	8.1	.5	3
320	240	17	440	340	10	350	24	3

the Upper Mississippi River—1977–1978 (μg/L)

Site 3			Site 4			Site 5		
$\overline{X}$	SD	N	$\overline{X}$	SD	N	$\overline{X}$	SD	N
6.8	4.2	17	7.8	3.1	10	3.1	2.8	3
20	15	17	39	27	10	13.8	19	3
86	61	17	101	45	10	15.9	5.3	3
3.2	.9	17	3.8	1.0	10	2.5	.2	3
.33	.34	17	.48	.22	10	.22	.18	3
.9	.8	17	.9	.6	10	.2	.1	3
880	1500	17	2700	3300	10	9900	8800	3
160	130	17	170	106	10	340	260	3
720	990	17	1700	1900	10	6200	6000	3
1.5	1.8	17	3.0	3.7	10	9.0	7.2	3
.23	.31	17	.48	.50	10	.27	.29	3
1.5	2.6	17	4.8	6.5	10	5.9	2.3	3
880	1500	17	2700	3300	10	9900	8800	3
180	120	17	210	92	10	360	280	3
810	1000	17	1800	1900	10	6200	6000	3
4.7	2.1	17	6.8	3.7	10	11.5	7.0	3
.55	.54	17	.96	.52	10	.49	.46	3
2.4	2.5	17	5.7	6.4	10	6.0	2.4	3

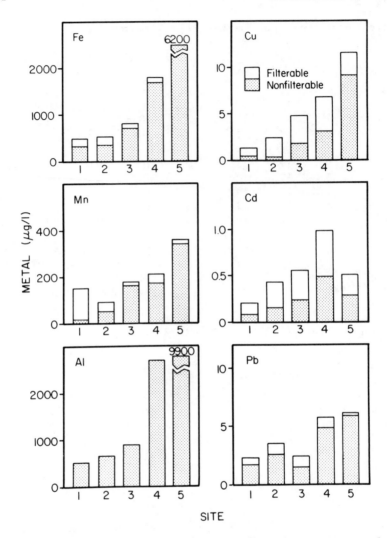

Figure 3.   *Mean dissolved and particulate metal concentrations in the upper Mississippi River, 1977–1978*

Colloidal and microparticulate matter can pass through 0.4-μm membrane filters and therefore bias metal distribution. A more detailed discussion of this is presented later.

The mean concentrations of total iron, aluminum, manganese, copper, cadmium, and lead increase with increasing distance downstream from Site 1–4. Total mean iron concentrations varied from 410 to 1800 μg/L. The relatively high concentrations of iron most likely were attributable to association with dissolved and colloidal organic matter which maintains it in solution or prevents precipitation as an amorphous ferric hydroxide (*4, 12, 25, 26*). Iron transported in the dissolved phase varies from 35% at Site 1, where the SS load is lowest, to 6% at Site 4, where the SS load is highest. Aluminum is transported more than 99% in the particulate phase, probably in an aluminum–silicate matrix.

Total manganese concentrations at Sites 1–4 ranged from 86 to 210 μg/L, with the value at Site 4 (150 μg/L) higher than at Site 2 (86 μg/L). The higher manganese concentration at Site 2 is attributable to increases in the dissolved fraction accounting for 93% of the total; manganese transported in the dissolved phase ranged from 11 to 43% for Sites 2–4. Dissolved manganese probably consists of organic and HCO_3^- complexes (*26*) with some contribution from microparticulate manganese (that is, MnO_2).

Total copper and cadmium concentrations increased from Sites 1–4, ranging from 1.2 to 6.8 μg/L for copper, and 0.2 to 1.0 μg/L for cadmium. Copper transported in the dissolved phase ranged from 56% at Site 4 to 68% at Sites 1 and 3, and 80% at Site 2. Dissolved cadmium accounted for more than 50% of the total concentrations at all sites. Thus, copper and cadmium transport in the Mississippi River was dominated by apparent solution-phase transport. Total mean lead concentrations ranged from 2.3 to 5.7 μg/L at Sites 1–4, with dissolved metal accounting for 16 to 38% of the total. The unexpectedly large contribution of dissolved lead indicates the potential role played by organic matter in maintaining lead in solution.

Wilson (*4*) has compiled available data on dissolved particulate metal distributions in rivers and in general, agrees with our observations on aluminum, iron, and manganese, but differs to varying degrees on copper, cadmium, and lead. Others have emphasized the importance of organic matter in mediating dissolved metal concentrations (*12, 26, 27*). However, Gibbs (*2, 27*) and Shuman et al. (*3*) have demonstrated the importance of particulate metal transport in major rivers. In contrast, Beck et al. (*25*) have found that dissolved metal is important in rivers of the southeastern U.S. where TOC is high, SS is low, and the pH is less than 7. Although conditions in the Mississippi River differ from those just described, the high TOC and low SS levels appear to emphasize the importance of dissolved metal.

Total metal concentrations in the Mississippi River are elevated somewhat over average river waters (Table IV), except for polluted areas. The elevated concentrations of copper, cadmium, and lead at Site 4 compared to Site 3 are in part attributable to the discharge of treated sewage effluent and urban runoff. The Minnesota River may be important for all metals studied with the exception of cadmium. Under low flow conditions, 10 to 50% of the trace metal load to the Mississippi

Table IV.   Concentrations of Metals in Average River Waters[a,b]

	Ocean		Rivers	
Element	Riley & Chester[c]	Bowen[d]	Turekian[e]	Upper Mississippi[f]
Li	180	1.1	3	—
Al	5	240	400	400–800
Sc	0.00015	—	0.004	—
Ti	1	8.6	3	—
V	1.5	1	0.9	—
Cr	0.6	0.18	1	5
Mn	2	12	7	80–360
Fe	3	670	—	400–1000
Co	0.08	0.9	0.2	1
Ni	2	10	0.3	10
Cu	3	10	7	1–7
Zn	5	10	20	15
Mo	10	0.35	1	—
Ag	0.1	0.13	0.3	—
Cd	0.05	0.08	—	0.2–1
Ba	30	54	10	—
Hg	0.05	0.08	0.07	0.1
Pb	0.03	5	3	2–6

[a] Modified from Ref. 4.
[b] All concentrations given in $\mu g \cdot L^{-1}$.
[c] Data from Ref. 77.
[d] Data from Ref. 78.
[e] Data from Ref. 79.
[f] Data from this study.

River downstream of the sewage discharge has been estimated to be attributable to the treatment plant (19). More recent data (28) suggest that 20 to 40% of the trace metal content at Site 4 results from sewage inputs under normal flow conditions, with the remaining amount contributed from upstream transport or from the Minnesota River.

The effects of seasonal and discharge variations in metal concentrations observed in rivers have been reviewed (4, 29, 30, 31). In general, no consistent relationship has been found between metal concentration and discharge, season, location, or geological environment. Intuitively, trace metal concentrations within a single river system must reflect the upstream geological environment, water chemistry, and the presence or absence of pollution discharges. Data from this study for Sites 3 and 4 reported elsewhere (32) showed little seasonal variation in the concentrations of TOC, TIC, and chloride, whereas the SS concentration reached a minimum during low flow periods (winter) and a maximum just before peak flow was achieved in spring. The SS load was responding to an increased flow velocity of the river, which moves sediment downstream, and to soil erosion.

Figures 4 and 5 show the seasonal variation in dissolved, particulate, and total iron, manganese, copper, and cadmium concentrations for Sites 3 and 4 in the upper Mississippi River. Maximum concentrations of iron and aluminum at Site 4 during maximum flows were three to five times

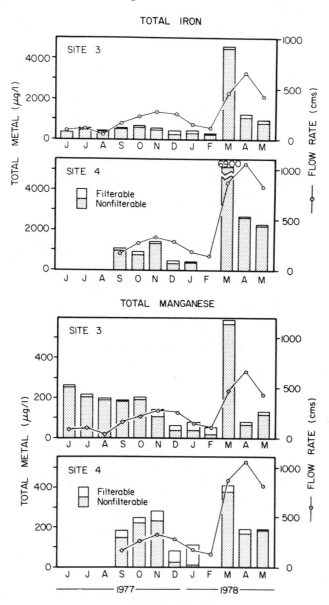

Figure 4. Seasonal variations in dissolved and particulate iron and manganese at Sites 3 and 4, 1977–1978

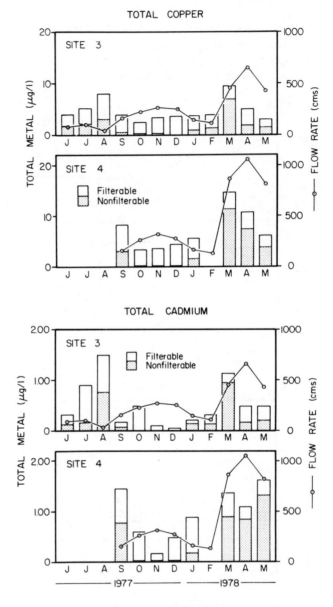

*Figure 5.   Seasonal variations in dissolved and particulate copper and
cadmium at Sites 3 and 4, 1977–1978*

the values observed at low flow. The increase in peak flow concentrations
of all metals was accounted for largely by increases in the particulate
phase. However, the increase in peak flow concentrations was significantly greater for aluminum, iron, and lead, which are dominated by

particulate-phase transport, but lesser for manganese, copper, and cadmium, which exhibit a high percentage of dissolved-phase transport.

The smaller increases of cadmium and copper concentrations during spring flow periods may mimic the lesser abundance of these metals in sediments. In general, total iron, aluminum, and manganese concentrations increased with increasing flow, suggesting particle transport, while copper, cadmium, and lead concentrations were inversely proportional to flow in summer, fall, and winter, but increased during spring flows. Wilson (4) has suggested that increasing flows should cause a decrease in the dissolved metal fraction by dilution, and an increase in the particulate phase, with discharge as a result of reentrainment of sediments. Dissolved metal did not decrease with increasing flow, but particulate and total metal concentrations did. It is apparent that the particulate and dissolved phases are not in equilibrium and that organic matter may influence trace metal associations in river water.

Ultrafiltration of Dissolved Metal. Specific analytical procedures for ultrafiltration were established expressly to maintain the integrity of the in situ metal–organic interactions. According to Buffle et al. (33), use of sequential ultrafiltration procedures minimizes the problems of inconsistent organic fractionations (34, 35), sorption (36), and leakage of organic material when the retentate concentration factor is large (37, 38). To avoid a large concentration gradient, either partial separation or repetitive washing of the retentate with distilled water of fixed ionic strength has been used. However, the latter procedure may result in irreversible alteration of the original metal–organic associations.

To maintain a maximum concentration factor less than or equal to 2 and to resolve the distribution of organic matter and organometallic complexes according to molecular size, a mass balance formalism was developed.

In sequential ultrafiltration successive volume reductions occur (Figure 6). Since volume is reduced by 50% at each stage, both the filtrates and retentates contain components of other molecular size fractions; consequently, the components contributing to each measured fraction must be taken into account in a mass balance.

The mass balance problem can be solved readily for organic carbon. The filtrates of each fraction filtered sequentially are missing organic matter of a size larger than the membrane cutoff. By measuring either the retentate or filtrate TOC in mg/L, the mass of TOC in each successive filtrate fraction can be determined if it is assumed that: (1) the TOC of molecular size less than the membrane cutoff passes in proportion to the volume; (2) the membrane behaves ideally as a molecular sieve at each cutoff; and (3) the volume is known. Given these conditions, a system of equations (Table V) for either retentate or filtrate fractions can be

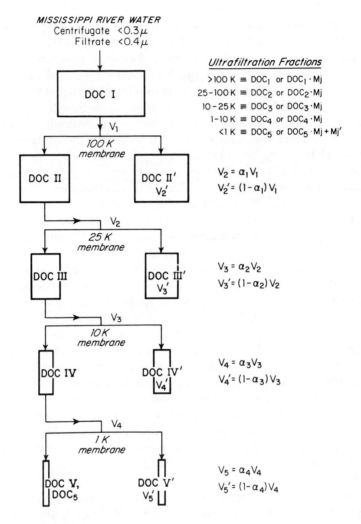

Figure 6. Mass balance approach to fractionation of dissolved organic matter and organometallic complexes by ultrafiltration (K = 1000 mol wt; v = volume (L); α = volume reduction factor)

written that accounts for the total mass of organic carbon in the original ($< 0.40\,\mu$) sample. The system of five simultaneous equations has five unknowns for which an exact solution can be obtained. For TOC in mass units, the solution to the mass balance problem is given in Equations 1–5, where α_1, α_2, α_3, and α_4 are successive volume reduction factors.

$$\text{TOC}_5 = \text{TOC V}/\alpha_1\alpha_2\alpha_3\alpha_4 \tag{1}$$

$$\text{TOC}_4 = \text{TOC IV}/\alpha_1\alpha_2\alpha_3 - \text{TOC}_5 \tag{2}$$

$$TOC_3 = TOC\ III/\alpha_1\alpha_2 - TOC_4 - TOC_5 \tag{3}$$

$$TOC_2 = TOC\ II/\alpha_1 - TOC_3 - TOC_4 - TOC_5 \tag{4}$$

$$TOC_1 = TOC\ I - TOC_2 - TOC_3 - TOC_4 - TOC_5 \tag{5}$$

Values for TOC I–TOC V in mass units were determined analytically by measurement of TOC (mg/L) in retentate or filtrate fractions and by knowing the total volume in each fraction. Alternatively, the measured retentate masses may be used directly to solve the same problem. The solution presented in Equations 1–5 can be expressed conveniently in the following series of equations for the filtrate fractions:

$$TOC_1 = TOC\ I - TOC\ II/\alpha_1 \tag{6}$$

$$TOC_2 = TOC\ II/\alpha_1 - TOC\ III/\alpha_1\alpha_2 \tag{7}$$

$$TOC_3 = TOC\ III/\alpha_1\alpha_2 - TOC\ IV/\alpha_1\alpha_2\alpha_3 \tag{8}$$

$$TOC_4 = TOC\ IV/\alpha_1\alpha_2\alpha_3 - TOC\ V/\alpha_1\alpha_2\alpha_3\alpha_4 \tag{9}$$

$$TOC_5 = TOC\ V/\alpha_1\alpha_2\alpha_3\alpha_4 \tag{10}$$

where TOC_1 is the $> 100K$ fraction, TOC_2 the 25–100K fraction, TOC_3 the 10–25K fraction, TOC_4 the 1–10K fraction, and TOC_5 the $< 1K$ frac-

Table V.   Two Systems of Simultaneous Equations Used to Solve
Sequential Ultrafiltration Mass Balance for Dissolved
Organic Carbon in Mass Units

$TOC = TOC_1 + TOC_2 + TOC_3 + TOC_4 + TOC_5$

TOC II' (retentate) $=$
  $TOC_1 + (1 - \alpha_1)[TOC_2 + TOC_3 + TOC_4 + TOC_5]$

TOC II (filtrate) $= \alpha_1[TOC_2 + TOC_3 + TOC_4 + TOC_5]$

TOC III' (retentate) $=$
  $\alpha_1 TOC_2 + (1 - \alpha_2)\alpha_1[TOC_3 + TOC_4 + TOC_5]$

TOC III (filtrate) $= \alpha_1\alpha_2[TOC_3 + TOC_4 + TOC_5]$

TOC IV' (retentate) $=$
  $\alpha_1\alpha_2 TOC_3 + (1 - \alpha_3)\ \alpha_1\alpha_2[TOC_4 + TOC_5]$

TOC IV (filtrate) $= \alpha_1\alpha_2\alpha_3[TOC_4 + TOC_5]$

TOC V' (retentate) $= \alpha_1\alpha_2\alpha_3 TOC_4 + (1 - \alpha_4)\alpha_1\alpha_2\alpha_3\ TOC_5$

TOC V (filtrate) $= \alpha_1\alpha_2\alpha_3\alpha_4\ TOC_5$

Table VI.   A System of Simultaneous Equations Used to Solve Sequential Ultrafiltration Mass Balance for Total Metal ($< 0.4 \ \mu m$)

$$MI = \sum_{i=1}^{5} TOC_i \cdot M_j + M_j'$$

$$MII = \sum_{i=2}^{5} \alpha_1 (TOC_i \cdot M_j + M_j')$$

$$MIII = \sum_{i=3}^{5} \alpha_1 \alpha_2 (TOC_i \cdot M_j + M_j')$$

$$MIV = \sum_{i=4}^{5} \alpha_1 \alpha_2 \alpha_3 (TOC_i \cdot M_j + M_j')$$

$$MV = \alpha_1 \alpha_2 \alpha_3 \alpha_4 (TOC_5 \cdot M_j + M_j')$$

tion. A similar set of simultaneous equations (Table VI) can be written for metal ions bound with the five different molecular weight fractions, although in each case the free aquated metal, M_j' must be taken into account. Unfortunately, this results in five equations and six unknowns. However, if ASV can be used to determine free and labile metal (cadmium, copper, lead), then an exact solution can be obtained for these metals. For example, solution to the metal retentate distribution is given in the following equations:

$$TOC_1 \cdot M_j = MI - MII/\alpha_1 \tag{11}$$

$$TOC_2 \cdot M_j = MII/\alpha_1 - MIII/\alpha_1\alpha_2 \tag{12}$$

$$TOC_3 \cdot M_j = MIII/\alpha_1\alpha_2 - MIV/\alpha_1\alpha_2\alpha_3 \tag{13}$$

$$TOC_4 \cdot M_j = MIV/\alpha_1\alpha_2\alpha_3 - MV/\alpha_1\alpha_2\alpha_3\alpha_4 \tag{14}$$

$$TOC_5 \cdot M_j = MV/\alpha_1\alpha_2\alpha_3\alpha_4 - M_j' \tag{15}$$

where $TOC_i \cdot M_j$ is the mass of organometallic complexes for metal M_j in the ith molecular size fraction and M_j' is the free unassociated metal.

The sequential ultrafiltration procedure was tested using known initial concentrations of the appropriate metal nitrate salt in Milli-Q water. Volume reduction factors were noted and metal concentrations for each retentate fraction were measured. From these values the mass of metal remaining in each discrete size range was determined. It follows from equations presented in Table VI that each retentate fraction will contain $[(1 - \alpha_{n+1})\ \alpha_n \alpha_{n-1} \alpha_{n-2} \ldots] M_j'$ of free metal M_j' where $4 \leq n \leq 0$, if each membrane acts as a molecular sieve. Ideally, as the total volume is reduced the mass of metal found in successive fractions will be reduced proportionately. The reduction in mass of each metal in successive retentate fractions can be predicted by knowing the volume reduction factors assuming a negligible adsorptive or desorptive interaction with the membrane or support surface. Results presented in Figure 7a for duplicate experiments agree with calculated values within 5%, which is well within the standard limit of precision using AAS. These results indicate that undesirable adsorptive or desorptive interactions with the clean membrane surfaces are inconsequential in the absence of complexation by high molecular weight organic matter and that free aquated metal ions are distributed proportionately according to volume.

Results of sequential ultrafiltration data for Site 3, summer and winter, and Site 4, summer, are presented in Figure 7(b–d). Although flow and temperature conditions were typical of late summer, the terrestrial ecosystem conditions were autumnal. Immediate conclusions that can be drawn are: (1) at the upstream site the organic carbon is found to be predominately in the < 10K fraction; (2) at the downstream site, which is approximately 8 km below the metropolitan wastewater treatment discharge, the metals appear to be distributed more evenly in approximate proportion to the distribution of organic carbon; and (3) the predominate fraction for the trace metals appear to be the < 10K fraction, which is consistent with the general concept of fulvic acid–metal associations. Sharp (39) has defined colloidal material to be in the size range 2.8–12.0 nm and microparticulate material to be in the size range 12.0–400 nm. Within the limits of these definitions, dissolved organic matter at Site 4 had a greater fraction present as colloidal and microparticulate material than Site 3. Copper, lead, and cadmium appeared to follow the mass distribution of organic carbon, while manganese and iron did not. Iron appeared most often in the larger molecular size fractions, which according to the aforementioned definition would be colloidal or microparticulate in nature. This observation is consistent with the aqueous-phase chemistry of iron(III) for which hydroxide or oxyhydroxide species are predicted to dominate at ambient pH and pE in the Mississippi River (26).

Results for Site 3 in winter are presented in Figure 7d. During this experiment calcium and magnesium were measured in each fraction, and both metals were found predominantly in the < 1K fraction. This result was somewhat surprising in view of the relatively high concentrations of these metals. Because of their high concentrations, calcium and mag-

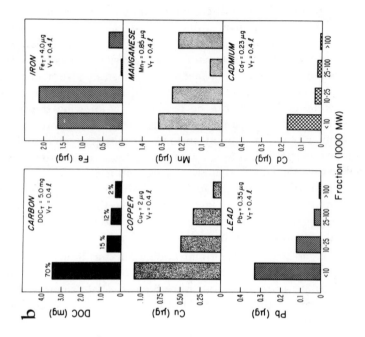

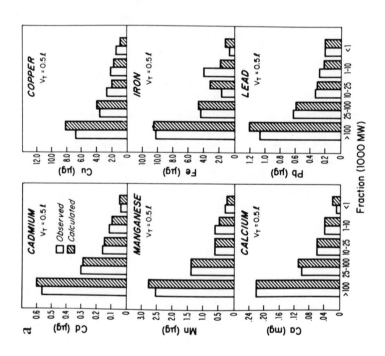

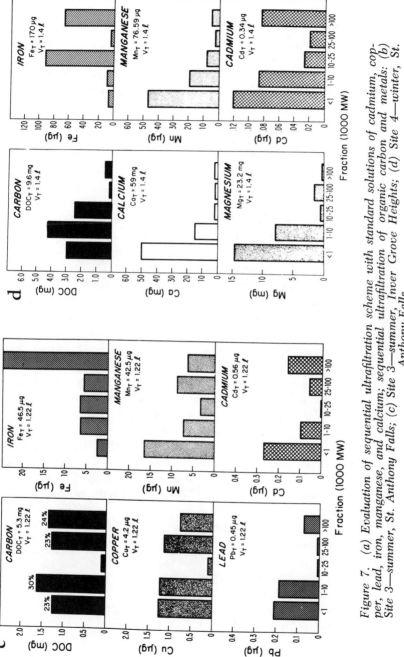

Figure 7. (a) Evaluation of sequential ultrafiltration scheme with standard solutions of cadmium, copper, lead, iron, manganese, and calcium; sequential ultrafiltration of organic carbon and metals: (b) Site 3—summer, St. Anthony Falls; (c) Site 3—summer, Inver Grove Heights; (d) Site 4—winter, St. Anthony Falls

nesium have been predicted to dominate (26) organometallic interactions in natural waters by simple mass action considerations assuming moderate formation constants for natural ligands. Trace metal–organic interactions, as a consequence, were thought to be relatively small in comparison to the major cationic metals. Current results do not support that notion. ASV measurements on the < 0.4-μm filtrate and on the $< 1K$ filtrate for the sequential ultrafiltrations indicate no detectable free copper, cadmium, or lead at pH 8.

Results of a spring 1978 sampling during a period of peak flow are presented in Figure 8 for Sites 1, 2, 4, and 5. Analytical procedures for trace metals were complicated by the fact that concentrations of copper, lead, and cadmium were extremely low. However, metal levels were consistent with an overall lower TOC level during spring runoff. Trends in metal distribution followed closely the distribution of organic carbon at each site with the exception of iron and manganese.

The general trend for spring is that the majority of metals are found in the lower molecular weight fractions and therefore can be considered to be soluble. Calcium was found for the most part in the lowest molecular weight fraction, whereas the major trace metals were found predominately in the 1–10K fraction. ASV measurements for copper, cadmium, and lead showed no detectable free or labile metal. These results, combined with the molecular size distribution of metals in proportion to organics, indicate that metal speciation in the upper Mississippi and Minnesota Rivers is dominated by organometallic interactions.

Further evidence for organometallic interactions was provided by complexation capacity measurements and pseudo stability constant determinations. Complexation capacity or the residual complexing ability of a natural water sample was determined by complexometric titration of the < 0.4-μm filtrate with soluble copper (II) (18, 40, 44) followed by ASV measurement of free and labile copper. Plots of ASV peak current vs. the cumulative copper concentration as shown in Figure 9 were made, and the intersection of lines drawn from the intitial and final linear response regions was defined as the residual complexation capacity (43, 44, 45). Complexation capacity measurements for fall 1978 are listed in Table VII. The average values for Sites 1–4 were 0.88, 0.99, 0.90, and 1.08 μM, respectively.

Stability constant determinations were based on the methods of Shuman and Woodward (43, 44) using ASV complexometric titration. The conditional formation constant for the formulation of a metal–ligand complex with an assumed stoichiometry of 1:1 can be determined from Equation 16

$$K_{\mathrm{ML}}' = \frac{(C_{\mathrm{M}} - (i_{\mathrm{a}/k}))}{(i_{\mathrm{a}/k})(C_{\mathrm{L}} - C_{\mathrm{M}} + i_{\mathrm{a}/k})} \tag{16}$$

where C_M is the total analytical metal concentration, C_L the total ligand concentration, i_a the anodic stripping current, and k an empirical proportionality constant. The numerator in Equation 16 corresponds to the total complex concentration, and the denominator is the product of the ASV free metal concentration $i_{a/k}$ and the equilibrium ligand concentration $(C_L - C_M + i_{a/k})$. During the initial stages of titration when C_L is greater than C_M, Equation 16 can be reduced to the following expression:

$$K_{ML}' = \frac{C_M}{(i_{a/k})(C_L - C_M)} \tag{17}$$

Rearranging Equation 17 and plotting i_a vs. $(C_M/(C_L - C_M))$ results in a linear relationship for which the slope is equal to k/K_{ML}'. From the latter portion of the complexometric titration curve where C_M is greater than C_L, a value of k can be obtained. In this region of the titration curve, i_a is a linear function of C_M, that is, i_a equals kC_M.

Since samples were buffered by a continuous flow of carbon dioxide gas during titrations and because the samples had a high alkalinity, a slight correction for competitive complexation of Cu^{+2} by carbonate and hydroxide was made according to the method proposed by Allen (46). This correction is based on the ASV current response when C_M is greater than C_L or, in the case of copper complexation, when $[Cu]_T$ is greater than $[L]$. In the absence of competitive ligands, $[Cu]_T \approx [Cu^{+2}]$ in this region. However, because of the constant level of $[CO_2]$ aq and high initial alkalinity, the ASV response will be reduced by a constant factor. If $\alpha_0 \equiv [Cu^{+2}]/[Cu]_T$ then

$$i_a = k\,[Cu^{+2}]/\alpha_0 \tag{18}$$

during the latter phases of titration where

$$\alpha_0 = (1 + \beta_1\,[OH^-] + \beta_2\,[CO_3^=] + \beta_3\,[CO_3^{-2}]^2)^{-1} \tag{19}$$

and β_1, β_2, β_3 are the formation constants for the monohydroxy, monocarbonato, and dicarbonato complexes, respectively. As reported by Smith and Martell (47) these values are $10^{6.3}$, $10^{6.75}$, and $10^{9.92}$, respectively. From the pH, alkalinity, and pK_a values for $H_2CO_3^*$, a value for α_0 can be readily obtained. Substitution of Equation 18 for $[Cu^{2+}]$ in Equation 17 followed by rearrangement will give the following

$$i_a = \frac{k}{\alpha_0 K_{ML}'}\left(\frac{C_M}{C_L - C_M}\right) \tag{20}$$

A linear regression fit of Equation 20 was used to determine K_{ML}' for copper where C_L was given by the complexation capacity measurement assuming the formation of a 1:1 complex. Average pseudo stability con-

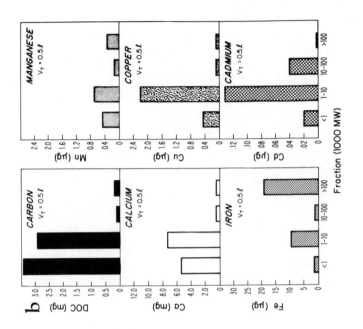

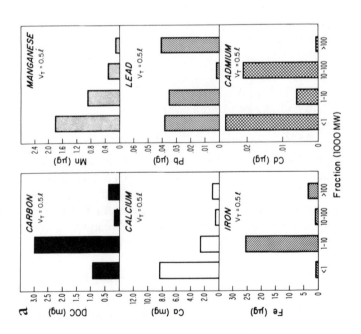

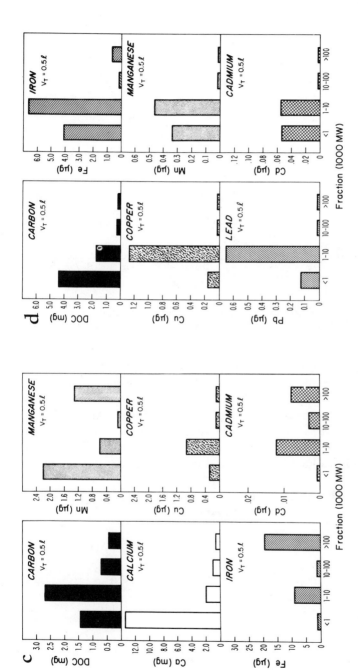

Figure 8. Sequential ultrafiltration of organic carbon and metals in spring 1978: (a) Site 1—Bemidji; (b) Site 2—Royalton; (c) Site 4—Inver Grove Heights; (d) Site 5—Minnesota River

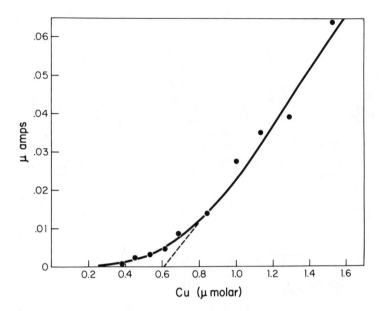

Figure 9. Measurement of copper complexation capacity in upper Mississippi River water—Site 4, 1978: current (μ amps) vs. copper added (μM)

stants for copper are given in Table VII. Log K_{ML}' ranged from 9.7 to 10.7 for Sites 1–4. To test this procedure, the formation constant for the complexation of Cu^{+2} by $HEDTA^{3-}$ at pH 8.1 was determined to be $10^{8.8}$, which is the same as the value reported by Smith and Martell (*47*).

Results of this study as illustrated in Figures 7 and 8 show that (1) the sequential ultrafiltration-mass balance approach is valid for known solutions; (2) cadmium, copper, and lead are found in the soluble phase

Table VII. Copper Complexation Capacity and Stability Constant Data—Fall 1978 [a]

Site	DOC (mg/L)	pH	pC_T [b]	CC (μM)	log K_{ML}'
1	18.4	8.05		0.88	10.4
2	13.2	8.23		0.99	9.9
3	24.3	8.29		0.90	10.7
4	17.6	8.16		1.08	10.0
EDTA [c, d]	—	8.10	5.0	—	8.8

[a] See Ref. *54*.
[b] C_T = total carbonate concentration.
[c] [EDTA] = $5 \times 10^{-6} M$.
[d] $Cu^{+2} + EDTA^{-4} \rightleftharpoons Cu(EDTA)^{-2}$ $pK = 18.8$
   $HEDTA^{-3} \rightleftharpoons EDTA^{-4} + H^+$ $pK_a = 10.0$

$Cu^{+2} + HEDTA^{-3} \rightleftharpoons Cu(EDTA)^{-2} + H^+$ $pK' = 8.8$

associated primarily with organic matter in the 1–10K mol wt range whereas calcium and magnesium are found predominately in the < 1K fraction; (3) iron and manganese are found primarily in the colloidal phase associated with organic matter in the 25–100K mol wt fraction; and (4) the distribution of trace metals correlates well with the distribution of dissolved organic carbon. ASV measurements summarized in Table VII show that (1) there is no detectable free or labile copper, cadmium, or lead; (2) the residual complexation capacity for copper is high; and (3) the pseudo stability constant for copper complexation is moderately high (that is, 10^9). Stability constants in the range 10^8–10^9 are indicative of metal complexes with multidentate ligands (47–50) containing a mixture of amino, sulfhydryl, carboxylate, or hydroxy functional groups.

A high level of residual complexation capacity, large pseudo stability constants, and metal molecular size distributions that closely parallel trends in organic carbon suggest that metal speciation is dominated by organometallic complexes with either soluble, colloidal, or microparticulate organic matter in the upper Mississippi River. This conclusion is consistent with the observations that trace metals are associated with an operationally defined organic fraction in water-borne particulates ($> 0.4\,\mu$) in the upper Mississippi River.

Ultrafiltration procedures and methods for data reduction by mass balance techniques developed as a part of this study have been shown to be useful for characterization of metal complexes by size fractionation. Ultrafiltration, coupled with ASV measurements for copper, cadmium, and lead allows the degree of metal association in discrete molecular size fractions to be determined. Common ultrafiltration problems such as membrane clogging, leakage, concentration polarization, ionic strength changes, and excessive concentration, which may lead to an alteration of the actual distribution of complexes, are avoided with these procedures. Flushing procedures with artificial waters of similar composition and ionic strength may effect changes in the degree of metal complexation by dilution or introduction of trace metal contaminants. However, it should be pointed out that size fractionation by ultrafiltration with these procedures is not problem free. Nominal molecular weight cutoffs are imprecise since the shape and degree of ionization of the molecule is an important factor in retention by or passage through a particular membrane. Ogura (37) has shown that cytochrome C (12.6K mol wt) was 95% retained by a 100K membrane and that cobalamin (1.357K mol wt) was 75% retained by a 0.5K membrane, although urea (0.06K mol wt) was not retained by either membrane. Macko et al. (35) have shown that the degree of ionization of low molecular weight organic acids is critical to the degree of retention by membranes with higher molecular weight cutoffs. They have shown also that retention is partially a function of flow rate and ionic strength when Ca^{+2}, Mg^{+2}, and $SO_4^=$ are primary solution components. Charged sulfonated membranes, which were developed to avoid fouling problems, are known to retain relatively high concentrations of of Ca^{+2} and Mg^{+2} (51). Flow rate problems can be overcome by passing sufficient water through new membranes until a constant flux has been obtained or, alternatively, by using scrupulously cleaned and washed used membranes that tend to have relatively constant fluxes (35). When metal distributions are of primary concern, results of the present study show that new, well-washed membranes with constant fluxes are preferable to clean, but used membranes.

Storage procedures are also important in ultrafiltration experiments. Unacidified field samples should be filtered as soon as possible to avoid sorptive losses to LPE bottle walls (52), especially when low levels of cadmium, lead, and copper are present initially.

Results of this study agree in general with the observations of other investigators. Benes et al. (36) concluded that transition metals were associated strongly with aquatic humus from results of ultrafiltration experiments and neutron activation analyses. Benes and Steinnes (53) found the majority of calcium, magnesium, and manganese to lie in the 10K mol wt fraction in the Glomma River. Schindler and Alberts (54) and Guy and Chakrabarti (56) reported that iron was found as microparticulate ion in the $>$ 100K fraction in the oxic waters of a Georgia reservoir and the Rideau River, respectively.

The observed copper complexation capacity of approximately 1.0 μM for the four Mississippi River sites is similar in magnitude to those reported by Chau and co-workers (18, 40, 41, 42) for Hamilton Harbor, the Niagara River, Lake Erie, and Lake Ontario. Allen (46) also reported comparable copper complexation capacities for the Mississippi River, Hidden Pond, and Saganashkee Slough. Higher copper complexation capacities have been reported by Shuman and Woodward (44) {20–126 μM} for some North Carolina rivers, and Guy and Chakrabarti (10) {23–49 μM} for Canadian freshwaters.

Moderately large pseudo stability constants for copper complexation in natural waters have been reported by other investigators. Allen (46) determined $-p\beta$ values of 9.0 for Lake Michigan and Maple Lake by ASV. Buffle et al. (33) used a copper ion specific electrode to determine $-p\beta$ values of 10.8 to 11.6 for Zaire riverwater, and van den Berg and Kramer (56) used a MnO_2 adsorption technique to determine $-p\beta$ values of some Canadian rivers and lakes that ranged from 7.2 to 9.5. Sunda and Hanson (57) reported a $-p\beta$ value for copper in Neuse River of 10. However, Shuman and Woodward (44) found apparent stability constants by ASV for copper complexation in natural waters to be near $10^{6.5}$ at pH 6.5. Bilinski et al. (58) have determined by ASV formation constants for copper, lead, and cadmium hydroxide or carbonate complexes and suggest that $PbCO_3^\circ$ and $CuCO_3^\circ$ are the preponderant lead and copper species in natural water. For cadmium, hydroxy or chloro complexes were thought to be predominant. Log K for $PbCO_3^\circ$ and $CuCO_3^\circ$ was found to be 6.4 and 6.0, respectively. By similar techniques Ernst et al. (59) found stability constants (log K) for the same complexes to be 6.2 and 6.1, respectively.

Results of this study and those just reported suggest that copper complexation in many waters is dominated by aquatic organics and that in certain waters, formation of carbonato complexes is of secondary importance. The magnitude of the observed stability constants for apparent 1:1 complexes indicates that the ligand involved is a multidentate organic chelate. The copper stability constants for simple organic chelates such as citrate are on the order of 10^6 under natural conditions. Stronger chelates containing amino or sulfhydryl groups, such as NTA or EDTA, more closely approximate the observed copper complexation characteristics in natural waters. Macrocyclic ligands such as porphyrins and phthalocyanines in which bonding takes place through pyrolle nitrogens may also play an important role in metal chelation in natural waters. The potential role of organic nitrogen and sulfur functionalities is strengthened by the experimental observation made by Perdue (60) that aquatic fulvic material isolated from Oregon river water contained aromatic carboxylic acids with two-thirds of the phenolic hydroxyl groups in meta and para positions. Only phenolic acids with ortho hydroxyl groups could participate in chelation via salicylate-like functional groups. The apparent strength of copper complexes in natural waters cannot be ascribed to carboxylate ligands alone.

Chemical Fractionation of Particulate Metal. Chemical fractionation schemes have been devised for determining metal transport phases in riverine particulates (2, 3), sediments (61–65), soils (66, 67, 68), and anaerobically digested sludge (22). Of these approaches, the fractionation schemes developed by Gibbs (2) and Shuman et al. (3) offer the best insight into the partitioning of metals between riverine dissolved and particulate forms. The modified fractionation scheme selected was the sequential leaching of isolated solids with $0.5N$ $MgCl_2$ at pH 7 for 7 hr (adsorbed/ion-exchangeable), $0.4N$ $Na_4P_2O_7$ at pH 7 for 10 hr (organic), $0.3N$ HCl at $90°C$ for 30 min (metal oxide), and conc. HCl–HNO_3–HF (crystalline). $MgCl_2$ was chosen over NH_4Cl or NH_4-acetate because it was more selective at removing sorbed metal, and oxide or organic phases were not attacked (2). Pyrophosphate at pH 7 was chosen over H_2O_2 and HOCl to remove organic-bound metal because released metal does not precipitate, and $Na_4P_2O_7$ does not attack metal oxide coatings, clay mineral structures, and metal carbonates and sulfides

($22, 69, 70$). Use of H_2O_2 as a chemical extractant was avoided since it was found to give poor results in the presence of clay minerals and MnO_2 (62). Metals released from organics by peroxide oxidation either were adsorbed on clays and MnO_2 and/or MnO_2 catalyzed the decomposition of H_2O_2 leaving organics intact. Pyrophosphate released less than 0.1% of standard Fe_2O_3 and MnO_2 under conditions identical to those mentioned previously. Although naturally occurring oxides are probably more amorphous in nature, less than 10% removal of metal oxide coating is expected with pyrophosphate. Hot $0.3N$ HCl was chosen over citrate–dithionate extraction at pH 3 for metal oxide coatings because solutions of lower salt content were obtained, and there is minimal disruption of the clay mineral structure (23).

Metal transport phases for iron, manganese, copper, cadmium, and lead are shown in Figures 10–14, with individual elemental data given in Tables VIII and IX. The fractionation scheme was applied to composite samples for Sites 1 and 2, and all fractionations were performed in duplicate.

Aluminum was transported in the crystalline phase ($> 80\%$) at Sites 1 and 2, followed in importance by solution- ($7-22\%$), oxide- ($2-7\%$), and organic- ($2-3\%$) phase transport. Surface-adsorbed or ion-exchangeable aluminum was unimportant. This distribution illus-

Table VIII. Particulate Metal Fractionation at Nonurban Sites in the Upper Mississippi River—1977–1978[a,b]

Transport Phase	Site	Al	Mn	Fe	Cu	Cd	Pb
				(%)			
Solution	1	22	60	19	94	88	62
	2	7	35	39	93	96	51
Adsorbed	1	.3	5	.1	.4	.2	~ 0
	2	.01	10	.07	.5	.08	~ 0
Organic solid	1	2	4	11	3	.2	5
	2	3	16	25	3	.4	26
Oxide coating	1	2	28	2	1	11	2
	2	7	36	6	1	3	11
Crystalline	1	75	3	68	2	1	31
	2	83	2	30	2	.2	11
Available metal	1	25	97	32	98	99	69
	2	17	98	70	97	~ 100	89
Total concentra-tion ($\mu g/L$)	1	15.8	265	1215	2.6	0.24	1.8
	2	216	116	437	2.1	0.24	1.4

[a] Fall and winter composite.
[b] Available metal = noncrystalline metal.

Table IX. Particulate Metal Fractionation at Urban Sites in the Upper Mississippi River—1977–1978

Transport Phase	Site	(%)					
		Al	Mn	Fe	Cu	Cd	Pb
Solution	3F	.4	3	3	64	62	16
	W	2	28	25	83	90	36
	4F	.4	3	5	59	86	9
	W	1	40	16	66	51	16
Adsorbed	3F	.3	27	.1	3.6	3	~0
	W	.8	11	.05	1.1	.2	~0
	4F	.2	22	.2	5	5	~0
	W	.1	19	.09	3	1	~0
Organic solid	3F	2	44	26	17	5	47
	W	2.4	46	23	11	8	46
	4F	2	56	27	22	6	83
	W	3	21	20	16	24	63
Oxide coating	3F	9	23	17	8	28	15
	W	9	13	10	2	23	6
	4F	9	14	12	7	2	5
	W	9	18	12	9	24	13
Crystalline	3F	88	3	54	6	1	22
	W	86	3	41	3	.3	12
	4F	88	5	56	7	.2	3
	W	87	3	53	6	.4	9
Available metal[a]	3F	12	97	46	94	99	78
	W	14	97	59	97	~100	88
	4F	12	95	44	93	~100	97
	W	13	97	47	94	~100	91
Total concentration (μg/L)	3F	568	235	576	7.0	.85	4.5
	W	399	72	555	3.5	.30	2.6
	4F	1480	100	1440	8.8	.85	17.3
	W	814	106	951	5.3	.83	5.1

[a] Available metal ≡ noncrystalline metal.

trates the influence that aluminum silicate minerals exert on aluminum transport in the river, not unlike conclusions drawn by others (*2, 3*).

Manganese transport at the background Site 1 was controlled primarily by the solution phase (60%) and metal oxide coatings (28%), while at site 2, solution and oxide coatings contributed equally (36%) to manganese transport, with solid organic phase contributing 10–16% for the two sites. In comparison, manganese transport in the Amazon and Yukon Rivers was dominated by the oxide fraction (> 50%) with lesser contributions from crystalline and solution phases. Equilibrium

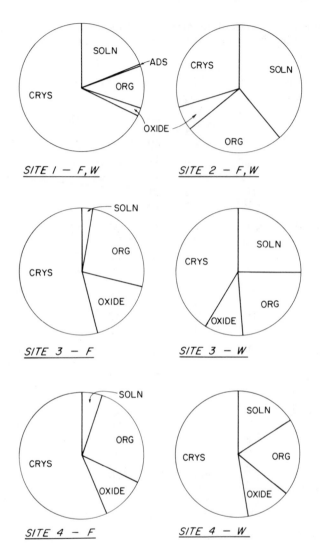

Figure 10. *Iron transport phases in the upper Mississippi River, 1977–1978*

models (26, 71) and field studies (72) suggest that manganese should exist as soluble (for example, $Mn(H_2O)_6^{+2}$, $MnHCO_3^+$) or insoluble complexes (for example, $MnCO_3$, MnO_2, mixed metal oxide). In the Mississippi River at Sites 1 and 2, solution and oxide phases dominate manganese transport as suggested by equilibrium modeling. Jenne (1) has suggested that manganese and iron oxides may control trace metal concentrations in natural waters.

Iron transport at Sites 1 and 2 was dominated by the crystalline fraction (68%) even though the SS load was lowest at this site. Solution and solid organic phases contributed approximately 19 and 11%, respectively. In contrast, iron at Site 2 was divided equally between solution, crystalline, and organic phases. The dominance of the crystalline phase at Site 1 is consistent with an apparent enrichment of iron in the mineral fraction (*18*) whereas the higher levels of solution and organic iron at Site 2 may be attributable to runoff of soil particles and iron oxides from

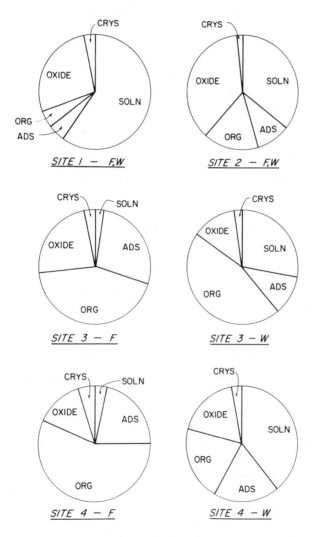

Figure 11.   Manganese transport phases in the upper Mississippi River,
1977–1978

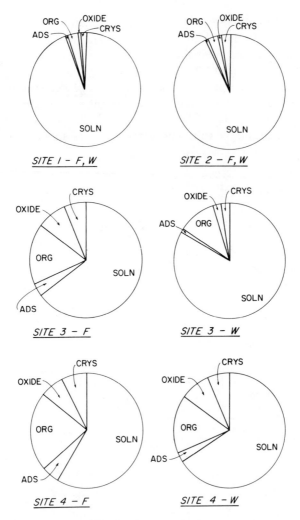

Figure 12. Copper transport phases in the upper Mississippi River, 1977–1978

agricultural lands. Greenland (73) has suggested that iron oxides complexed with organic macromolecules form amorphous gels on soil clay particles that retain metal scavenging ability. The pyrophosphate reagent should liberate the amorphous iron oxide, releasing it to solution.

The transport of copper and cadmium at the unpolluted Sites 1 and 2 was dominated by the solution phase ($\geq$ 90%). This is in direct contrast to data for the Yukon and Amazon Rivers where approximately 80% was bound in the crystalline phase and derived from a high SS load from upland areas (2). In the upper Mississippi River, the SS load is low with relatively high TOC, implicating natural organic matter interactions.

Lead transport was dominated by the solution phase (> 50%) but had significant contributions from the oxide, organic, and crystalline phases. Hem (74) has suggested that lead levels are held below equilibrium solubility levels by adsorption processes, altered in this case by low solid surface areas and high TOC levels.

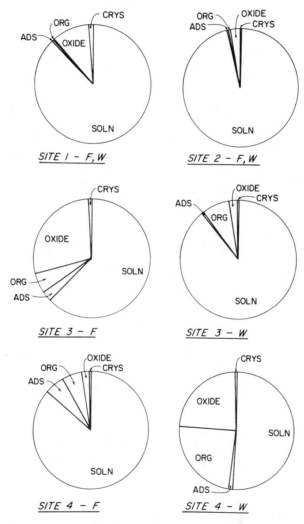

Figure 13.   Cadmium transport phases in the upper Mississippi River,
1977–1978

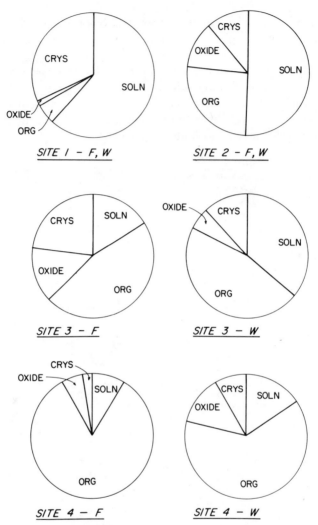

Figure 14.   Lead transport phases in the upper Mississippi River, 1977–1978

Sites 3 and 4 represent a direct contrast to the transport phases at the background sites. Site 3 receives upstream flow and runoff from a portion of the urban Twin Cities area, while Site 4 is influenced additionally by urban runoff, treated sewage effluent, and tributary flow from the Minnesota River.

Solution-phase transport of metals increased generally between fall and winter 1977, at Sites 3 and 4, although the overall contribution to total metal was still less than for Sites 1 and 2. The most significant increase in solution metal transport from fall to winter occurred for manganese and iron (at 4 to 10 times), with smaller percent increases noted for copper, cadmium, and lead. This trend was coincident with a decrease in SS concentration. On an absolute basis, copper, cadmium, and lead concentrations in the solution phase were reduced by approximately 50%.

Adsorbed/ion-exchangeable metal transport was significant only for manganese, representing about 20% of the total at Sites 3 and 4. The minor role of the adsorbed phase agrees with the findings of Gibbs (2) and Shuman et al. (3). Seasonally, the relative contribution of adsorbed metal decreased from fall to winter for iron, manganese, copper, and cadmium, coincident with a decrease in the SS load, suggesting that changes in particle surface area or properties exert only a small effect on metal transport.

Manganese and lead were transported primarily in association with solid organic matter at Sites 3 and 4, while the solid organic matter had a secondary influence on iron and copper. The relative importance of solid organic metal did not change appreciably for any metal at Site 3 between fall and winter. In contrast, the amount of manganese and cadmium transported in the solid organic phase increased from 22 to 56% and from 6 to 24%, respectively, from fall to winter. The amount of iron, copper, and lead in the organic phase decreased from fall to winter. No apparent explanation exists for the behavior of manganese, lead, and cadmium at Sites 3 and 4 since the relative quantities of carbon, hydrogen, oxygen, and nitrogen in the pyrolyzable fraction did not differ (75). The solid organic-phase transport of copper, cadmium, and lead was greater at Site 4 than at Site 3, suggesting that the nature of the organic matter may differ. Hullett and Eisenreich (76) have found elevated concentrations of sewage-derived fatty acids at Site 4, and the SS isolated from Site 4 exhibited visual and odorous qualities typical of activated sludge. The increased importance of solution and solid organic transport of trace metals at Site 4 suggests that either metals bound to sewage sludge are being added to the river or metals in the river bind to sewage sludge organics. In any case, the sewage effluent alters both the concentrations and speciation of the particulate metals.

Metal oxide coatings are most important for manganese, iron, and cadmium transport at Sites 3 and 4, but represent less than 25% of the total metal transport in all cases. Crystalline-bound aluminum (> 80%) and iron (> 50%) represented the dominant transport mode, which was unimportant for copper, cadmium, and manganese, and crystalline lead contributed about 20 and 6% at Sites 3 and 4, respectively, to total lead transport. The decrease in crystalline lead at Site 4 may result from dilution of lattice-bound lead by solid organic lead emanating from the sewage effluent.

The transport modes of iron, aluminum, manganese, copper, cadmium, and lead can be summarized as in the following list, which represents the relative ranking of each metal in each phase at the unpolluted and polluted site. In each case, the ranking refers to the percent metal transport relative to total metal transport.

Unpolluted

Al	Cry >	Soln >	Ox >	Org >	Ads
Mn	Soln >	Ox >	Org >	Ads >	Cry
Fe	Cry >	Soln >	Org >	Ox >	Ads
Cu	Soln >	Org >	Cry >	Ox >	Ads
Cd	Soln >	Ox >	Cry ~	Org ~	Ads
Pb	Soln >	Cry >	Org >	Ox >	Ads

Polluted

Al	Cry >	Ox >	Org >	Soln >	Ads
Mn	Org >	Ox >	Ads >	Soln >	Cry
Fe	Cry >	Org >	Ox >	Soln >	Ads
Cu	Soln >	Org >	Ox >	Cry >	Ads
Cd	Soln >	Ox >	Org >	Ads >	Cry
Pb	Org >	Soln >	Cry >	Ox >	Ads

At the unpolluted sites, solution-phase transport dominates for manganese, copper, cadmium, and lead, while aluminum and manganese are dominated by crystalline-phase contributions. In contrast, at the polluted sites the solution and organic phases dominate for manganese, copper, cadmium, and lead, while the crystalline phase remains dominant for iron and aluminum transport. The effect of urban and sewage discharges to the river is to increase the total metal concentrations and more importantly, the particulate trace metal speciation to geochemically available forms. With the exception of iron and aluminum, more than 90% of the manganese, copper, cadmium, and lead occurred in transport phases thought to be accessible to further aqueous chemical and biotic interactions.

Summary

Trace metal transport phases in the upper Mississippi River were investigated during 1977–1978 at sites representative of natural bog/marsh drainage, agricultural drainage, urban runoff, and municipal and industrial discharges. The metals iron, aluminum, manganese, copper, cadmium, and lead were found to partition between the dissolved ($< 0.4\,\mu m$) and particulate ($> 0.4\,\mu m$) phases according to the water chemistry of each element, its source, and the degree of interaction with dissolved and particulate organic matter in the river. Aluminum, iron, and manganese were transported primarily in the particulate phase at all sites, although a significant fraction (10–20%) was associated with the dissolved phase. At the background site, manganese, and to a lesser extent iron, showed greater dissolved concentrations. The transport of copper and cadmium, and to a lesser extent lead, was controlled by the dissolved phase (30–70%). The fraction of copper, cadmium, lead, iron, and manganese in the dissolved phase increased with decreasing SS concentrations in the river, showing the importance of lattice-bound and surface-associated metal transport. The concentrations of all metals increased with increased river flow, and the fraction in the dissolved phase was highest in the winter months, corresponding to minima in the flow and SS concentration.

Particulates isolated from the river were subjected to a chemical fractionation scheme that separated iron, aluminum, manganese, copper, cadmium, and lead phases into solution, adsorbed, solid organic, oxide coating, or crystalline fractions. Aluminum was transported in the crystalline mode ($\sim 90\%$), probably associated with aluminosilicate minerals. Iron also had the crystalline phase as an important transport mode but had contributions from the solution, solid organic, and oxide phases. Manganese was transported primarily in the solution and oxide phases at unpolluted sites, with greater contributions of solution, adsorbed, and solid organic phases at polluted sites. Copper, cadmium, and lead exhibited significant solution-phase transport at all sites. The solid organic phase of metal transport increased at the site receiving sewage input. Copper, cadmium, and lead in the river discharged as soluble complexes in the sewage effluent may associate with solid organic matter being discharged. This has the effect of increasing metal transport and the overall quantity of metal accessible to the biota. With the exception of iron and aluminum, more than 90% of the manganese, copper, cadmium, and lead occurred in transport modes thought to be available to aqueous and biotic interactions.

The interaction of trace metals with dissolved, colloidal, and microparticulate organic matter appears to control metal transport in the dissolved ($< 0.4\,\mu m$) fraction of river water. Ultrafiltration and ASV were used to determine the distribution and degree of association of copper, cadmium, and lead with organic matter. A novel mass balance approach was developed for fractionation of dissolved organic matter and organometallic complexes into discrete molecular size ranges. Mass balance and analytical procedures used showed that the highest concentrations of trace metals occurred in the $< 10K$ mol wt range. Calcium and magnesium were found predominantly in the $< 1K$ molecular size range, and appear to have minimal impact on the extent of interaction of trace metals with organic matter. Mississippi River water routinely exhibited residual copper complexation capacities of approximately 1.0 μM and a pseudo stability constant of about 10^{10}. The results suggest that (1) trace metal complexation is dominated by aquatic organics; (2) formation of copper–carbonato complexes is of secondary importance; and (3) organic ligands forming the complexes are likely multidentate and contain nitrogen and sulfur ligand atoms.

Acknowledgment

This work was supported by a National Science Foundation Grant (ENV 77-04496) awarded to W. J. Maier, M. R. Hoffmann, and S. J. Eisenreich.

Literature Cited

1. Jenne, E. A. "Controls on Mn, Fe, Co, Ni, Cu and Zn Concentrations in Soils and Waters: The Significant Role of Hydrous Mn and Fe Oxides," In "Trace Organics in Water," Adv. Chem. Ser. 1968, 73, 337–387.
2. Gibbs, R. J. Geol. Soc. Am. Bull. 1977, 88, 829.
3. Shuman, M. S.; Haynie, C. L.; Smock, L. A. Environ. Sci. Technol. 1978, 12, 1066.
4. Wilson, A. L. "Concentration of Trace Metals in River Waters: A Review," Water Research Centre Technical Report TR 16, Medmenham Laboratory, Buckinghamshire, England, 1976, 60 p.
5. Kharkar, D. P.; Turekian, K. K.; Bertine, K. K. Geochim. Cosmochim. Acta 1968, 32, 285.
6. Turekian, K. K.; Scott, M. R. Environ. Sci. Technol. 1967, 1, 940.
7. Turekian, K. K. Geochim. Cosmochim. Acta 1977, 41, 1139.
8. Siegel, A. "Metal Organic Interactions in the Marine Environment," In "Organic Compounds in Aquatic Environment"; Faust, S. D., Hunter, J. V., Eds.; Marcel Dekker: New York, 1971.
9. Stumm, W.; Brauner, "Chemical Speciation," In "Chemical Oceanography"; Riley, J. P., Skirrow, G. D., Eds.; Academic: London, 1975; Vol. 1.
10. Guy, R. D.; Chakrabarti, C. L. Can. J. Chem. 1976, 54, 2600.
11. Florence, T. M.; Batley, G. E. Talanta 1977, 24, 151.
12. Reuter, J. H.; Perdue, E. M. Geochim. Cosmochim. Acta 1977, 41, 325.
13. Barber, R. T.; Ryther, J. H. J. Exp. Mar. Biol. Ecol. 1969, 3, 191.
14. Swallow, K. C.; Westall, J. C.; McKnight, D. M.; Morel, N. M. L.; Morel, F. M. Limnol. Oceanogr. 1978, 23, 538.
15. Jackson, G. A.; Morgan, J. J. Limnol. Oceanogr. 1978, 23, 268.
16. Davey, E. W.; Morgan, M. J.; Erickson, S. J. Limnol. Oceanogr. 1973, 18, 993.
17. Giesy, J. P., Jr.; Lawrence, G. J.; Williams, D. R. Water Res. 1977, 11, 1013.
18. Gachter, R.; Lum-Shue-Cham, K.; Chau, Y. K.; Schweiz, Z. Hydrol. 1973, 35, 252.
19. Vanderboom, S. A. "The Fate of Trace Metals in Wastewater Effluents Discharged to the Mississippi River," M.S. Thesis, University of Minnesota, Minneapolis, MN, 1976, 69 pp.
20. Eggiman, D. W.; Betzer, P. R. Anal. Chem. 1976, 48, 886.
21. Standard Methods for Analysis of Water and Wastewaters, U.S.P.H.S., 1975.
22. Stover, R. C.; Sommers, L. E.; Silviera, D. J. J. Water Pollut. Control Fed. 1976, 48, 2165.

23. Malo, B. A. *Environ. Sci. Technol.* **1977,** *11,* 277.
24. Minnesota Pollution Control Agency, Water Quality Report, 1978.
25. Beck, K. C.; Reuter, J. H.; Perdue, E. M. *Geochim. Cosmochim. Acta* **1974,** *38,* 341.
26. Stumm, W.; Morgan, J. J. "Aquatic Chemistry"; John Wiley: New York, 1970.
27. Gibbs, R. J. *Science* **1973,** *180,* 71.
28. Eisenreich, S. J.; Hoffmann, M. R.; Maier, W. J. "A Mass Balance of Trace Elements in Natural Fresh Waters," In "Trace Metals and Metal Organic charges," Env. Eng. Prog., Univ. of Minn., 1979.
29. Andelman, J. B. "Incidence, Variability and Controlling Factors for Trace Elements in Natural, Fresh Waters," In "Trace Metals and Metal-Organic Interaction in Natural Waters"; Singer, P. C., Ed.; Ann Arbor Science: Ann Arbor, 1973; p. 57–88.
30. Angino, E. E.; Magnuson, L. M.; Waugh, T. C. *Water Resour. Res.* **1974,** *10,* 1187.
31. Hellman, H. *Dtsch. Gewaesserkd. Mitt.* **1970,** *14,* 42.
32. Rastetter, D., M.S. Thesis in progress, University of Minnesota, Minneapolis, MN, 1979.
33. Buffle, J.; Deladoey, P.; Haerdi, W. *Anal. Chim. Acta* **1978,** *101,* 339.
34. Wilander, A.; Schweiz, A. *Hydrol.* **1972,** *34,* 190.
35. Macko, C. A.; Maier, W. J.; Hoffmann, M. R.; Eisenreich, S. J. *Proc. AIChE,* Feb. (**1978**).
36. Benes, P.; Skinnes, E. *Water Res.* **1974,** *8,* 947.
37. Ogura, N. *Mar. Biol.* **1974,** *24,* 305.
38. Wheeler, J. R. *Limnol. Oceanogr.* **1976,** *21,* 846.
39. Sharp, J. H. *Limnol. Oceanogr.* **1973,** *18,* 441.
40. Chau, Y. K. *J. Chromatogr. Sci.* **1973,** *11,* 579.
41. Chau, Y. K.; Lum-Shue-Chan, K. *J. Fish. Res. Board. Can.* **1974,** *31,* 1515.
42. Chau, Y. K.; Lum-Shue-Chan, K. *Water Res.* **1974,** *8,* 383.
43. Shuman, M. S.; Woodward, G. P., Jr. *Anal. Chim.* **1973,** *45,* 2032.
44. Shuman, M. S.; Woodward, G. P., Jr. *Environ. Sci. Technol.* **1977,** *11,* 809.
45. Rosenthal, D.; Jones, G. L., Jr. *Anal. Chim. Acta* **1971,** *53,* 141.
46. Allen, H. L. "Biological Consequences of Chemical Speciation of Heavy Metals in Natural Waters," Final Report, U.S. DOE, Washington, D.C., 1978.
47. Smith, R. M.; Martell, A. E. "Critical Stability Constants—Inorganic Complexes"; Plenum: New York, 1976; Vol. 4.
48. Martell, A. E.; Smith, R. M. "Critical Stability Constants—Amino Acids"; Plenum: New York, 1974; Vol. 1.
49. Smith, R. M.; Martell, A. E. "Critical Stability Constants—Amines"; Plenum: New York, 1975; Vol. 2.
50. Smith, R. M.; Martell, A. E. "Critical Stability Constants—Other Organic Ligands"; Plenum: New York, 1977; Vol. 3.
51. Kesting, R. E. "Polymeric Membranes"; McGraw–Hill: New York, 1971.
52. Batley, G. E.; Gardner, D. *Water Res.* **1977,** *11,* 745.
53. Benes, P.; Steinnes, E. *Water Res.* **1974,** *8,* 947.
54. Schindler, J. E.; Alberts, J. J. *Arch. Hydrobiol.* **1974,** *74,* 429.
55. Guy, R. D.; Chakrabarti, C. L. *Proc. Int. Conf. Heavy Metals in the Environment, Toronto, Ontario,* 1975, p. 275.
56. van den Berg, C. M. G.; Kramer, J. R. *Anal. Chim. Acta* **1979,** *106,* 113.
57. Sunda, W. G.; Hanson, P. J. "Chemical Speciation of Copper in River Water," In "Chemical Modeling in Aqueous Systems," *ACS Symp. Ser.* **1979,** 93.
58. Bilinski, H.; Huston, R.; Stumm, W. *Anal. Chim. Acta* **1976,** *84,* 157.
59. Ernst, R.; Allen, H. E.; Mancy, K. H. *Water Res.* **1975,** *9,* 969.
60. Perdue, E. M. *Geochim. Cosmochim. Acta* **1978,** *42,* 1351.

61. Gupta, S. K.; Chen, K. Y. *Environ. Lett.* **1975,** *10,* 129.
62. Guy, R. D.; Chakrabarti, C. L.; McBain, D. C. *Water Res.* **1978,** *12,* 21.
63. Agemian, H.; Chau, A. S. Y. *Environ. Contam. Toxicol.* **1977,** *6,* 69.
64. Filipek, L. H.; Owen, R. M. *Can. J. Spectrosc.* **1978,** *23,* 31.
65. Deuer, R.; Förstner, U.; Schmoll, G. *Geochim. Cosmochim. Acta* **1978,** *42,* 425.
66. Jackson, M. L. "Soil Chemical Analysis"; Prentice–Hall: Englewood Cliffs, 1958.
67. McKeague, J. A. *Can. J. Soil Sci.* **1967,** *47,* 95.
68. McLaren, R. G.; Crawford, D. V. *Soil Sci.* **1973,** *24,* 172.
69. Lahann, R. W. *J. Environ. Sci. Health, Part A* **1976,** *11,* 639.
70. Smith, R. G., Jr. *Anal. Chem.* **1976,** *48,* 76.
71. Hoffmann, M. R.; Eisenreich, S. J. "Chemical Equilibrium Model for the Variation of Fe and Mn in the Hypolimnion of Lake Mendota," *Env. Sci. Technol.,* in press.
72. Krom, M. D.; Sholkovitz, E. R. *Geochim. Cosmochim. Acta* **1978,** *42,* 607.
73. Greenland, D. J. *Soil Sci.* **1971,** *111,* 4.
74. Hem, J. D. *Geochim. Cosmochim. Acta* **1976,** *40,* 599.
75. Yendell, K., M.S. Thesis, in progress, Univ. of Minnesota, Minneapolis, MN, 1979.
76. Hullet, D. "Biochemically-Important Naturally Organic Acids in the Upper Mississippi River," M.S. Thesis, Univ. of Minnesota, Minneapolis, MN, 1979.
77. Riley, J. P.; Chester, R. "Introduction to Marine Chemistry"; Academic: London, 1971.
78. Bowen, J. H. M. "Trace Elements in Biochemistry"; Academic: London, 1966.
79. Turekian, K. K. "Rivers, Tributaries and Estuaries," In "Impingement of Man on the Oceans"; Wiley–Interscience: New York, 1971; pp. 9–73.

RECEIVED November 13, 1978.

Chemical Associations of Heavy Metals in Polluted Sediments from the Lower Rhine River

ULRICH FÖRSTNER and SAMBASIVA R. PATCHINEELAM[1]

Institut für Sedimentforschung, Universität Heidelberg Im Neuenheimer Feld 236, D-69 Heidelberg, West Germany

Differentiation of sedimentary metal phases was performed on grain size fractionated samples from the lower Rhine River by successive chemical leaching (review). Pollution affects the significant increase of nonresidual associations of chromium, copper, lead, and zinc. Except for manganese the metal contents in most of the extracted phases decrease as the grain size increases. Phase concentration factors (PCF; relative enrichment of metal content in major carrier substances) are high for chromium in moderately reducible phases (20-fold increase in clay-sized particles), for manganese and zinc in the easily reducible sediment fraction (30- and 55-fold enrichment), and for copper and zinc in the carbonates (15 or 25 times compared with total sediment).

The availability of trace metals for metabolic processes is closely related to their chemical species both in solution and in particulate matter. The type of chemical association between metals and sediment constituents has therefore become of interest in connection with problems arising from the recovery and disposal of contaminated dredged material. Chemical extraction procedures from soil studies (1) and sedimentary geochemistry (such as the differentiation of detrital and authigenic phases in limestones (2), shales (3), pelagic Mn/Fe-concre-

---

[1] Present address: Instituto de Geociências Universidade Federal da Bahia, 40.000-Salvador, Brazil.

0-8412-0499-3/80/33-189-177$05.00/0

tions (4), and recent deposits of less polluted aquatic environments (5, 6,7)) have been developed enabling the estimation of the toxicity potential of a sediment (8–13). The basic rationale of these studies is that the nonlithogenic fractions of the sediment, dredged material, or sewage substances constitute "the reservoir for potential subsequent release of contaminants into water columns and into new interstitial waters" (14), thus being predominantly available for biological uptake (15, 16, 17, 18). Criteria to predict the pollution potential of these substances must reflect the sediment fraction that has a detrimental effect on water quality and the associated biota (19–23).

In that respect, the differentiation of oxidizable, reducible, and residual phases on polluted sediments near a sewer outfall of the County of Los Angeles was made by Bruland et al. (24). Studies by Gupta and Chen (25) on sediment from the Los Angeles harbor and by Brannon et al. (26) on contaminated harbor sediments from Lake Erie, Mobile Bay, Alabama and Bridgeport, Connecticut, were performed by applying successive extraction techniques to include the determination of the metal contents in interstitial water, and in ion-exchangeable, easily reducible, and moderately reducible organic and residual sediment fractions. It has been established by these and other investigations (27, 28) that the surplus of metal contaminants introduced into the aquatic system from man's activities usually exists in relatively unstable chemical associations and is, therefore, predominantly accessible for biological uptake (29).

Direct investigations on the effect of metal partitioning in different sediment components on metal availability to organisms were initially performed by Luoma and Jenne (19, 20, 21). Using deposit-feeding clams on various types of sedimentary substrates, which were labeled with heavy metal nuclides (for example, cadmium-109), these studies indicate that the bioavailability of the heavy metals was inversely related to the strength of metal–particulate binding in the sediments. Regression analyses (availability indices) were evaluated from sediment fractions and metal concentrations in related deposit-feeding bivalves (30), indicating, for example, that the uptake of zinc by *Macoma baltica* in San Francisco Bay is controlled by the competitive partitioning of zinc between extractable iron and manganese forms in the substrate. Despite many unanswered questions, this new approach may be useful as "a statistical interface between more sophisticated chemical and biological models" (30).

Further insight into the processes of metal accumulation in aquatic sediments and estimations of these possible effects on biota could be expected from an analysis of the metal phases within different grain size intervals and from the correlation of metal associations with the percentages of their respective carrier substance in the sediment samples.

Results of studies on recent mud deposits of the heavily polluted lower Rhine River (*31, 33*) near the German/Dutch border (*13*) are compared with the data from a sediment study on Lake Constance (*33*), where the contamination by heavy metals is still relatively low (*34*).

Extraction Procedures

Grain size fractionation was performed by sieving, separating in settling tubes, and by centrifuging (laboratory centrifuge; 100-mL test tubes) in the range <0.2 μm to >63 μm. The latter two procedures were performed with distilled water and each step was repeated 5–10 times. The solids were recovered by evaporation of the suspension in a porcelain bowl at 70°C. Other than in the case of filtration no metal components are lost. However, it cannot be excluded that metal will be stripped off from coarser grained particles, which would enable accumulation in finer grained fractions. The finest grained centrifuge fraction (<0.06 μm) could not be used for further experimentation because of the enrichment of salts (for example, sodium chloride and gypsum). Six grain size fractions of each sample were analyzed by AAS following sequential extraction procedures (Table I): only very small amounts of heavy metals were measured in the water-soluble fraction, a fact which can be expected with the grain size fractionation procedure. From bulk samples of Lake Constance, however, only 2% copper content and less than 0.3% for the other metals (Fe, Mn, Zn, Cr, Ni, Co) were found in the water-soluble portion of the sediment (Deurer (*33*)). This indicates that in most cases—at least for oxidized materials—the effect of grain separation in water is negligible. The cation-exchange procedure using BaCl$_2$–triethanolamine was favored over ammonium acetate (for

Table I.   Extraction Procedures Used in the Present Study[a]

Extraction Method	Major Extractable Phase	Ref.
H$_2$O	Easily soluble fraction	
0.2M BaCl$_2$–triethanolamine, pH 8.1	Easily extractable fraction (e.g., exchangeable cations)	(*1*)
0.1N NaOH	Humates, fulvates	(*36*)
Acidic cation exchange	Carbonates	(*37, 38*)
0.1N NH$_2$OH · HCl + 0.01N HNO$_3$	Easily reducible fraction (Mn-oxides, amorphous Fe-oxides)	(*39*)
30% H$_2$O$_2$ + 1N NH$_4$OAc	Organic residues + sulfides	(*25*)
1N NH$_2$OH · HCl + 25% acetic acid	Moderately reducible fraction (hydrous Fe-oxides)	(*4*)
HF/HClO$_4$ digestion	Inorganic residues	

[a] Modified after Refs. *9, 25, 35.*

example, (7)), since control experiments have indicated that the release of metals such as zinc and cadmium is enhanced significantly by the latter agent, probably because of the dissolution of carbonate minerals (40).

On the other hand, sorption/desorption experiments with copper and clay minerals have shown that true cation exchange (with $BaCl_2$) is relatively insignificant and that the triethanolamine reagent (added to raise the pH of the solution) effects the release of additional metal compounds, which may be more appropriately designated as easily extractable phases (41). Extraction of humate phases was performed with 0.1N NaOH; the same reagent was used for the determination of the concentrations of humic substances by comparison with standard humic acids (42).

From recent sediments of the Blyth River in Great Britain, Cooper and Harris (43) have indicated a further differentiation of the organic fraction into more or less reactive phases, that is, lipids, asphalts, and organic residues. The latter extraction was performed with 30% H_2O_2 and ammonium acetate after separation of the easily reducible metal fractions. Extraction of carbonate-associated metals was done by shaking the suspension of the sample with acidic cation exchangers at pH 5 in the solution (37) after the separation of water-soluble, exchangeable, and humate-associated heavy metals. The use of more acidic solvents in combination with hydroxylamine hydrochloride leads to the successive extraction of easily reducible (mainly MnO_2) and moderately reducible phases (predominantly hydrous iron oxides). Both procedures were tested extensively by Engler, Chen, and colleagues (9, 14, 25, 26); Malo (44) has suggested a treatment with 0.3N HCl as an especially suitable extraction method for reducible metal phases. The residual inorganic fraction was then digested with hydrofluoric/perchloric acid.

It must be clearly pointed out that most of the extraction steps described here are not "selective" as stated in some publications. Repeated treatment often shows a further release of metals, especially in reducible fractions (45). Readsorption of metals can occur, for example, after extraction with dilute acids and H_2O_2 (46). It has been suggested by Burton (47) that while a solution of sodium pyrophosphate and sodium hydroxide powerfully attacks humic materials, the reagent as generally used also attacks silicates to an appreciable degree. Formation of soluble basic metal oxides may take place during initial high pH conditions, for example, for zinc (see next section). Reactions are influenced by the duration of treatment (48) and by the ratio of solid matter to volume of extractant. However, the separation of carbonate phases with acid cation exchanger has a certain selectivity; that the buffering capacity of the carbonates is eliminated before the other extraction steps is the major advantage of this procedure. One of the major disadvantages of this kind of work is presently in the partitioning of metals in anoxic deposits, since no extraction procedure is yet available to differentiate organic and sulfidic metal associations (*49*).

Sample Characteristics

Some properties of the sediment samples from Lake Constance and the Rhine River, possibly relevant for the study of the accumulative effects on heavy metals, are summarized in Figure 1. The concentrations of the metals in both samples (*13, 33*) indicate an increase from coarser to fine grained fractions, as exemplified by zinc. Elevated concentrations of metals in the sand-sized material probably originate from heavy minerals or from corrosion products. An even greater grain size effect occurs for phosphorus; approximately 0.35% (dry weight) has been

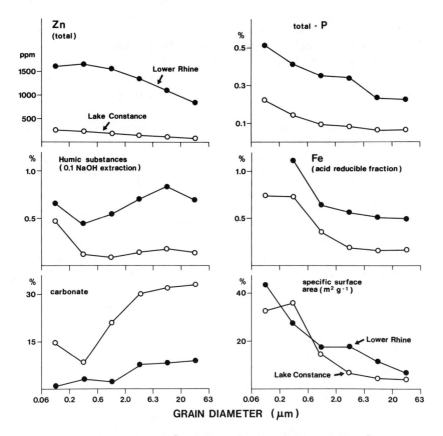

Figure 1.   Percentages and concentrations of zinc, various sedimentary phases, and specific surface area in different grain size intervals of sediments from the Rhine River and Lake Constance

Table II.   Metal Concentrations and Average Percentages of Metals
Samples from Lake Constance (L.C.)

	HF/HClO$_4$ Digestion (total sediment)		Nonresidual Fraction	
	L.C.	R.R.	L.C.	R.R.
Fe	2.4 %	4.0 %	23	51
Mn	545 ppm	830 ppm	80	60
Cr	71 ppm	388 ppm	6	23
Cu	29 ppm	268 ppm	54	89
Pb	40 ppm	482 ppm	54	92
Zn	79 ppm	1240 ppm	37	97

analyzed from the bulk sample of the Rhine and 0.1% for the sediment
from Lake Constance. The concentration of iron associated with hydrous
oxides (reducible phases) also indicates a distinct increase towards finer
grained material, particularly in the pelitic fraction (< 2 μm); calculated
as FeOOH, approximately 1% occurs in the Rhine sediment and approximately 0.4% in the bulk sample from Lake Constance. The grain size
dependency curves for the humic substances (extracted with 0.1N NaOH)
indicate an initial maximum with very small particles, but then an
unexpected increase with medium silt-sized substances. Since the concentrations of extracted humic acids are approximately 0.3% and 1%,
respectively, in the bulk samples, there is an obvious loss of humic
substances during grain size separation by settling and centrifugation.
And in materials smaller than 0.06 μm (which were not studied further) a
strong accumulation of humic substances was found. The carbonate contents of the materials from Lake Constance are significantly higher than
those from the lower Rhine sediments; the increase within the most fine-
grained particles from Lake Constance is probably attributable to the
precipitation of calcite from biogenous chemical processes (50). No distinct difference is indicated for the specific surface areas of both samples in
the fine-grained fractions; however, the silt material from the Rhine exhibits significantly higher surface areas than the respective grain size fractions
from the Lake Constance samples. The difference in the specific surface
areas of the bulk samples (7.2 m^2/g in the Lake Constance sediment,
24.3 m^2/g in the Rhine sediment) corresponds to the smaller mean size
of the Rhine sediment (5.5 μm) as compared to the sample from Lake
Constance (9 μm). The material from Lake Constance indicated a pH
of 7.5 and Eh + 180 mV (the original sample had an in situ pH of 6.9
and Eh from +115 mV at the surface to −75 mV at a 10-cm depth).
Experiments with the Rhine sample began at a pH of 6.8 and Eh ~ +50
mV (wet sample).

in Nonresidual Fractions (Percent of Total Contents) of Sediment and the Lower Rhine River (R.R.)

0.2N $BaCl_2$ (easily extractable)		0.1N $NaOH$ (humic acids)		Acid-Reducible Fraction	
L.C.	R.R.	L.C.	R.R.	L.C.	R.R.
1	2	2	25	20	24
11	18	1	4	68	38
0	0	2	2	4	21
27	17	8	22	19	50
5	3	0	55	49	34
0	1	1	15	36	81

Results and Discussion

The ratio of residual and nonresidual associations in a sediment sample will allow early estimations of the bioavailability of trace metals. In Table II, we compare the metal concentrations in the total sediment samples and within the nonresidual phases, with the latter as average values from the grain size fractionated samples. It is shown that as the metal concentrations in the Rhine River sample are enriched because of pollution influences (except for manganese where diagenetic effects are probably involved), there is a distinct increase in the nonresidual metal fraction. This is mainly valid for copper, lead, and zinc, of which more than 90% in the Rhine River sample can be considered as being potentially remobilizable under natural conditions. Most readily available for biological processes are the metal cations in water-soluble and easily extractable forms; copper is typically enriched in these chemical phases in both samples.

Of the other forms of nonresidual associations, some metal–organic compounds, for example, fulvic and humic acids, have been shown to be particularly effective in the transfer of (toxic) metals from inorganic matter into organisms (51). According to Table II, where data of humate extractions with 0.1N NaOH are compared, these effects should be more relevant for iron, copper, zinc, and lead in the sample from the Rhine River than in the sediment material from Lake Constance. The other nonresidual metal associations (easily reducible, carbonates, moderately reducible forms) partly indicate higher percentages in Lake Constance sediments (Mn and Pb), whereas others (chromium, copper, and zinc) are enriched in the Rhine sample.

Grain-Size Effects

Leaching experiments on selected heavy metals (Fe, Mn, Zn, Pb, Cu, and Cr) from different grain-size fractions in the sample from the strongly

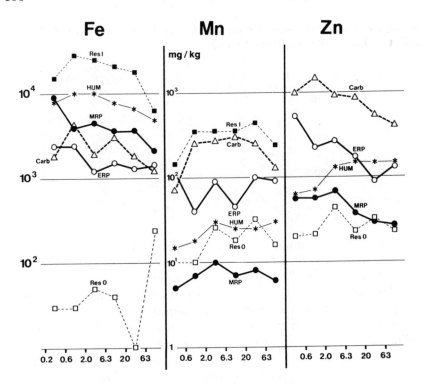

*Figure 2. Concentration of heavy metals (mg/kg) in different grain size
intervals of Rhine sediments (52)*

polluted lower Rhine River (Figure 2) indicate a distinct reduction of
the metal content as grain size increases for most of the extracted phases.
These effects are particularly evident for the zinc and lead associations
with carbonates ("Carb"), easily reducible (ERP), and moderately reducible (MRP) fractions, but can also be observed in the residual organic
(ResO) and inorganic (ResI) fractions such as for chromium, or in the
humate (Hum) fractions, as for iron, lead, and copper. Such effects
should be even more pronounced if the mechanical fractionation would
more accurately separate individual particles according to their grain
size. However, this is not the case and "coatings", for example, of iron/
manganese oxides, carbonates, and organic substances on relatively inert
material with respect to heavy metals, such as quartz grains, act as
substrates of heavy metals in coarser grain size fractions (52).

In Figure 3 the metal concentrations in the various extracts are
related to the percentages of the extracted carrier material in the grain
size fractionated samples (humic substances, hydrous iron oxides, manganese oxides, carbonate minerals, organic and inorganic residues). For
lead, copper, and chromium, the moderately reducible fraction dominates

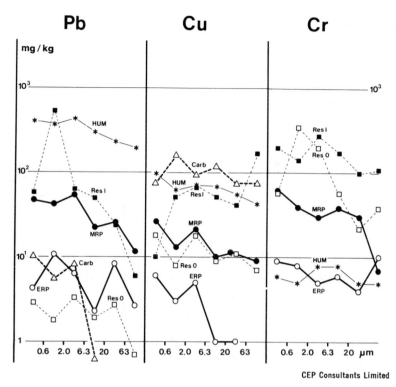

Figure 2.   Continued

among the acid-reducible fractions. Zinc is significantly associated with the easily reducible fraction and with carbonate (the latter is also valid for copper, manganese, and iron). In the resistant fraction, the inorganic associations dominate for iron, manganese, and lead, whereas zinc, copper, and chromium show a closer tendency to organic residues.

Residual Fraction

Generally, there is a decrease in metal concentrations in both residual forms as the grain size increases. An exception is manganese, which tends to be incorporated into larger inorganic particles. The large amount of copper in sand-sized inorganic residues probably originates in scrap materials; the enrichment of chromium in fine-grained organic residues could be explained as resulting from the input of chromium-rich organic fibers from the leather tanning industry, for example, from the Weschnitz tributary of the upper Rhine area (10). Apart from the example of chromium in the Rhine sediment sample, all other metals studied are more or less diluted by the organic and inorganic residues.

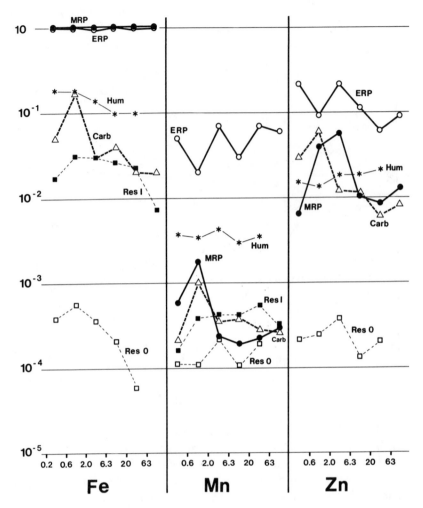

Figure 3. Relative concentrations of heavy metals associated with major carrier phases: mg/kg metal vs. mg/kg carrier phase (ordinate: absolute values)

This becomes even more evident from the data in Table III where the phase concentration factors (PCF) are calculated as the percentage of a certain metal association vs. the percentage content of the respective carrier material (phase is defined here as one of the major sediment fractions, for example, Fe/Mn hydroxides, humic acids, carbonates). If the computed value lies above one, an enrichment has occurred in the studied phase; if the value is less than one (as for all metals in the residual phases except chromium), the metal concentration in the sample

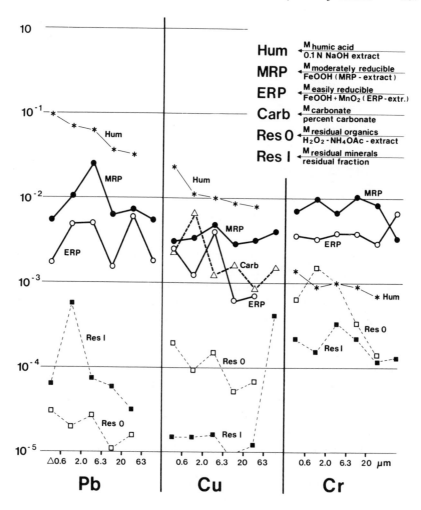

Figure 3.   Continued

has been reduced by the presence of that phase. The PCF is a relative value that is not influenced by the total content and is especially suited for comparison of samples with large differences in the total metal concentrations.

Reducible Fractions

PCF values of chromium, copper, lead, and zinc in the moderately reducible sediment fraction are found to be significantly higher for the silt-sized particles than for the fine-grained substances. Similar effects

Table III. Phase Concentration Factors Indicating the Relative
Carrier Substances of Clay- and Silt-Sized

	Organic Residues		Inorganic Residues		Moderately Reducible Fraction	
	Clay	Silt	Clay	Silt	Clay	Silt
Fe	< 0.01	0.01	0.5	0.7	—	—
Mn	0.2	0.2	0.4	0.5	2	2
Cr	3	1	0.5	0.7	16	22
Cu	0.5	0.3	0.1	0.1	8	10
Pb	0.03	0.05	0.2	0.2	8	15
Zn	0.1	0.2	< 0.1	< 0.1	4	7

[a] Percent metal phase of total sediment/percent carrier.

occur for chromium and lead in the easily reducible sediment fractions (PCF values of reducible fractions from Lake Constance sediment are 15 and 87 for copper, and 5 and 9 for chromium in the clay and silt particles, respectively). This unexpected development has also been reported by Gibbs (53) for the contents of manganese, nickel, cobalt, chromium, and copper in suspended solids from the Amazon and Yukon Rivers. It was suggested that these effects are related to the differences in the thickness of iron hydroxide coatings on mineral grains. Thick layers on coarse material are considered, among other factors, to be the result of the weathering environment: the coarse material in the soil would have a higher permeability, bringing about a greater supply of the precipitating (and coprecipitating) ions. Since the metal concentrations in the Rhine River example are largely affected by man-made contamination, it would seem that these mechanisms likewise take place in polluted environments, for example, on sewage particles in treatment plants. Our data confirm the suggestions made by Lee (54), Jenne (55), and Gibbs (53) that under aerobic conditions the incorporation of metals into hydrous Fe/Mn oxides represents an important sink for potentially toxic heavy metals. As for the carbonate phases (see next section), the possible presence of discrete heavy metal associations in the acid-reducible extraction step has to be considered.

Carbonate Fraction

The association of trace metals with carbonate has been given little or no attention, although it has been shown in laboratory experiments that coprecipitation with calcium carbonate is an effective elimination

Enrichment (or Reduction) of Metal Concentrations in Major Sediment Particles from the Rhine River[a]

Easily Reducible Fraction		Carbonate Fraction		Humic Acid Fraction	
Clay	Silt	Clay	Silt	Clay	Silt
—	—	2	< 1	40	30
30	30	10	4	6	4
5	7	0	0	3	3
6	4	15	5	50	35
3	5	< 1	< 1	100	100
55	55	25	8	8	17

mechanism for heavy metals from aqueous solutions (56). Recent investigations by Salomons and Mook (57) in the IJssel Sea in Holland show that a decrease in the dissolved loads of zinc, cadmium, and nickel depends directly on the precipitation of carbonate minerals, which is chiefly a result of an increase in pH. Measurements on polluted Rhine sediment samples by Patchineelam (13) using a CO_2-extraction step (CO_2 gas was introduced into a bottle containing a suspension of the sample material) led to a considerable remobilization of zinc and cadmium. However, while it could not be excluded that with the latter method hydroxidic phases (that is, hydrous iron oxides) will precipitate, the acidic cation-exchange procedure might also influence other phases, for example, labile phosphorus compounds. In the Rhine sediment sample, carbonate associations of heavy metals such as manganese, copper, and zinc are more concentrated in the fine-grained sediment fractions than in silt-sized materials. The enrichment of iron in the pelitic (< 2 μm) carbonate fraction is probably attributable to the formation of siderite under slightly reducing conditions (siderite mineralizations have been found in the harbor of Neuss/Rhine, where they led to river-bed solidification (10)). In the sample from Lake Constance, PCF values ranged from 0.8 to 1.5 for zinc, 0.4 to 0.9 for copper, 0.3 to 0.5 for iron, and 2 to 4 for manganese for the silt-sized and pelitic carbonate fractions, respectively. Because of the large differences in the carbonate-associated metal contents between the sediment samples from Lake Constance and the Rhine River, it seems possible that in the latter example, part of the observed concentrations (mainly for copper and zinc) result from discrete heavy metal carbonate precipitates, which could have been formed, for example, in sewage treatment plants (through the addition of lime, etc. (58)).

Humic Acid Fraction and Phosphates

In the 0.1N NaOH extractable fraction (which is assumed to reflect the humic acid phases) of the Rhine sediment, PCF values are particularly high for iron, copper, and lead, exhibiting a slight increase in the fine-grained particles compared with the silt-sized material. For the sample from Lake Constance, the concentration factors in the humate phase have been found to be 11 and 40 for copper, 9 and 12 for iron, and 5 and 9 for chromium, in the pelitic and silty sediment fractions, respectively. Similar concentration factors of trace metals in the NaOH-extractable matter have been observed by Chen et al. (14) for chromium, lead, copper, and zinc, whereas iron and manganese showed no significant association with this fraction.

The strong enrichment of lead, copper, and iron in the sodium hydroxide fraction of the Rhine sediment sample led us to consider if there might be other extractable metal phases in addition to humic and fulvic acids, for example, phosphates. Stumm and Morgan (59) have suggested that in the pH range 4.5–6.5, phosphate tends to be bound to the solid phase by $Fe(III)$ and $Al(III)$, either by precipitation or by adsorption ($FePO_4$, clays). For metals such as manganese, zinc, and copper, complex-formation reactions of inorganic phosphates might significantly affect distribution of the metal ion, the phosphates, or both (59). From theoretical phase relationships together with several field observations, Nriagu (60) concluded that lead phosphate-like pyromorphite and plumbogummite (with very low solubility constants compared to lead carbonate, hydroxide, sulfide, and sulfate) are important in the fixation of lead in the sediments. For characterization of phosphate compounds in lake sediments, Williams et al. (61) used—together with other solvents—0.3N NaOH to extract various inorganic phosphorus phases. It was shown that the phosphates associated with iron and aluminum were extracted up to 90% with sodium hydroxide, but not calcium-bound phosphate. Our data from the grain-size fractionated Rhine sediment indicate a close relationship between the 0.1N NaOH extractable concentrations of lead, iron, and copper, and the total phosphorus contents. Direct evidence for these associations might be obtained from the simultaneous determination of phosphorus and heavy metals in the various leachates, which, however, still pose problems in some instances. To improve the present leaching sequence, it is proposed that the NaOH extraction step be completed after treatment of the easily reducible fraction.

Cation Exchange

Contrary to the metal enrichments in the more or less well-defined phases just described, sorbed metal values cannot be attributed to a specific carrier substance. Clay minerals, fine-grained carbonate, humic substances, and hydrous oxides all contribute to the total cation-exchange capacity of the sediment sample. A generalized sequence of the capacity of the solids to sorb heavy metals was established by Guy and Chakrabarti (62):

$$MnO_2 > \text{humic acids} > \text{hydrous iron oxides} > \text{clay minerals}$$

It has been shown that clay minerals through sorption of heavy metals contribute only little to the equilibria between aqueous and solid phases, but are important as carrier substances for organic compounds, hydrous oxides, carbonates, and phosphates (55). These coatings still play a characteristic role in the bonding of metals in coarser grain size fractions as well (see previous sections). Furthermore, it must be considered that the sorptive interactions between heavy metals and organic substances, hydrous oxides, carbonates, and phosphates depend on grain size, molecular weight, functional groups, aging of the material, etc.

Bioavailability of Metal Fractions

Of the various chemical fractions differentiated by sequential leaching, the metals extracted with $0.2M$ $BaCl_2$–triethanolamine should be available primarily for biological uptake, followed by trace metals in association with humic substances. Additional work is needed to develop more specific extraction techniques for the latter compounds, since the NaOH method used in this study also remobilizes metals from other phases such as oxides, phosphates, and silicates. Metal associations with phosphorus compounds could be evaluated by measuring heavy metals in the extractants obtained in the phosphorus-leaching sequence (61) and—vice versa—the phosphorus contents in the heavy metal leachates. Two more aspects studied in the present investigation should be tested with respect to biological uptake: (1) which grain size fractions (together with their chemical phases) are ingested by various species of deposit-feeding organisms and (2) to what extent can the metals from carbonate phases become available to plants under different pH conditions of the substrate.

Acknowledgments

We wish to express our thanks to R. Deurer for supplying us with data from his research on sediments of Lake Constance. The German Research Foundation supported our study of chemical leaching sequences on natural and polluted sediments. We also acknowledge the assistance of D. Godfrey in preparing the English version of this manuscript.

192   PARTICULATES IN WATER

Literature Cited

 1. Jackson, M. L. "Soil Chemical Analysis"; Prentice–Hall: Englewood Cliffs, NJ, 1958.
 2. Hirst, D. M.; Nicholls, G. D. *J. Sediment. Petrol.* **1958,** *28,* 468.
 3. Gad, M. A.; LeRiche, H. H. *Geochim. Cosmochim. Acta* **1966,** *30,* 811.
 4. Chester, R.; Hughes, M. J. *Chem. Geol.* **1967,** *21,* 143.
 5. Nissenbaum, A. *Isr. J. Earth Sci.* **1972,** *21,* 143.
 6. Gibbs, R. J. *Science* **1973,** *180,* 71.
 7. Nissenbaum, A. *Isr. J. Earth Sci.* **1974,** *23,* 111.
 8. Keeley, J. W.; Engler, R. M. Dredged Material Research Program, Misc. Pap. *D-74-14* **1974.**
 9. Engler, R. M.; Brannon, J. M.; Rose, J.; Bigham, G. Presented at the 168th Meeting, ACS, 1974.
10. Förstner, U.; Müller, G. "Schwermetalle in Flüssen und Seen"; Springer–Verlag: Berlin, Heidelberg, New York, 1974.
11. Lee, G. F.; Plumb, R. H. Dredged Material Research Program, Contract Rept. *D-74-1* **1974.**
12. Burrows, K. C.; Hulbert, M. H. In "Marine Chemistry in the Coastal Environment," *ACS Symp. Ser.* **1975,** *18,* 382–392.
13. Patchineelam, S. R., Ph.D. Dissertation, Universität Heidelberg, Heidelronmen. *Effects, Mobile, Ala.* ASCE **1976,** 435–454.
14. Chen, K. Y.; Gupta, S. K.; Sycip, A. Z.; Lu, J. C. S.; Knezevic, M.; Choi, W.-W. *DMRP,* Contract Rept. *D-76-1* **1976.**
15. Chen, K. Y.; Lu, J. C. S.; Sycip, A. Z. *Proc. Spec. Conf. Dredging Environmen. Effects, Mobile Ala.* ASCE **1976,** 435–454.
16. Gambrell, R. P.; Khalid, R. A.; Patrick, W. H., Jr. *Proc. Spec. Conf. Dredging Environmen. Effects, Mobile, Ala.* **1976,** 418–434.
17. Gambrell, R. P.; Khalid, R. A.; Verloo, M. G.; Patrick, W. H., Jr. Dredged Material Research Program, Contract Rept. *D-77-4* **1977,** *2.*
18. Khalid, R. A.; Gambrell, R. P.; Verloo, M. G.; Patrick, W. H., Jr. Dredged Material Research Program, Contract Rept. *D-77-4* **1977,** *2.*
19. Luoma, S. N., Jenne, E. A. In "Radioecology and Energy Resources," Cushing, C. E., Ed.; Dowden, Hutchinson and Ross, Inc.: Stroudsburg, PA, 1976.
20. Luoma, S. N.; Jenne, E. A. In "Biological Implications of Metals in the Environment," Wildung, R. E., Drucker, H., Eds.; NTIS-CONF-750929, Springfield, VA, 1977; 213–230.
21. Luoma, S. N.; Jenne, E. A. In "Trace Substances in Environmental Health—X," Hemphill, D. D., Ed.; University of Missouri Press: Columbia, MO, 1977; 343–351.
22. Jackson, T. A. *Proc. Int. Conf. "Management and Control of Heavy Metals in the Environment,"* London, 1979, p. 457.
23. Prosi, F. *Proc. Int. Conf. "Management and Control of Heavy Metals in the Environment,"* London, 1979, p. 288.
24. Bruland, K. W.; Bertine, K.; Koide, M.; Goldberg, E. D. *Environ. Sci. Technol.* **1974,** *8,* 425.
25. Gupta, S. K.; Chen, K. Y. *Environ. Lett.* **1975,** *10,* 129.
26. Brannon, L. M.; Engler, R. M.; Rose, J. R.; Hunt, P. G.; Smith, I. Dredged Material Research Program, Misc. Pap. *D-76-18* **1976.**
27. Campbell, P. G. C.; Tessier, A.; Bisson, M. *Proc. Int. Conf. "Management and Control of Heavy Metals in the Environment,"* London, 1979, p. 453.
28. Förstner, U.; Wittmann, G. "Metal Pollution in the Aquatic Environment"; Springer–Verlag: Berlin, Heidelberg, New York, 1979.
29. Lee, C. R.; Engler, R. M.; Mahloch, J. L. Dredged Material Research Program, Misc. Pap. *D-76-5,* **1976.**

30. Luoma, S. N.; Bryan, G. W. In "Chemical Modelling in Aqueous Systems," *ACS Symp. Ser.* **1979**, *93*, 577–609.
31. Banat, K.; Förstner, U.; Müller, G. *Naturwissenschaften* **1972**, *59*, 525.
32. Förstner, U.; Müller, G. *Geoforum* **1973**, *14*, 53.
33. Deurer, R. Ph.D. Dissertation, Universität Heidelberg, Heidelberg, West Germany, 1978.
34. Förstner, U.; Müller, G. *Tschermaks Mineral. Petrogr. Mitt.* **1974**, *21*, 145.
35. Förstner, U. In "Origin and Distribution of the Elements," Ahrens, L. H., Ed.; Pergamon: Oxford, 1979; 849–866.
36. Volkov, I. I.; Fomina, L. S. *Am. Assoc. Pet. Geol. Bull.* **1974**, *20*, 456.
37. Deurer, R.; Förstner, U.; Schmoll, G. *Geochim. Cosmochim. Acta* **1978**, *42*, 425.
38. Schmoll, G., Förstner, U. *Neves Jahr. Mineral.* **1979**, *135*, 190.
39. Chao, L. L. *Soil Sci. Soc. Am. Proc.* **1972**, *36*, 764.
40. Förstner, U.; Patchineelam, S. R. *Chem.-Zt.* **1976**, *100*, 49.
41. Förstner, U.; Stoffers, P., unpublished data.
42. Eloff, J. N. *S. Afr. J. Agric. Sci.* **1965**, *8*, 673.
43. Cooper, B. S.; Harris, R. C. *Mar. Pollut. Bull.* **1974**, *5*, 24.
44. Malo, B. A. *Environ. Sci. Technol.* **1977**, *11*, 277.
45. Heath, G. R.; Dymond, J. *Geol. Soc. Am. Bull.* **1977**, *88*, 723.
46. Rendell, P. S.; Batley, G. E.; Cameron, A. J. *Environ. Sci. Technol.* **1980**, *14*, 314.
47. Burton, J. D. In "Biogeochemistry of Estuarine Sediments," Goldberg, E. D., Ed.; Unesco: Paris, 1978; 33–38.
48. Bowser, C. J.; Mills, B. A.; Callender, E. In "Marine Geology and Oceanography of the Central Pacific Manganese Nodule Province," in press.
49. Calvert, S. E.; Batchelor, C. E. Deep-Sea Drilling Project Initial Rept. **1978**, *42*, 527.
50. Wagner, G. *Z. Hydrol.* **1976**, *38*, 191.
51. Singer, P. C. In "Fate of Pollutants in the Air and Water Environment," Suffet, I. H., Ed.; Wiley: New York, 1977; 155–182.
52. Förstner, U.; Patchineelam, S. R.; Schmoll, G. "Management and Control of Heavy Metals in the Environment," Proc. Int. Conf.: London, 1979; 316.
53. Gibbs, R. J. *Bull. Geol. Soc. Am.* **1977**, *88*, 829.
54. Lee, G. F. In "Heavy Metals in the Aquatic Environment," Krenkel, P. A., Ed.; Pergamon: Oxford, 1975; 137–147.
55. Jenne, E. A. In "Molybdenum in the Environment," Chappel, W., Peterson, K., Eds.; M. Dekker: New York, 1976; 425–553.
56. Popova, T. P. *Geochemistry* **1961**, *12*, 1256.
57. Salomons, W.; Mook, W. G. *Abstr. Intern. Congr. on Sedimentology, 10th, Jerusalem, July 1978*, pp. 569–570.
58. Netzer, A.; Bowers, A.; Norman, J. D. *Proc. Conf. Great Lakes Res., 16th, Huron, Ohio, 1976*, pp. 260–265.
59. Stumm, W.; Morgan, J. J. "Aquatic Chemistry," Wiley: New York, 1970, p. 518.
60. Nriagu, J. O. *Geochim. Cosmochim. Acta* **1974**, *38*, 887.
61. Williams, J. D. H.; Syers, J. K.; Harris, R. F.; Armstrong, D. E. *Soil Sci. Soc. Am. Proc.* **1971**, *35*, 250.
62. Guy, R. D.; Chakrabarti, C. L. *Abstr. Intern. Conf. Heavy Metals in the Environment, Toronto, Oct. 1975*, D-29.

RECEIVED October 10, 1978.

Mathematical Model for Simulation of the Fate of Copper in a Marine Environment

GERALD T. ORLOB

University of California, Davis, CA 95616

DAVORIN HROVAT

Wayne State University, Detroit, MI 48202

FLORENCE HARRISON

Lawrence Livermore Lab, Livermore, CA 94550

A mathematical model is formulated to describe the first-order kinetics of ionic copper released into a marine environment where sorption on suspended solids and complexation with dissolved organic matter occur. Reactions are followed in time until equilibrium between the three copper states is achieved within about 3 hr (based on laboratory determinations of rate and equilibrium constants). The model is demonstrated by simulation of a hypothetical slug discharge of ionic copper, comparable to an actual accidental release off the California coast that caused an abalone kill. A two-dimensional finite element model, containing the copper submodel, was used to simulate the combined effects of advection, diffusion, and kinetic transformation for 6 hr following discharge of 45 kg of ionic copper. Results are shown graphically.

Nuclear power stations, situated along a coastline and using once-through cooling for main condensers, may have a significant impact on the marine environment in the vicinity of the discharge. Because flows are large, roughly 45 m^3/sec (1588 cfs) for a 1000 MW(e) unit, the zone influenced by the discharge plume also may be extensive, a function of the degree of dispersion and mixing with ambient seawater. Depending on the design and operation of such power stations, and on

0-8412-0499-3/80/33-189-195$05.00/0

the metals and alloys used in heat exchangers, there is a potential hazard to the marine ecosystem from emissions of toxic metals and from the products of their interactions with the indigenous constituents in seawater. One such metal of particular concern, because of its common use in fabrication of condenser tubes and because of its high toxicity to aquatic organisms, is copper. The prospect of copper entering solution as a corrosion product and reaching critical concentration levels, perhaps in the form of the acutely toxic cupric ion Cu^{++} (2), is enhanced when the flow of coolant water is initiated again through the condensers after a period of shutdown.

In the summer of 1974, intermittent testing of the cooling water system was performed at the Diablo Canyon Nuclear Power Station, a facility located on the central California coast. In July, dead and dying red and black abalones were discovered in Diablo Cove, the site of discharge of effluent waters from the station (15). This discovery prompted an investigation of copper concentrations in the discharge waters during start-up. During one test, a transient "slug" of copper with a peak of 7700 µg/L was introduced into the near-shore area after start-up (20). Subsequent to this investigation, the copper–nickel (90–10 alloy) heat-exchange tubing was replaced by noncorrosive titanium alloy tubing. Field sampling by one of the authors at this same station at start-up in the summer of 1977 revealed concentrations of approximately 30 µg/L of copper which declined gradually in a few hours toward seawater background levels of about 1 µg/L. The source of this small pulse was copper present in part of an auxiliary pumping system. (Following this incident copper heat-exchange tubing was replaced by noncorrosive titanium alloy tubing.)

Laboratory experience indicates that some forms of copper can be toxic to sensitive marine biota in natural seawater at concentrations less than 10 µg/L (9, 14, 18). Inasmuch as copper may be complexed with dissolved organic matter (5, 19) and sorbed onto the surface of naturally occurring suspended particles (3, 5), it is reasonable to presume that in these forms it may find alternative pathways through the marine ecosystem, perhaps reaching toxic levels because of accumulation in indigenous biota. Given the acute toxicity of some forms of copper and the prospects for exceeding toxic levels in the discharge from power stations, there is a compelling need to mitigate damage through improved design and/or operation procedures. The research reported here is aimed at developing a methodology to predict the fate of transient copper emissions in the vicinity of power station discharges and to evaluate mitigation measures. Specific attention is given to the several forms of copper and the kinetic relationships that determine the relative concentrations of each within the zone of potentially damaging impact.

Chemical Considerations

The physiochemical forms of copper that are present in seawater may be numerous. Copper released into the water column is partitioned between the soluble and particulate fractions. The form in the soluble fraction is related to the inorganic and organic constituents in the water; the form in the particulate fraction is related to the kinds of sinks present on the particles.

The copper species in the soluble fraction has been classified as "labile" or "bound" (6). Definition of the two groups is determined by the experimental conditions under which the measurement is made. In this report we shall refer to the following as labile species: ions, ion pairs, readily dissociable (labile) inorganic and organic complexes, and easily exchangeable copper adsorbed on either colloidal inorganic or organic matter. Inorganic anions to which the copper may be complexed are hydroxides, carbonates, chlorides, sulfates, phosphates, and nitrates; organic anions are amino acids, amino sugars, alcohol, urea, etc. Equilibrium between these forms would be established very rapidly—so fast, in fact, that differentiation of specific forms is not possible with conventional analytical techniques.

Bound species include stable metal–organic complexes, metals bound to high molecular weight organic material, some inorganic complexes, and metals occluded in, or sorbed tightly on, highly dispersed colloids. Included in this group of bound species is copper complexed to humic substances, the refractory organic material that comprises approximately 90% of organic material in water.

Laboratory experience in evaluating the partitioning of ionic copper injected into natural seawater, containing dissolved organic matter (DOM) and suspended particles (SS), indicates that the complexing to DOM and the sorption of copper to particles from labile forms occur at rates much slower (relatively) than those of the pure chemical reactions between free ionic copper and inorganic constituents (5, 10). Designating labile copper by Cu^{++}, the characteristic equilibria with DOM and SS are given by

$$Cu^{++} + DOM \underset{k_{21}}{\overset{k_{12}}{\rightleftharpoons}} Cu^{++} \cdot DOM \tag{1}$$

and

$$Cu^{++} + SS \underset{k_{31}}{\overset{k_{13}}{\rightleftharpoons}} Cu^{++} \cdot SS \tag{2}$$

where $Cu^{++} \cdot DOM$ is the copper complexed with high molecular weight DOM naturally occurring in seawater, $Cu^{++} \cdot SS$ the copper sorbed on the

surface of SS, and the k's the kinetic coefficients governing the rates of
reaction. The possibility also exists, at least conceptually, for exchange
of copper between the compartments $Cu^{++} \cdot DOM$ and $Cu^{++} \cdot SS$ such that

$$Cu^{++} \cdot DOM + SS \underset{k_{32}}{\overset{k_{23}}{\rightleftharpoons}} Cu^{++} \cdot SS + DOM \tag{3}$$

The conceptual model for disposition of copper among these three
compartments is illustrated in Figure 1. (Note: For the present the model
is restricted to these three compartments, although it is recognized that
it will be necessary ultimately to include interactions with additional
compartments such as the sediment bed and indigenous biota, and
perhaps others.)

In such a system as illustrated in Figure 1, with a given complexing
capacity determined by the concentration of DOM and a sorption capacity
determined by the concentration of SS, equilibrium eventually will be
reached so that

$$C_s = \sum_{i=1}^{n} C_{i,\,eq} \tag{4}$$

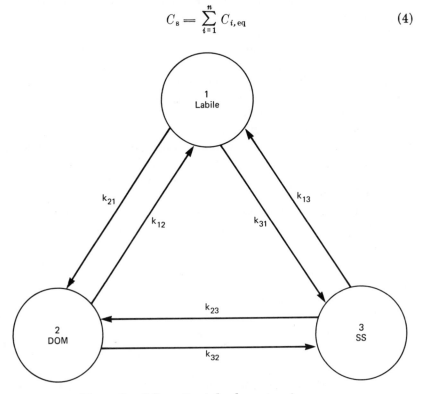

Figure 1.   Schematic of the dynamics of copper

where C_s is the total concentration of copper in n compartments (three in our case) and $C_{i,eq}$ is the equilibrium (steady state) concentration in compartment i. The distribution of copper between compartments is described by a distribution coefficient $K_{d,i}$ that can be evaluated experimentally:

$$K_{d,i} = \frac{C_{i,eq}}{C_s - C_{i,eq}} \left[\frac{V}{M_i} \right], \text{mL/g} \qquad (5)$$

where V is the volume of solution and M_i is the mass of material with which copper is transferred to compartment i by sorption or complexation.

The relative proportion of copper in the ith compartment to the copper contained in the others is given by

$$\beta_i = \frac{C_{i,eq}}{C_s - C_{i,eq}} = K_{d,i} \left[\frac{M_i}{V} \right] \qquad (6)$$

It follows that for DOM and SS, the particular compartments we are considering in addition to copper in the ionic state, the corresponding β's, are

$$\beta_2 = K_{d,2} \times \text{DOM}$$

$$\beta_3 = K_{d,3} \times \text{SS} \qquad (7)$$

where DOM and SS are ambient concentrations.

The rate of sorption of copper to suspended particles was determined for samples collected at the discharge area of a coastal power station (5). Steady-state conditions of sorption were approached within 10 hr after spiking with ionic copper using ^{64}Cu as a tracer. Distribution coefficients range from 11,000 to 52,000 and k values from 0.2 to 0.8 hr^{-1}. There was some evidence from the data that part of the copper was sorbed in a very short period (less than about 10 min), while the remainder was sorbed onto the particles at an exponential rate over the next 10 hr as steady-state conditions were approached. However, for present purposes, considering sorption on particles as a single compartment, the transfer of copper from labile forms to those sorbed on suspended particles is fairly well represented as a first-order process in which k_{13} approximately equals 0.75 hr^{-1}.

Additions of organics, particularly in the form of high molecular weight DOM naturally occurring in seawater (5), substantially reduced the distribution of copper; the ionic and suspended sediment compartments indicate the high complexing capacity of these substances. Naturally occurring DOM in water discharged from coastal power plants gave

$K_{d,2}$ values ranging from 100,000 to 600,000 mL/g at DOM levels of 1 to 2 mg/L (see Appendix I for sample calculations). No information is available on the number of binding sites per molecule for the DOM present, but it appears that the sites were more than adequate for complexing the copper. The limited data on the kinetics of complexation indicate that the reactions can be described by first-order kinetics if the values $k_{1,2}$ are in the range of 0.5 to 1.0 hr^{-1} (10). The concentration of DOM must be much greater than that of copper because of the limited number of sites per molecule of DOM (molecular weight = 10,000 to 100,000, or more). This is true in our case where we are dealing with copper concentrations of a few μg/L (less than 40 μg/L), and DOM concentrations of 1000–3000 μg/L. The assumption of first-order kinetics appears to be a reasonable approximation of actual behavior in such a case.

Mathematical Considerations

If for conditions in which the number of copper atoms is very much smaller than the number of binding sites on SS or DOM, that is, where the rates of complexation and sorption are independent of the concentrations of complexing or sorption agent, then the rates of reaction can be characterized mathematically as pseudo first order. Accordingly, the physiochemical transformations are described by the following rate equations

$$\frac{\partial C_1}{\partial t} = -k_{12}\widetilde{C}_1 + k_{21}\widetilde{C}_2 - k_{13}\widetilde{C}_1 + k_{31}\widetilde{C}_3 \tag{8}$$

$$\frac{\partial C_2}{\partial t} = -k_{21}\widetilde{C}_2 + k_{12}\widetilde{C}_1 - k_{23}\widetilde{C}_2 + k_{32}\widetilde{C}_3 \tag{9}$$

$$\frac{\partial C_3}{\partial t} = -k_{31}\widetilde{C}_3 + k_{13}\widetilde{C}_1 - k_{32}\widetilde{C}_3 + k_{23}\widetilde{C}_2 \tag{10}$$

where $\widetilde{C}_i = C_i - C_{i,eq}$, $i = 1, 2, 3$. It is noted that the processes of desorption, decomplexing, and exchange between the DOM (C_2) and SS (C_3) compartments are included for completeness, although the kinetics of these changes in state are not yet fully evaluated.

The general statement of these equations is of the form

$$\frac{\partial C_i}{\partial t} = \sum_{j=1}^{3} k_{ij}\widetilde{C}_j \qquad i = 1, 2, 3 \tag{11}$$

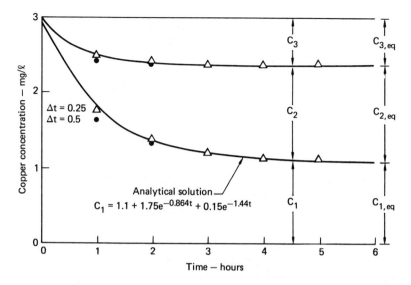

Figure 2. Test case for copper kinetics

In matrix form

$$[\dot{C}] = \{K\}[C] \tag{12}$$

the analytical solution of which, for spatially homogeneous copper concentrations, is

$$[C] = \sum_{l=1}^{3} \alpha_l \, [\epsilon_l] \, \exp \, [\lambda_l t] \tag{13}$$

where $[\epsilon_l]$ is the eigenvector; λ_l is the eigenvalue; t is time; α equals $[T]^{-1} [C]_{t=0}$; and $[T]$ equals $\{[\epsilon_1], [\epsilon_2], [\epsilon_3]\}$.

A test problem comparing the analytical and numerical solutions of the same problem using a finite element model (*12*) is illustrated in Figure 2. The solution corresponds to the case of a continuously stirred tank reactor (CSTR) in which first-order kinetics are assumed, and the rates of reaction are comparable to those we have observed in the laboratory (*5, 10*).

Simulation of Prototype

For purposes of simulation of the fate of copper in a realistic environmental situation, the kinetic model just described was imbedded in a previously developed two-dimensional finite element model (*11*).

The model, referenced in its latest version as RMA-4 (12), is based on the advection diffusion equation, which for a shallow vertically mixed system is written for the two coordinate axes as

$$z\left(\frac{\partial C_i}{\partial t}\right) = -uz\frac{\partial C_i}{\partial x} - vz\frac{\partial C_i}{\partial y} + z\frac{\partial}{\partial x}\left(E_x\frac{\partial C_i}{\partial x}\right) +$$

$$z\frac{\partial}{\partial y}\left(E_y\frac{\partial C_i}{\partial y}\right) + R_i(\widetilde{C_i}) \pm S_o \qquad (14)$$

where u and v are mean velocities over the depth z corresponding to the axes x and y, E_x and E_y are effective diffusion coefficients, $R_i(\widetilde{C_i})$ is a vector of first-order rate coefficients describing the kinetics of C_i, and S_o are sources and sinks of C. The first two terms of the right side of the equation describe transport with the flow and the second two terms characterize mixing processes associated with secondary currents and ambient turbulence. The coefficients E_x and E_y are usually determined empirically.

Solution of Equation 14 is obtained by the finite element model RMA-4 for any stipulated flow field $u(x,t)$ and $v(y,t)$ over a continuum of elements, usually of triangular or rectangular shape (although isoparametric elements with quadratic functions defining the sides are allowed). Depths may be fixed or variable, both in time and space. RMA-4 has been applied successfully to a wide variety of practical problems (13). The version used here is a modification, RMA-4A, that is designed specifically to deal with the kinetics of copper as described.

Model Setup for Case Study

Demonstration of the copper model on a case of practical interest was undertaken by considering a hypothetical slug release of ionic copper through the outfall of Unit One of the San Onofre Nuclear Power Station on the Southern California coast.

The receiving water zone modeled is shown as an inset in Figure 3. It encompassed an area of 2.7 × 5.4 km adjacent to the station site with the discharge situated near the center of the field about 800 m offshore at a depth of about 7.3 m. Water depths ranged from 0 to 15.5 m with bottom contours roughly paralleling the shore in a north–south direction.

The finite element model for the receiving water zone, shown in Figure 4, is comprised of 124 elements of both rectangular and triangular shape, varying in size with proximity to the outfall. The dispersion zone, immediately adjacent to the outfall, was modeled with triangular elements arranged symmetrically around the outfall terminus. The innermost group of four triangles, forming a diamond-shaped area, was considered as a

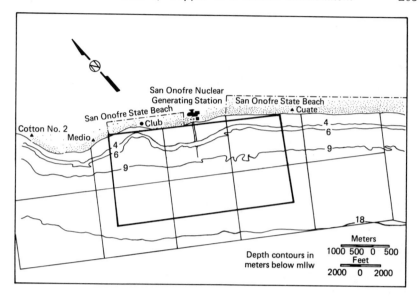

Figure 3.   Site of San Onofre Nuclear Power Station (area modeled shown in inset)

"mixing zone" within which the slug discharge was assumed to be mixed continuously with ambient seawater during the 1 hr discharge period as if by an efficient diffusion system.

Field current patterns imposed on the model were approximated from data collected by Southern California Edison (*17*), owners of the station. For purposes of the demonstration simulation, a changing sequence of currents was selected to represent the history of local transport over a half tidal cycle of 6 hr. It consisted of a 0.3 knot current downcoast (south) for the first hour while copper discharge was occurring, a null period of zero current velocity for 2 hr, and a 3-hr period of 0.3 knot current upcoast. Current vectors were directed parallel to the coastline and were considered to represent average velocities through the full depth of the water column. Dispersion characteristics of the nearshore waters were represented by values of $E_x = 3$ m^{2}sec^{-1} and $E_y = 2$ m^{2}sec^{-1} in the mixing zone, and were increased to 4 m^{2}sec^{-1} and 3 m^{2}sec^{-1} for all other elements (*16*). They were maintained steadily at these values throughout the 6-hr period simulated.

Copper emitted from the outfall was assumed to enter in labile state (C_1) at a rate corresponding to a steady flow of 21.25 m^3/sec (750 cfs) at 600 μg/L for a period of 1 hr. This concentration was reduced immediately by dilution so that by the end of the first hour, at which time discharge ceased, the maximum concentration of ionic copper in the

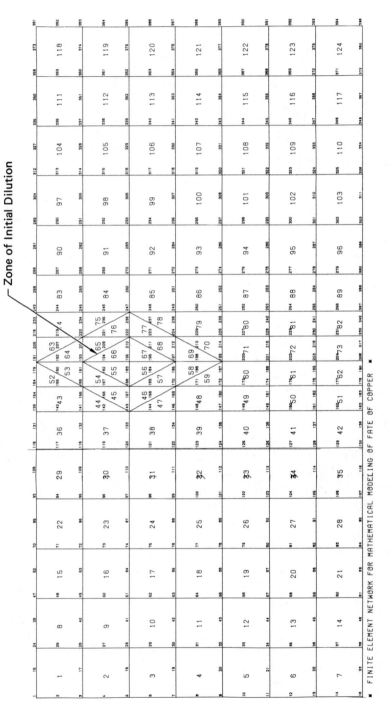

Figure 4.   Finite element network for simulation of fate of copper in the marine environment

mixing zone was less than 20 μg/L. The total emission of about 45.9 kg of copper was estimated to approximate that of the Diablo Canyon incident (20).

The equilibrium states for labile, bound, and sorbed copper, estimated from field and laboratory measurements, were set as follows:

$$\frac{C_{1,\,\text{eq}}}{C_s} = 0.62 \qquad \frac{C_{2,\,\text{eq}}}{C_s} = 0.31 \qquad \frac{C_{3,\,\text{eq}}}{C_s} = 0.07$$

In this test $C_s = C_1(0)$, $C_2(0) = 0$, and $C_3(0) = 0$. The corresponding values of β_i are 1.632, 0.449, and 0.075. If $k_{d,2}$ and $k_{d,3}$ are assumed to be 200,000 and 12,000 mL/g, respectively, the corresponding values of DOM (C_2) and SS (C_3) would be 2.2 mg/L and 6.3 mg/L. These are in reasonable agreement with field experience at the San Onofre site.

In this demonstration we assumed that there was no exchange between the DOM and SS species, that is, $k_{23} = k_{32} = 0$. From laboratory experiments (5, 10) we estimated the remaining k_{ij}'s as follows:

$$k_{12} = k_{21} = 0.684 \text{ hr}^{-1}$$
$$k_{13} = 0.230 \text{ hr}^{-1}$$
$$k_{31} = 0.173 \text{ hr}^{-1}$$

corresponding to half-lives of unreacted constituents of 1, 3, and 4 hr, respectively.

Simulation of Prototype

Simulation of the fate of copper after injection into the marine environment is illustrated by Figures 5, 6, 7, and 8.

During the first hour, when the entire slug of copper is introduced, dispersion, dilution, and first-order kinetic reactions result in a two-dimensional pattern of labile copper of the shape illustrated in Figure 5. The pattern shows a distribution of concentration about the peak value as a result of initial mixing and subsequent dispersion. The peak is translated downcoast by the advective process and the distribution is skewed somewhat in this same direction as a result of the kinetic transformation of labile copper to bound and sorbed states. A peak value of 18.4 μg/L is indicated which may be compared to a maximum possible value of about 40 μg/L if the entire slug were simply mixed with the entire volume of the dispersion zone (the diamond-shaped area around the outfall).

Figure 6 depicts the temporal history of labile copper in the first 3 hr after the slug is discharged as it is subjected to further translation, dispersion, and decay. This history may be contrasted with that of bound copper (C_2) shown in Figure 7 where the same mechanisms are operative, but the mass transferred to this compartment increases as the

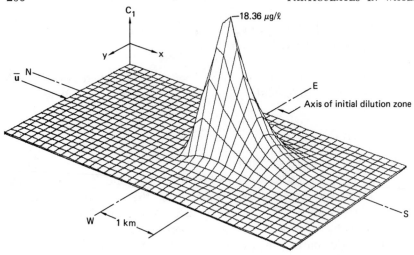

Figure 5.   Dispersion of labile copper 1 hr after initial injection

system progresses toward physiochemical equilibrium. Note, for example, that the peak concentration of C_2 occurs about 2 hr following discharge, declining thereafter primarily because of the mechanism of dispersion. A similar pattern is noted for copper sorbed on SS, as illustrated in Figure 8 and summarized with the histories of the other copper species in Table I.

If one compares the ratios of $C_2:C_1$ and $C_3:C_1$ it appears that the system is in a state of rapid change for the first several hours; equilibrium conditions are achieved virtually after about 3 hr for copper complexed with DOM and after about 5 hr for copper sorbed on SS. Subsequently, concentration changes are attributable entirely to the physical mechanisms of transport and dispersion.

Table I.   Temporal Variation in Peak Copper Concentrations in Plume Following Discharge: Results of Simulation with RMA-4A

Time t (hr)	Copper Concentration ($\mu g/L$)			C_2/C_1	C_3/C_1
	C_1	C_2	C_3		
1	18.36	2.42	0.41	0.1318	0.0223
2	9.15	4.24	0.76	0.4634	0.0831
3	6.45	3.26	0.60	0.5054	0.0930
4	4.88	2.47	0.48	0.5061	0.0984
5	3.94	1.99	0.40	0.5051	0.1015
6	3.28	1.65	0.34	0.5030	0.1037
∞				0.5000	0.1129

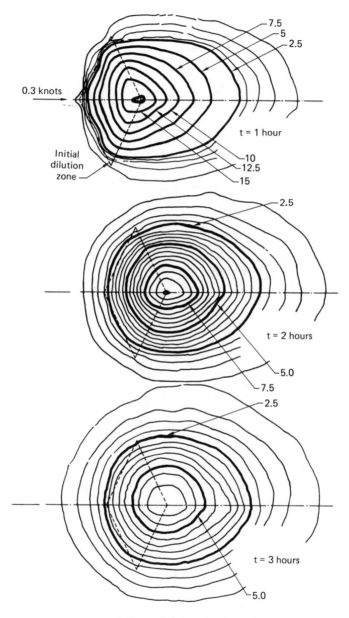

Figure 6.   Advection and dispersion of labile copper during first 3 hr following discharge (contours show lines of equal concentration, μg/L)

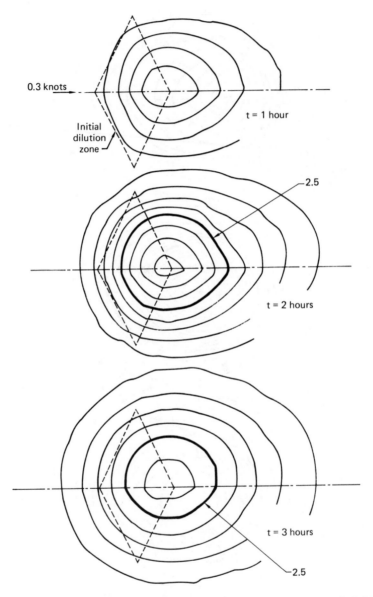

Figure 7.   *Advection, dispersion, and complexation of copper with DOM
during first 3 hr following discharge*

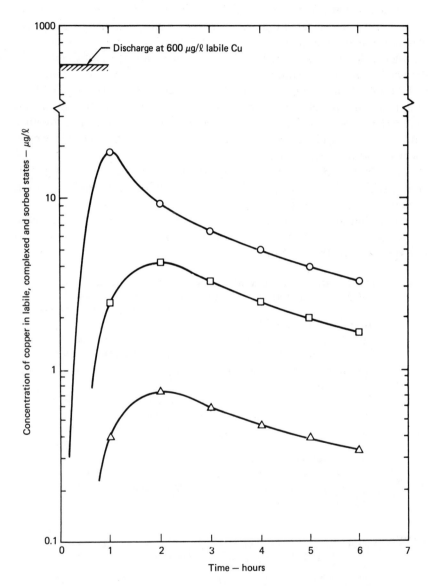

Figure 8. Fate of copper in labile (○), complexed with DOM (□), and sorbed on SS (△) states following discharge to a marine environment

Conclusions

A mathematical model capable of describing the fate of copper species in a marine environment following a slug discharge of copper was developed. The model in its present form is limited to three compartments: labile copper, including free copper ions, ion pairs, and easily dissociated inorganic complexes; copper complexed by high molecular weight DOM indigenous to seawater; and copper sorbed on SS. It is unique among models thus far developed for metal species in that kinetic interactions between the three compartments can be simulated together with transport and mixing processes as the concentrations of copper species tend toward equilibrium.

Within the limits of the assumptions on which the model is predicated, the results are reasonable, and the model could be used as a preliminary tool in evaluation of discharge episodes. However, there is a need to improve capabilities and to test the model rigorously. Among the areas where future development and research are needed the following appear to be the most pressing:

1. The sensitivity of the model's response to kinetic coefficients and K_d's estimated from laboratory experiments should be assessed. Particular limitations in the assumptions of first-order kinetics should be determined.
2. The desirability of adding additional compartments, for example, deposited sediments and biota, should be explored.
3. The effects of other environmental factors, for example, pH, salinity, and temperature, should be included.
4. The effects of background concentrations of DOM and SS on the fate of copper should be evaluated.
5. The model should be tested under more realistic conditions, including actual current structure, bathymetry, and influences of thermal discharge.
6. The model should be extended to predict partitioning of copper in fresh and brackish water, for example, streams, lakes, and estuaries.

Appendix I. Calculation of $K_{d,2}$ Values for DOM at San Onofre Nuclear Power Station[a]

Date	Soluble[b] Organic Carbon (mg/L)	Organically[c] Bound Copper (%)
7/21/77	1.6	14
10/12/77	1.6	49
1/ 9/78	1.5	13

1) $K_{d,2} = \dfrac{(0.14)}{(0.86)} \dfrac{(1000)}{(0.0016)} = 101{,}700$

2) $K_{d,2} = \dfrac{(0.49)}{(0.51)} \dfrac{(1000)}{(0.0016)} = 600{,}500$

3) $K_{d,2} = \dfrac{(0.13)}{(0.87)} \dfrac{(1000)}{(0.0015)} = 99{,}600$

[a] Data from (7).
[b] It is assumed that all soluble organic carbon is involved in the binding of copper. If only part of it is involved, the $K_{d,2}$ value will increase in proportion to the decrease in mass.
[c] Values obtained from ultrafiltration studies.

Literature Cited

1. Ariathurai, R.; Krone, R. B. "Finite Element Model for Cohesive Sediment Transport," *J. Hydraul. Div., Am. Soc. Civ. Eng.* **1976,** 323–338.
2. Chen, C. W.; Orlob, G. T. "Ecologic Simulations for Aquatic Environments," in "Systems Analysis and Simulation in Ecology"; Academic: New York, 1975; Vol. III, pp. 475–588.
3. Cooper, B. S.; Harris, R. C. "Heavy Metals in Organic Phases of River and Estuarine Sediment," *Mar. Pollut. Bull.* **1974,** *5,* 24–26.
4. Davey, E. W., Morgan, M. J.; Erickson, S. J. "Biological Measurement of the Copper Complexation Capacity of Sea Water," *Limnol. Oceanogr.* **1973,** 993–997.
5. Emerson, R. R.; Harrison, F. L. "Partitioning of Copper in Sea Water and Sediment near Nuclear Power Plants at Diablo Canyon and San Onofre, California," NUREG Rept. 1967, in press.
6. Florence, T. M.; Batley, G. E. "Determination of the Chemical Forms of Trace Metals in Natural Waters, with Special Reference to Copper, Lead, Cadmium and Zinc," *Palanta* **1977,** *24,* 151–158.
7. Harrison, F. L.; Bishop, D.; Emerson, R. R.; Rice, D. W., Jr. "Concentration and Speciation of Copper in Waters in the Vicinity of the San Onofre and Diablo Canyon Nuclear Power Stations," NUREG CR-073, UCRL-52706, Lawrence Livermore Laboratory, Livermore, CA, 1979.
8. Jackson, T. A. "Humic Matter in Natural Waters and Sediments," *Soil Sci.* **1975,** *119,* 56–64.
9. Knezovich, J. P.; Harrison, F. L.; Tucker, J. S. "The Influence of Organic Chelators on the Toxicity of Copper to Embryos of the Pacific Oyster," *EOS, Trans. Am. Geophys. Union* **1977,** *58,* 1166.
10. Katekaru, J. Y.; Pullen, P. E.; Harrison, F. L. "Determining the Copper-Complexing Capacity of Known Organic Compounds and Soluble Organic Matter in Seawater by Differential Pulse Anodic Stripping Voltametry Peak Potential Shift," unpublished data.
11. King, I. P. "Operating Instructions for the Computer Program RMA-4"; Resource Management Associates: Lafayette, CA, March 1977.
12. King, I. P.; Norton, W. R. "Recent Application of RMA's Finite Element Models for Two-Dimensional Hydrodynamics and Water Quality," *Int. Conf. on Finite Elements in Water Resources, 2nd, London, England,* 1978.
13. King, I. P.; Norton, W. R.; Orlob, G. T. "A Finite Element Model for Two-Dimensional Stratified Flow," Report to Office of Water Resources Research, March, 1973.

14. Lewis, A. G.; Whitfield, P. H.; Ramnarine, A. "Some Particulate and Soluble Agents Affecting the Relationship Between Metal Toxicity and Organisms Survival in the Calanoid Copepod *Euchaeta japonica*," *Mar. Biol.* **1972,** *17,* 215–221.
15. Martin, M. "A Summary of the Cause and Impact of an Abalone Mortality—Diablo Cove, San Luis Obispo County, California," Summary Report to U.S. Nuclear Reg. Comm. by Calif. Dept. of Fish and Game, Dec. 1974.
16. Orlob, G. T. "Eddy Diffusion in Homogeneous Turbulence," *J. Hydraul. Div., Am. Soc. Civ. Eng.* **1959,** $HY9(2150)$, 75–101.
17. "Ocean Current Information, San Onofre Nuclear Generating Station," Southern California Edison Company: April, 1973.
18. Sylvia, R. N. "Environmental Chemistry of Copper in Aquatic Systems," *Water Resources* **1976,** *10,* 789–792.
19. Sunda, W.; Guillaid, R. R. L. "The Relationship Between Cupric Ion Activity and the Toxicity of Copper to Phytoplankton," *J. Mar. Res.* **1976,** *34,* 311–329.
20. Warrick, J. W.; Sharp, S. C.; Friedrich, S. J. "Chemical, Biological and Corrosion Investigations Related to the Testing of the Diablo Canyon Unit 1 Cooling Water System," Report No. 733-129-75, Pacific Gas and Electric Company, San Ramon, CA, 1975.
21. Williams, P. M.; Baldwin, R. J. "Cupric Ion Activity in Coastal Seawater," *Mar. Sci. Commun.* **1976,** *2,* 161–181.
22. Zitko, P.; Carson, W. V.; Carson, W. G. "Prediction of Incipient Lethal Levels of Copper to Juvenile Atlantic Salmon in the Presence of Humic Acid by Cupric Electrode," *Bull. Environ. Contam. Toxicol.* **1973,** *10,* 265–271.

RECEIVED September 8, 1978.

Modeling Particulate Transport in Impounded Rivers

H. H. HAHN, F. KÄSER, and R. KLUTE

University of Karlsruhe, 75 Karlsruhe 1, Postfach 6380, Am Fasanengarten, West Germany

Phenomena affecting particle distribution are dispersion, convection, aggregation, sedimentation, and erosion. An attempt to formulate kinetic models for each process on the basis of existing concepts and data leads to a complex model that defies practical application because of numerical difficulties and a lack of input data. A closer inspection of this more analytical model indicates possibilities for a permissible simplification of the conceptually derived model. In analogy to other water quality models it is proposed to represent all major phenomena that affect particle concentration as a function of time and location by first-order decay or regeneration terms, controlled by a flow structure-dependent equilibrium concentration and superimposed by convection and/or dispersion terms. First attempts to use this simplified model complex in practice have shown that observed particulate distribution can be described with a satisfactory degree of precision by this model. Furthermore, changes in particulate transport, especially in terms of altered sedimentation behavior, can be predicted when solution or temperature effects on the settling behavior are defined.

A description and prediction of particulate matter distribution in water are of great direct importance for the assessment of water quality in the broadest sense. The term "particulates" is used to summarize suspended inorganic and organic particles in that order of size where interparticle forces become important. For all practical purposes particulate matter in natural water consists mainly of the different clay fractions (kaolinite, montmorillonite, illite, etc.) of metal oxides and of

0-8412-0499-3/80/33-189-213$05.00/0

bacteria and algae as well as metabolic intermediaries, all on the order of magnitude from less than 1 μm to 100 μm. While their chemical structure and even their surface chemical properties may differ (1), their transport and deposition behavior is to some degree similar because of the dominance of electrostatic and van der Waals particle interaction forces (2).

Particulate matter in this sense affects water quality from the viewpoint of a potential user or from an ecological point of view in a direct way, such as causing turbidity (3), or through limiting light penetration (4), or by producing obstructing and/or degrading sediments (5). Particulates also have a very marked, indirect influence on water quality through adsorption–desorption reactions with the dissolved phase (6). The close association of heavy metals with suspended inorganic or organic solids and with sediments illustrates this phenomenon. Exact mechanisms of sorption and in particular desorption are still unknown. Likewise the rates of sorption and desorption are determined for very specific situations only. However, the overall significance of this phenomenon is documented by findings that often the ratio of dissolved metal ion or complex to sorbed metal compound is in the order of 10^1 to 10^3 or more (7, 8, 9, 10, 11). It must be pointed out that some of these effects are thought of as positive, depending on the point of view, such as fixation of toxic heavy metals in sedimented material. Other effects, such as particulate deposition in waterways or degradation of sedimented material, are, from most any viewpoint, negative.

The preceding discussion has shown that it is important for sound ecological use of natural waters to anticipate changes in particulate transport and deposition attributable either to shifts in relevant milieu factors or to specific management measures taken. Furthermore, the various reactions controlling particulate distribution have been indicated, that is, aggregation, floc-shearing, sedimentation, and erosion as processes that are controlled by physical and chemical factors, and convective and turbulent diffusive transport as governed by hydromechanic parameters. It becomes apparent that such complex reactions can only be studied for very specific conditions in laboratory-type models; description and prediction of particulate behavior in natural systems under most diverse conditions might be accomplished more efficiently by mathematical modeling.

Such a mathematical model must fulfill the following requirements: (1) It must include all hydrodynamic transport phenomena and all physicochemical aggregation, sedimentation, and entrainment reactions that have been found significant, yet it should not attempt impractical completeness (such as including the still unknown factor of consolidation of sedimented material). (2) It should require input data—including global parameters that reflect less known reaction steps—that are readily obtainable, if necessary by simple additional analysis, such as milieu-dependent sedimentation rates, as opposed to time and location variant colloid-stability values plus aggregation controlling velocity gradients. (3) The mathematical structure must be such that any shift in model parameters will lead to identifiable and assignable variations in the output and not be clouded by a nondefinable host of partly compensating, partly enhancing reactions; furthermore, it should allow ready use from the viewpoint of computer expenses. These demands may appear redundant for mathematical models; however, in the case of modeling particulate transport, they seem to be neglected at times, as will be seen in the following section.

Previous Investigations

Transport and deposition phenomena of particulates in the previously described sense have received widespread interest from workers in different fields. From the viewpoint of changes in water quality, Fair, Moore, and Thomas (5) were among the first to describe the effects of hydrography on the degree of deposition, as well as the scour rate of degradable settled solids for actual rivers; they used a simple method to describe these effects and listed the order of magnitude of these phenomena. Hjulström (12) elucidated the problem by pointing out that deposition and scour of particulates occurred in different ways and by quantifying the relationship between average flow velocity and particle size. Postma (13) carried these studies further and described deposition and scour in terms of hydraulic parameters for an estuarine system.

Detailed quantitative investigations on settling and re-entrainment of particulates in complex, real-world flow patterns either in laboratories or in situ followed much later.

Mehta and collaborators (14) showed for a number of particulates, such as kaolinite and naturally occurring sediments containing different clays and organic matter from shallow estuaries, the deposition and erosion behavior in a circular laboratory flume. They found that characteristic ratios of an equilibrium particulate concentration to the initial concentration existed for specified shear values and that the depth of flow had no influence on this equilibrium. Conversely, the rate of attaining this equilibrium was found to depend on flow depth and particle concentration. Although the authors found that physicochemical parameters, such as the cation-exchange capacity of clays, had an effect on the depositional behavior, still no conceptual model was derived. Similarly, Ariathurai and co-workers (15) investigated deposition and erosion in a laboratory flume and arrived at data that are basically in agreement with Mehta. Contrary to the former, however, they used these data to confirm a conceptual model they had developed for particulate transport, thus deriving some mechanistic insights.

In situ observations on deposition (and where applicable on erosion) of particulates are numerous. To indicate the broad spectrum of interest, the work by Whitehouse et al. (16), Förstner (17), Sakamoto (18), and Kudo et al. (19) will be mentioned. Whitehouse and collaborators focused mainly on a detailed analysis of deposited material and, leaving hydromechanic aspects aside, found evidence for the effects of differential settling of different clays. Förstner and his co-workers dealt mainly with a detailed survey of heavy metals in lake and river sediments. The survey contains extensive analyses of the distribution and characteristics of heavy metals associated mainly with inorganic particulates, giving evidence for the effect of physicochemical parameters, but presenting only global hydraulic aspects. Sakamoto investigated the fate of particulates in a river–estuary system, reporting not only physicochemical parameters but also **fluid-mechanical characteristics** which allow the testing of existing notions on the role of aggregation and sedimentation within the phenomenon of particulate transport. Kudo and associates initiated one of the most ambitious studies on the fate of mercury in a Canadian river, monitoring and controlling as closely as possible all variables that might be significant. While they seem to be able to shed some light on the interactions of solid and dissolved phases, their conclusions with respect to more hydrodynamic effects seem lacking. With the exception of Sakamoto and a few others, in situ studies usually are deficient in assessing hydraulic effects while, on the contrary, laboratory studies seem to neglect taking into account physicochemical reactions.

Modeling particulate transport, or various process phases, has been attempted only relatively recently. Sayre (20) gave a very sound basis for further work by using a momentum solution of the two-dimensional convection–diffusion equation characterizing particle transport when additional terms for sedimentation, bed adsorbance, and re-entrainment (erosion) are included. He showed, with extensive hypothetical calculations, which hydrodynamic parameters are important and how they could be quantified. He was also able to show that his concept of bed adsorbance and re-entrainment requires further elucidation and indicated that there might be a turbulence effect on the sedimentation step. Hahn et al. (21) approached the problem from physicochemical viewpoints and developed a quantitative concept for aggregation and subsequent sedimentation in systems where convection and turbulent diffusion can be neglected. Laboratory and in situ data confirm the postulated mechanisms and models, thus providing a basis for formulation of those reactions that Sayre pointed out as still undefined. Subsequently, a large number of publications appeared dealing with the modeling of particulate transport. With the exception of one of the most recent by Somlyody (22), which appears to be a most ambitious attempt to incorporate all known phenomena in a conceptual way, most models, such as those of Nihoul et al. (*23*), Smith et al. (*24*), and Ariathurai et al. (*15*), concentrate on the exact formulation of particulate convection and diffusion in a two-dimensional way (both depth-integrated or width-integrated). It seems as if none of these researchers accepted the basis developed by Sayre along with his recommendations to focus on more physicochemical aspects. Therefore, practicality of these modeling concepts depends on the availability of input data that represent all physicochemical characteristics of the modeled system as far as they are relevant to particulate transport.

A Physicochemical and Hydromechanic Concept for Modeling Particulate Transport and Deposition

It becomes apparent from the discussion of existing models that further efforts should concentrate on the development of a descriptive and predictive model, that is, a model that goes beyond the reproduction of observed hydrodynamic laboratory data or more chemical in situ data. This implies that all pertinent reactions affecting particulate transport have been identified, formulated, and included in the model. Besides the phenomena of convective and diffusive transport, which appear to be well understood, the model should take into account particle aggregation, particulate sedimentation, and erosion of recent as well as aged and consolidated sediments. Implicitly contained in this is the notion that these processes might affect each other (for example, increase of floc size through aggregation and therefore changed sedimentation characteristics). This means that the effect on the process steps of pertinent physicochemical and hydraulic systems, such as composition of the dissolved phase (pH value, total salt content, concentration of aggregation-causing ions, etc.) and the flow-characterizing parameters (such as the distribution of flow velocity and dissipation), must be known.

The mixing of particulates following a discharge will be described on the basis of the continuity equation, the momentum equation, and the equation stating the conservation of species. The equations will be solved for depth- or width-integrated two-dimensional cases using the boundary layer approximation. The effects of aggregation are noticeable floc formation and subsequent enhancement of the sedimentation tendency. These effects are formulated as changes in particle/floc size including the phenomenon of the entrainment of water into the aggregate. Finally, the notion of an equilibrium concentration is introduced, quantifying the carrying capacity of the aqueous system, and controlling sedimentation and erosion as a function of time and location. These individual building blocks will be discussed in detail in subsequent paragraphs (see Figure 1).

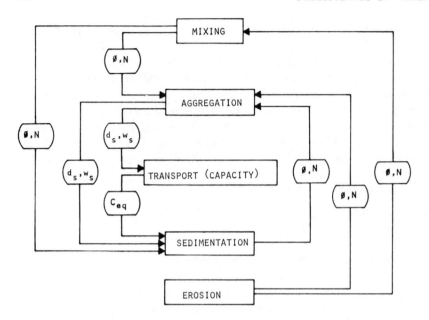

Figure 1. Building blocks of the particulate transport model representing the most important physical and chemical processes. Control variables and interrelationships are indicated.

To solve this modeling problem, systems where only one of the listed physical or chemical phenomena dominates are studied in detail. Actual laboratory and/or in situ data are collected to verify a conceptually derived mathematical model. The data used in this study are taken in part from the literature, in particular from Ref. *14,* and partly stem from our own investigations.

The Mixing Process

Reactions in natural systems are controlled by the degree to which the equilibrium is disturbed. In this case nonequilibrium conditions have to be considered separately for the dissolved phase and the suspended solid phase. Disturbances of natural equilibria, for example, in a river, are most frequently found downstream of wastewater discharges, etc. To quantify concentration-dependent reaction rates the two concentration fields resulting from mixing must be described. This is done using known and tested models to describe turbulent diffusion of dissolved and undissolved suspended substances (*26*).

The three-dimensional phenomenon is described in this case by a two-dimensional depth-integrated model assuming immediate and complete mixing in the vertical direction. The gradient of flow parameters and concentration values in the vertical axis was considerably lower than that within the cross-sectional axis. Flow velocities in the direction of flow usually are much higher than those orthogonal to the flow; however, gradients in the concentration values along the axis of flow appear negligible as compared to those across the width of the river. This leads to the following assumptions for the boundary layer:

$$u \gg v \text{ and } \frac{\delta}{\delta y} \gg \frac{\delta}{\delta x}$$

where u and v are the respective flow velocities in the direction of flow (x-axis) and orthogonal to this axis (y-axis). Furthermore, the system is considered to be stationary ($\delta/\delta t = 0$).

With these assumptions and definitions the following equations are derived and solved as described in (26):

For the conservation of mass:

$$\frac{\delta u}{\delta x} + \frac{\delta v}{\delta y} = 0 \tag{1}$$

For the conservation of momentum:

$$u \frac{\delta u}{\delta x} + v \frac{\delta u}{\delta y} = \frac{1}{\rho} \frac{\delta}{\delta y} \left(\mu_t \frac{\delta u}{\delta y} \right) \tag{2}$$

For the conservation of species:

$$u \frac{\delta c}{\delta x} + v \frac{\delta c}{\delta y} = \frac{1}{\rho} \frac{\delta}{\delta y} \left(\frac{\mu_t}{\sigma_c} \frac{\delta c}{\delta y} \right) + \frac{S}{\rho} \tag{3}$$

where c is the species concentration; ρ the density; S the source/sink; σ_c the turbulent Schmidt number; and μ_t the eddy viscosity (described by the kinetic energy and its dissipation according to Rastogi and Rodi (27)).

The expression S allows the introduction of generation or reduction terms for any species within the finite volume under consideration. In the case of conservative substances this will disappear, while in the case of particulates there will be rate laws reflecting aggregation, sedimentation, and erosion (see the following sections). Figure 2 describes schematically velocity and concentration profiles as computed by the model (Equations 1–3).

The Aggregation Process

The most significant effect of particulate coagulation or flocculation in this context is the resulting increase in overall particle size and conse-

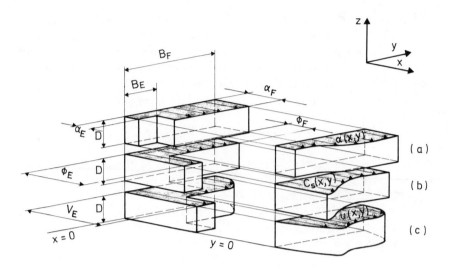

Figure 2. Schematic of the concentration profiles of the dissolved (a) and the suspended (b) phases, and the flow velocity profiles (c). (The symbols correspond to Table I.)

quently in the sedimentation velocity. The aggregation rate of particles in a flow system is described as follows (28):

$$\frac{dN}{dt} = -\frac{1}{2}\alpha(N\pi d_s)^2 u_r N \tag{4}$$

where α is the collision efficiency factor; $u_r = \sqrt{1/15}\ (\epsilon/\nu)^{1/2}d_s$ (a function of dissipation (ϵ), kinematic viscosity (ν), and particle diameter (d_s)); and N the particle concentration.

During the aggregation, excluding sedimentation, the floc volume ratio remains constant and allows one to relate change in particle number concentration to particle diameter,

$$\frac{N_i}{N_{i-1}} = \left(\frac{d_{s,i-1}}{d_{s,i}}\right)^3 \tag{5}$$

and, in agreement with the law of Stokes, a formulation of the changing sedimentation velocity:

$$\frac{w_{s,i-1}}{w_{s,i}} = \left(\frac{N_i}{N_{i-1}}\right)^{2/3} \tag{6}$$

where i, $i-1$ indicate subsequent reaction steps and w_s is the settling velocity of the sedimentous part of the suspension in a quiescent medium.

Usually the particle size distribution of natural suspensions can be described by a log-normal distribution. But this form cannot be integrated to obtain cumulative frequencies. The latter one is needed for subsequent mathematical manipulation, that is, to divide the suspension into a sedimentous and a nonsedimentous part. For this reason an empirical expression of simpler mathematical form is introduced for the cumulative distribution function:

$$Q = \frac{1}{1 + A \exp(- B \cdot \log w_s)} \qquad (7)$$

The constants A and B describe characteristic properties of the aggregating suspension: A is a measure of the mean settling velocity and B reflects the degree of dispersion (analogous to the standard deviation). The constants are determined by regression analysis from experimental data.

Figure 3 shows data observed for an aggregating and sedimenting system; comparing the data and the cumulative distribution function it is seen that this empirical model reproduces the observations to a satisfactory degree.

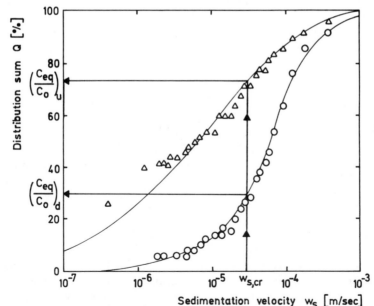

Figure 3. A typical variation of settling velocity distribution of a natural suspension underlying aggregating phenomena in a flowing system. Data are observed with a photosedimentometer in an aggregating kaolinite suspension, where the suspending medium was a water sample from a river, upstream (△) and downstream (○) from a waste discharge.

The Sedimentation Process

The vertical movement of particulates in natural waters is determined by gravitational and turbulent forces. This sedimentation process in a nonquiescent system is controlled by three groups of parameters: fluid parameters, particle properties, and characteristics of the hydrodynamic system. A dimensional analysis leads to the following:

$$\frac{s \cdot D}{u} = f\left(\frac{w_s}{u}, \frac{d_s}{D}, (C - C_{eq}), \frac{u \cdot D}{\nu}, \frac{B}{D}, \frac{u^2}{g \cdot D}\right) \tag{8}$$

where $s = dC/dt$ is the sedimentation rate; w_s the settling velocity of the sedimentous part of the suspension in a quiescent medium; $C - C_{eq}$ the amount of material that could be deposited (where C_{eq} is determined from a critical settling velocity—Equation 13); ν the kinematic viscosity; D the flow depth; B the flow width; and u the flow velocity.

For a natural river system (such as the Neckar) the variable groups d_s/D, B/D, and $u^2/g \cdot D$ are of no significance and can be omitted. Furthermore, the dimensionless group $u \cdot D/\nu$ corresponds to the Reynolds number of the main flow. This reflects the influence of the turbulent flow system on the sedimentation of a single particle. One can also omit the Reynolds number in Equation 8, if the settling velocity in quiescent medium w_s is translated into a settling velocity in a turbulent flow system w_{st}. Generally, the settling velocity can be calculated:

$$w_s = \left(\frac{4}{3}\left(\frac{\gamma_s}{\gamma} - 1\right)\frac{g \cdot d_s}{C_D}\right)^{1/2} \tag{9}$$

where γ_s is the specific weight of suspensa; γ the specific weight of water; g the gravitational acceleration; and C_D the coefficient of drag.

If w_s should be replaced by w_{st}, the drag coefficient must be replaced analogously:

$$\frac{w_{st}}{w_s} = \sqrt{\frac{C_D}{C_{Dt}}} \tag{10}$$

There is little information only on C_{Dt}. Thus in first approximation it is postulated that the drag of a settling particle depends on the internal friction of the surrounding fluid, which is described by the viscosity, eventually leading to:

$$\frac{w_{st}}{w_s} = \sqrt{\frac{\mu}{\mu_t}} \tag{11}$$

With this, Equation 8 can be simplified further when replacing w_s by w_{st}. Assuming a first-order rate equation for the deposition process (as has proven a reasonable hypothesis in many cases), one obtains for the rate of change of particle concentration attributable to deposition:

$$\frac{dC}{dt} = - \frac{w_{st}}{D} (C - C_{eq}) \tag{12}$$

The rate of sedimentation depends directly on the available amount of sedimenting particulates and leads to an exponential rate law. This conceptually derived rate law explains satisfactorily the asymptotic behavior of the concentration curve of the observed data (*14*).

In this context the number of particles depositing for given hydraulic conditions $(C - C_{eq})$ must be determined. This means that the characteristics of the particulates must be correlated to local parameters describing the turbulent flow field. This leads to a critical sedimentation velocity $w_{s,cr}$ for a particle. It is derived on the basis of an energy balance: the potential energy loss attributable to settling in a non-turbulent system must equal the turbulent kinetic energy that must be imparted on the particle in a turbulent flow system to prevent sedimentation (*29*).

$$w_{s,cr} \propto d_s \sqrt{\frac{k}{18\nu}} \sqrt{\frac{u}{D}} \tag{13}$$

where k is the turbulent kinetic energy of the system.

Knowing $w_{s,cr}$, and the actual distribution of sedimentation velocities of a suspension, one can determine the fraction of particulates with $w_s < w_{s,cr}$. This reflects the transport capacity C_{eq}/C_0 of a system defined by hydrodynamic and particle characteristics. The actual settling velocity of the sedimenting fraction corresponds to the average of this fraction (Figure 3).

This concept was tested on the basis of experimental data for a one-dimensional system. Figure 4 shows the time-dependent decrease of the particulate concentration in an annular flume; observed values and those computed according to Equation 12 show good agreement. In the experiments neither the hydrodynamic system nor the characteristics of the suspended matter were varied. One can see that the transport capacity C_{eq}/C_0 is constant (always about 50% of the initial concentration C_0 is sedimenting) and independent of the absolute value of C_0. This supports the hypothesis that there exists a critical settling velocity $w_{s,cr}$, which divides the entire amount of particulates into sedimentous and nonsedimentous parts.

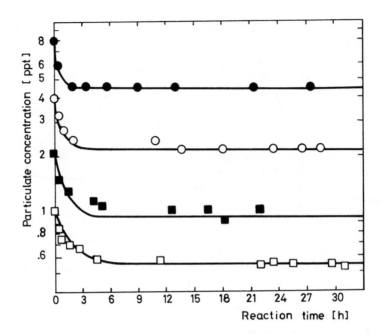

Figure 4. Particulate profiles as observed in a laboratory flume and as computed for a sedimenting environment. (Observed data as described in Ref. 14.) (●) $C_o = 7,875$ ppm, $C_{eq} = 4,471$ ppm; (○) $C_o = 3,983$ ppm, $C_{eq} = 2,155$ ppm; (■) $C_o = 2,037$ ppm, $C_{eq} = 996$ ppm; (□) $C_o = 1,014$ ppm, $C_{eq} = 584$ ppm.

In situ data from the River Neckar were likewise juxtaposed to model data, assuming that the river can be represented initially as a one-dimensional system. Here again the concept of a critical settling velocity, determined by particle and flow system characteristics (leading to a transport capacity C_{eq}/C_o), leads to a satisfactory reproduction of the observed data.

Application and Validity of Model

The described differential equations were solved by the finite difference method of Patankar and Spalding (26). The boundary-layer nature of the problem permits us to solve the equations by moving in the direction of flow for discrete grid points, starting with known initial conditions.

Input Data. As indicated previously, two concentration fields and one flow field must be computed to describe the transport of particulates. One concentration field defines dissolved substances that control particle

Table I. Input and Output Data for the Model

Input Data	Intermediary Results[a]	Output
Main stream		
mean flow velocity u_F	profiles of collision	velocity profiles $u(x,y)$
discharge rate Q_F	efficiency factor	concentration profiles
collision efficiency factor	$\alpha(x,y)$	of dissolved sub-
α_F	profiles of the turbu-	stances $C_d(x,y)$
volume concentration of	lent kinetic energy	concentration profiles
the suspended phase	$k(x,y)$	of the suspended
ϕ_F	profiles of dissipation	phase $C_s(x,y)$
density ρ	$\epsilon(x,y)$	sediment contours
kinematic viscosity ν		$S(x,y)$
coefficient of roughness		
(Manning) n		
river width B_F		
river depth D_F		
energy-line slope S_o		
Waste discharge (or other		
input)		
collision efficiency factor		
α_E		
mean particle diameter		
d_s		
volume concentration of		
the suspended phase		
ϕ_E		
discharge rate Q_E		
width B_E		
settling velocity		
distribution of the		
suspended particles		
$Q(w_s)$		

[a] Aggregation and transport parameters.

aggregation, while the other defines the particle concentration itself. The flow field affects particle aggregation and sedimentation as well as transport in all directions (see Table I).

A brief description of the course of computation and the necessary input indicates how suitable such a model is for illustrating very basic phenomena and/or solving very practical problems.

The effect of dissolved substances on particle aggregation is reflected in all presently used kinetic models in the collision efficiency factor. Since a direct correlation between constituents of the dissolved phase and particle stability can only be established for very specific situations (30), the computation of the concentration field is modified such that a quasiefficiency-factor field is calculated on the basis of observed efficiency factors for the discharge and river stream. The error introduced by the assumption of direct additivity of aggregation-causing phenomena is small compared to the possible variations resulting from uncertainties in assumed or precalculated coagulant collision efficiency curves.

The initial concentrations in both flows as well as the respective average diameters and distributions of settling velocities of the suspended particles together with fluid viscosity and fluid density have to be known. Furthermore, the critical bottom shear characterizing the deposited sediments has to be known if erosion is to be included explicitly, and the geometry and the discharge of the flow systems must be known for the computation of the flow field (depth and width). In addition, the energy-line slope and the coefficient of roughness (Manning) must be determined for the system so that the turbulent kinetic energy and the dissipation (according to the turbulence model (27)) can be computed.

Output Data. Basically, concentration maps of all dissolved substances can be computed. Yet, because of the previously described simplifying assumptions, the computation will proceed routinely via the direct calculation of collision efficiency factors as functions of location. Transport parameters, such as dissipation and kinetic energy, are likewise computed as functions of location at the predetermined grid points. These parameters, too, are on the one hand, intermediary results in the computation of particle aggregation and aggregate transport, but on the other hand, the basis for computed velocity profiles (see Table I).

The actual particulate transport will be described by both concentration profiles of the suspensa or by sediment contours. These profiles are computed for limited time spans only, since the model is based on the assumption of stationary conditions. It appears conceptually feasible to combine several different stationary computations to simulate longer time periods (such as over the course of a day).

Validity. To prove the applicability of the model, the first in situ measurements were performed in the Neckar, an impounded river, under almost stationary flow conditions. A tracer (Rhodamine B) was dosed into the effluent of a sewage treatment plant. The outlet of the channel was situated at the right bank of the river with an angle of 35° between the river axis and the channel. The concentration profiles in a transverse direction were measured with a fluorometer installed on a boat. Preliminary measurements showed that the river is well-mixed along the depth, so that a sample taken at a depth of 1 m could be considered representative. The discharge of the river was 51 m^3/sec and the discharge of the treatment plant was 1.2 m^3/sec.

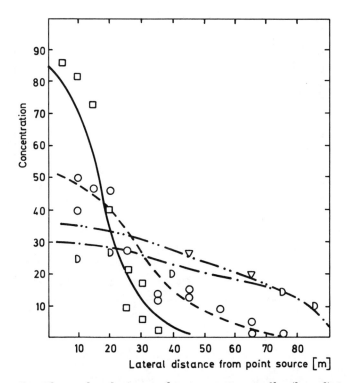

Figure 5.   Observed and computed concentration profiles (lateral) for an inert substance discharged into the River Neckar. Data are from an in situ investigation of a Rhodamine B plume in the Neckar. Rhodamine was discharged into a wastewater discharging channel. (——) computed, (□) observed at 400 m from point source; (– – –) computed, (○) observed at 900 m from point source; (– · – ·) computed, (▽) observed at 1900 m from point source; (– · · – · ·) computed, (D) observed at 2900 m from point source.

In Figure 5, observed (*31*) and computed Rhodamine concentration profiles are compared for a lateral discharge. The agreement between these data is satisfactory and shows the validity of the model for conservative constituents (Equations 1–3).

Next, the distribution of particulates was measured in the same section of the River Neckar under conditions similar to those just described; the concentration of the particulates at the discharge was ten times as high as in the river. Observed data were then compared to data computed with the model (that is, Equations 1–4, 6, and 12). Figure 6 shows the results of the computation. Part A (one-dimensional consideration) testifies to the reasonable agreement between observed and calculated data and Part B (two-dimensional consideration) exemplifies the type

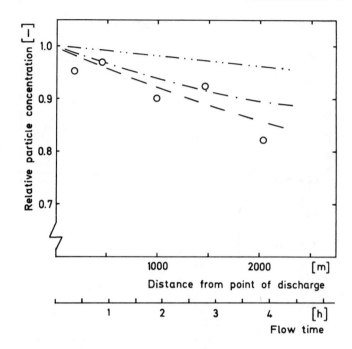

Figure 6A. Observed and computed particulate profiles in a sedimenting environment (River Neckar). Data stem from an in situ analysis of suspended particles in the River Neckar, downstream from a wastewater discharge with relatively high particle concentration. (– · · –) aggregation only; (– · –) sedimentation only; (– – –) aggregation and sedimentation.

of concentration profiles that can be computed. Mixing as well as (aggregation enhanced) sedimentation lead visibly to a reduction of particulate concentration downstream from the point of discharge. Closer inspection of the data shows that the effects of mixing are dampened faster than those of particle sedimentation.

Summary

Particulate transport controlling water quality through effects on dissolved and suspended water constituents has been treated until now either as an exclusively hydrodynamic problem or as a predominantly surface-chemical phenomenon. This study attempted to introduce tested colloid concepts, that is, aggregation and sedimentation kinetics, quantitatively into a complex, but practically solved, hydrodynamic transport model. The proposed model for particulate transport and sedimentation consists of several submodels, each describing the mixing process (mixing of colloid stability-controlling substances as well as particles), the aggre-

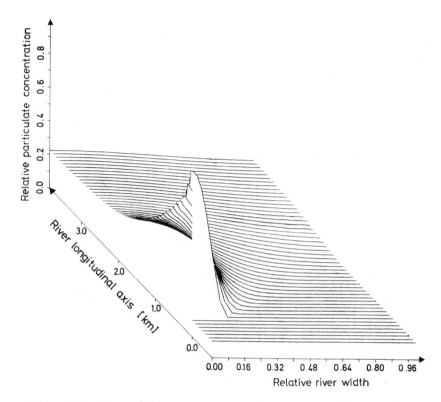

Figure 6B.   Computed concentration profiles of particulate matter discharged into a river.

gation process (as it is determined by chemical and hydrodynamic parameters), the aggregation-controlled sedimentation process, and its counterpart, erosion. The necessary input consisted of the usual hydrographic and hydrologic parameters, a predetermined turbulence model, and laboratory-determined suspension characteristics such as particle size distribution and particle stability for specific boundaries. This type of limited yet feasibly available input makes the model suitable for practical situations. Preliminary tests with respect to the agreement of simulated and observed data show the validity of the underlying concept.

Acknowledgments

One of the authors (F. Käser) is grateful to DFG and the Sonderforschungsbereich 80 for financial support. The computations were performed on the UNIVAC 1108 computer of the University of Karlsruhe.

Glossary of Symbols

A, B = constants in Equation 7

B_F = river width (m)

B_E = width of the waste discharge (m)

C = local concentration of species (i.e., floc volume ratio, if the suspended phase is modeled, and α-value, if the dissolved phase is modeled)

C_{eq} = equilibrium concentration of the suspended matter

C_o = initial concentration of the suspended matter

C_D = coefficient of drag

D = depth of the flow (m)

d_s = diameter of particle resp. floc (m)

g = gravitational acceleration (m/sec^2)

k = turbulent kinetic energy per unit mass (m^2/sec^2)

N = particle number per volume (1/m^3)

n = Manning's coefficient of roughness (sec/m$^{1/3}$)

Q = cumulative distribution function of suspensa's settling velocity

Q_E = waste discharge (or other input) (m^3/sec)

Q_F = river discharge (m^3/sec)

S = volumetric source resp. sink (kp sec/m^4)

S_o = energy-line slope

s = sedimentation rate (sec^{-1})

t = time (sec)

u_r = parameter describing hydrodynamic influences in the aggregation law (sec^{-1})

u = flow velocity in main flow direction (x-axis)(m/sec)

v = flow velocity orthogonal to the main flow direction (y-axis) in horizontal plane (m/sec)

w_s = settling velocity of particles in a quiescent medium (m/sec)

w_{st} = settling velocity of particles in a turbulent flowing system (m/sec)

$w_{s,cr}$ = critical settling velocity of particles (m/sec)

α = collision efficiency factor

γ = specific weight of water (kp/m^3)

γ_s = specific weight of suspensa (kp/m^3)

ϵ = dissipation of turbulent kinetic energy per unit mass (m^2/sec^3)

μ = dynamic viscosity (kp sec/m^2)

μ_t = eddy viscosity (kp sec/m^2)

ν = kinematic viscosity (m^2/sec)

ρ_c = density of water (kp sec^2/m^4)

σ_c = turbulent Schmidt number

ϕ = volume concentration

Literature Cited

1. Hahn, H. H.; Eppler, B. "Colloid and Interface Science"; Academic: New York, 1976; Vol. 4, p. 125–139.
2. Hahn, H. H.; Stumm, W. *Am. J. Sci.* **1968,** *268,* 354–368.
3. Am. Water Works Assoc. "Standard Methods for the Examination of Water, Sewage and Industrial Wastes"; APHA: New York, 1955; p. 207.
4. Cains, J., Jr.; Lanza, G. R. "Water Pollution Microbiology"; Wiley–Interscience: New York, 1972; p. 245–273.
5. Fair, G.-M.; Moore, E. W.; Thomas, H. A., Jr. *Sewage Works J.* **1941,** *13,* 270–278.
6. Bilinski, H.; Schindler, P.; Stumm, W.; Zobrist, J. "Jahrbuch vom Wasser"; Verlag Chemie: Weinheim, 1974; Vol. 43, p. 107–116.
7. Gardener, J. *Water Res.* **1974,** *8,* 157–164.
8. Förstner, U.; Müller, G. "Schwermetalle in Flüssen und Seen als Ausdruck der Umweltverschmutzung"; Springer–Verlag: Heidelberg, 1974.
9. Gibbs, R. J. *Science* **1973,** *180,* 71–73.
10. Hellmann, H. "Deutsche Gewässerkundliche Mitteilungen," Bundesanst. Gewässerkunde, Koblenz **1970,** *13,* 108–113.
11. Ibid, *14,* 42–47.
12. Hjulström, F. *Bull. Geol. Inst. Univ. Uppsala* **1935,** *25,* Chap. 3.
13. Postma, H. "Estuaries"; American Association for the Advancement of Science: Washington, DC, 1967; Vol. 83, pp. 267–281.
14. Mehta, A. J.; Partheniades, E. *J. Hydr. Res. Int. Assoc. Hydraul. Res.* **1975,** *13*(4), 361–381.
15. Ariathurai, R.; Krone, R. B. *J. Hydr. Division Am. Soc. Civil Eng.* **1976,** *102*(HY3), 323–338.
16. Whitehouse, U. G.; Jeffrey, L. M.; Debbrecht, J. D. "Clay Minerals in Saline Waters"; *Clays Clay Miner.* **1958,** *7,* 1–79.
17. Förstner, U. *Neues Jahrb. Mineral. Abh.* **1973,** *118,* 268–312.
18. Sakamoto, W. *Bull. Ocean Res. Inst. Univ. Tokyo* **1972,** *5.*
19. Kudo, A.; Miller, D. R.; Akagi, H.; Mortimer, D. C.; Nagase, H.; Townsend, D. R.; Warnock, R. C. *Preprints of the 9th International Conference* Int. Assoc. Water Pollut. Res. Stockholm **1978,** 329–341.
20. Sayre, W. W. *J. Hydr. Division* Am. Soc. Civil Eng. **1969,** *95*(HY3), 1009–1037.
21. Hahn, H. H.; Klute, R. *Proc. Intern. Conf. Environm. Sens. Assessment,* Las Vegas 1976, Vol. 1, p. 14–25.
22. Somlyody, L. *Preprints of International Symposium on Modeling the Water Quality of the Hydrological Cycle, Baden, 1978.*
23. Nihoul, J. C. J.; Adam, Y. *J. Hydr. Res.* Int. Assoc. Hydraul. Res. **1975,** *13*(2), 171–186.
24. Smith, T. J.; O'Connor, B. A. *Preprints of the Seventeenth Congress of the IAHR Baden-Baden, 1977,* Vol. 1, p. 79–86.
25. Jobson, H. E.; Sayre, W. W. *J. Hydr. Division* Am. Soc. Civil Eng. **1970,** *96*(HY3), 703–724.
26. Patankar, S. V.; Spalding, D. P. "Heat and Mass Transfer in Boundary Layers"; Intertext: London, 1970.
27. Rastogi, A. K.; Rodi, W. Proc. 2nd World Cong., Int. Water Res. Assoc., New Delhi, India, 1975, Vol. 2, p. 143–151.
28. Delichatsios, M. A.; Probstein, R. F. *J. Water Pollut. Control Fed.* **1975,** *47*(5), 941–949.
29. Käser, F. Dissertation, Universität Karlsruhe, Karlsruhe, West Germany, 1980.
30. Klute, R.; Neis, U. "Colloid and Interface Science"; Academic: New York, 1976; Vol. 4, p. 113–123.
31. Küpper, L. *Abschlußbericht an das BMFT*—Institut für Siedlungswasserwirtschaft, Universität Karlsruhe, Karlsruhe, West Germany, **1977.**

RECEIVED December 14, 1978.

10

Interactions Between Marine Zooplankton and Suspended Particles

A Brief Review

MICHAEL M. MULLIN

Institute of Marine Resources, A-018, University of California, San Diego, La Jolla, CA 92093

Zooplanktonic organisms remove particles from seawater either by entrapping them in a flow of mucus (salps, pteropods) or filtering them on a meshwork of fibers (larvaceans) or setae (many crustaceans). Removal of particles from suspensions may be determined by microscopic or electronic counting of the uneaten particles, and ingestion of radioisotopically labeled particles may be measured as short-term uptake of radioactivity by the animals. The filter-feeding of crustaceans on single classes of particles follows saturation kinetics, and in mixtures of particles of different sizes, the largest are often removed preferentially. The degree of control of the filtration in response to changes in the size spectrum of available particles and the extent of discrimination against detrital or inorganic particles are as yet unresolved. Zooplankters produce particles as fecal pellets, cast exoskeletons or abandoned feeding structures, and corpses; vertical transport of material is thereby accelerated.

Qualitative Aspects of Feeding

Many species of zooplankton feed by removing phytoplankton and other small particles from suspension. Some pelagic tunicates—salps and doliolids—trap particles on a mucous net through which water is driven by the swimming motions of the animal, and then ingest the mucus (1, 2, 3). Zooplanktonic molluscs—the thecosomatous pteropods—also produce mucus which is ingested together with entrapped particles (4). The appendicularians (larvaceans) pull water by swimming motions through a coarse filter which excludes large particles and then through a very fine filter from which particles are sucked by ciliary action (5, 6, 7, 8). The crustaceans that are filter-feeders—especially copepods and euphausiids—have a meshwork of setae on which particles are strained from a current set up by the swimming appendages, or a meshwork which itself functions as a rake on a moving appendage (1, 9, 10). The crustaceans have been studied extensively because they survive capture and laboratory manipulation more readily.

Methods of Study of Feeding

Microscopic observation of the gut contents of animals and visual determination of the change in concentration of suspended particles on which the animals graze are the classical techniques for determining the types of food eaten. The former method is still very useful for several types of qualitative studies (11, 12), particularly for animals too large or too fragile to be experimentally confined, but can be biased if organisms feed while within the net used to capture them (13) or by the fact that different types of ingested food remain distinguishable for variable periods. The latter difficulty might be alleviated by immunological identification of gut contents using fluorescent antibodies (cf. 14).

The ingestion of radioisotopically labeled food can be determined in much shorter experiments than are possible when ingestion is estimated from the change in concentration of particles in a known suspension; even the latter type of experiment has been greatly facilitated through the use of electronic particle counters (see Ref. 15 for a comparison of these techniques). Labeling different foods with different isotopes (16) can provide information on preferential feeding, and a device that releases labeled food while entrapping animals in situ has been used, so the animals are not removed from the environment until the experiment is completed (17). Most electronic particle counters are also able to enumerate different sizes of particles, thus permitting the study of differential removal by grazing. However, these counters do not distinguish differences in taxonomic status or nutritional value of food particles, or even whether or not the particles are organic.

Feeding on Phytoplanktonic Cells

The mucous feeders can retain cells as small as 1 μm in diameter, and are relatively unselective as to the particles they retain, except perhaps in the smallest categories (2, 3, 4). The crustaceans generally capture cells greater than or equal to 5 μm, though some cladocerans can remove finer material (18), as can appendicularians (1, 8). Studies of copepods that were fed suspensions of a single type of cell have generally indicated saturation kinetics. The animal filters a large amount of water when phytoplankton is scarce, and this amount decreases when cells are more concentrated so that ingestion tends to increase asymptotically to some maximal level. The exact nature of this function and the underlying mechanism are still unclear (*19*). In nature, the concentration of food is often so low that the maximal rate of ingestion may never be approached (*20*). Many studies indicate that some nonzero concentration of food particles must be present to stimulate feeding behavior, though this may be an artifact of data presentation (*19, 20, 21*).

Within the size range of phytoplankton that can be ingested at all, many copepods remove large cells at higher rates than they do small ones (*22, 23*). Some investigations have indicated that this is simply a property of the mesh sizes created by the setae of the mouthparts (*24, 25*), so selective feeding is relatively unvarying and mechanical. Selection might also result from a shift from relatively passive filtering to active searching for and seizing specific types of food (*26, 27*). Several studies (*28, 29, 30*) suggest that copepods can adjust their preferential feeding in response to changes in the size spectrum of particles available as food, though the mechanism of this adjustment is unclear.

Feeding on Other Particles

Phytoplankton often is not the major component of suspended particulate matter (seston), and the degree to which zooplankters can feed selectively on a basis other than size, or discriminate between organic and inorganic particles, is still unclear. Since copepods will ingest plastic beads (*31, 32*), the effect of particle size on feeding can be studied independently of other differences between particles (*23, 28*). Some investigators have found that the beads are eaten less readily than phytoplanktonic cells, perhaps because the beads must be swallowed whole. The fact that copepods will ingest microcapsules has been used advantageously in studies of nutrition (*33*) and chemosensory stimuli to grazing (*34*).

Some copepods have been found to ingest small oil particles from a fuel oil spill, and to be potentially important in sedimenting the oil in fecal pellets (*35*). The "red mud" produced in industrial extraction of aluminum from bauxite is also ingested, and at high concentrations causes nutritional stress (*36*). However, there is evidence of partial discrimination against inorganic particles in a natural suspension (*37*), and chemosensory stimuli (*34, 38*) may facilitate this. Copepods have been observed to backflush the filtration mechanism when it becomes clogged with unwanted material (*27*), and appendicularians can clean the exclusion filter (*7*), though salps apparently lack this ability (*3*).

The use of detritus (nonliving, particulate organic matter) as food by zooplankton seems to depend primarily on its source. Natural detritus from deep water apparently is not eaten, or at least is not assimilated readily, but fecal pellets are ingested and provide some nutritional value (*31, 39*). Detritus from near-shore macrophytes is ingested, but its main nutritional value seems to lie in the ciliates and bacteria growing on the particles (*40, 41*), which provide protein and decrease the carbon:nitrogen ratio of the detrital–microbial assemblage. Detritus may therefore be quantitatively important in the diet during much of the year in near-shore areas (*42*).

Production, Modification, and Sedimentation of Particles

The capture and ingestion of particles is not a perfectly efficient process; chains of phytoplankton may be broken into smaller chains (*43*), and production of small particles is sometimes reported in experiments in which grazing is measured by electronic particle counting and sizing (*23, 44*).

Assimilation is also incomplete; in crustaceans, the fecal material is packaged in a peritrophic membrane, and the fecal pellets contain unassimilated organic matter and inorganic remains. Living phytoplanktonic cells sink at a rate of 0–3 m/day in still water when actively growing; the largest cells might sink ten times as fast when senescent (*45*). Fecal pellets sink at a rate of 30–1000 m/day depending on size (*45, 46*), which is correlated with the size of the animal producing them (*39*). The pellets are important in moving bomb- and reactor-produced radioisotopes (*47*) and the skeletal remains of phytoplankton (coccoliths, diatom frustules, etc.) to the deep sea (*48, 49, 50, 51*). Indeed, accelerated sinking in fecal pellets is probably the reason why biogeographic patterns in surface waters are recorded rather accurately on the deep-sea floor as microfossils. If the microfossils sank as individual particles, they would be greatly displaced from their origin by oceanic currents.

Fecal pellets are a numerically small component of the total seston in the sea and are rarely caught by conventional sampling with water bottles; their importance is attributable to their size and sinking rate. The downward flux of material in fecal pellets depends not only on the rate of production of feces near the surface (which in turn depends on the abundances of zooplankters and their food), but also on the temperature-dependent rate of bacterial disintegration of the surrounding membrane (*50*) and the probability of re-ingestion (coprophagy) by other zooplankters (*39*). Small zooplankters are more numerous and have a higher rate of feeding per unit bodily size than do large ones, but small fecal pellets sink more slowly than large ones, stay longer in the warm surface layers, and may be ingested by a greater variety of animals. Some studies (*52*) indicate that most fecal pellets do not reach great depths intact, but sediment traps below 1000 m catch quantities of recognizable pellets (53, 54, 55). Indirect evidence (56, 57, 58) also suggests that fecal pellets and other relatively large particles account for much of the downward flux of chemical elements in the ocean. However, the most complete study to date (54) showed that recognizable fecal pellets contributed less than 10% to the sedimentation of carbon, nitrogen, and phosphorus in the central Pacific, and 10–65% of the sedimentation in coastal waters during upwelling. Seasonal variations in the importance of fecal pellets relative to other components has also been reported (59).

Though quantitatively of less importance than fecal pellets, the exuviae (cast exoskeletons) of crustaceans represent another form of particle production (60); these have sinking rates on the same order of magnitude as fecal pellets, and transport radioisotopes as well (46, 47, 61). The abandoned houses of larvaceans (62) and gelatinous pseudoconchs of pteropods (4), together with entrapped particles, form mucoid aggregates that develop microbial microcosms of their own.

From scatology and exuviology to necrology, the role of zooplanktonic corpses must finally be mentioned as a source of particles (63, 64). These sink rapidly (46) but also decompose rapidly (65), and are of less quantitative importance than fecal pellets (47), yet they may represent a source of high-quality food for deep-sea creatures.

Some Final Comments

Sediment traps are intended to measure downward flux, but technical difficulties may affect interpretation of the results. Loss of trapped material during recovery is still a problem (53, 55), and replication is not always good (55) because the traps affect the flow of water around them (cf. Ref. 66). Also, trapping of what is really resuspended benthic material may create artifacts (for example, 67), even a few hundred meters off the bottom of the deep sea (55).

The sinking rates of fecal pellets in still water have been measured frequently (see previous sections), but how these rates are modified by the natural, turbulent environment is not clear. Also, in areas with high concentrations of organic aggregates many fecal pellets may settle onto the aggregates and sink not as single particles, but as part of a heterogeneous assemblage (68).

Clearly, the role of zooplankton in sedimentation of material is of great interest beyond biology (for example, 69), and there are interesting questions concerning the effects of the behavior of organisms on geochemical processes. For example, work on particulate radionuclides indicates that carbonate is sedimented in larger (or at least faster sinking) particles than is iron (70); one would like to know whether selective feeding of zooplankton might be a contributing factor. Some of the kinds of experiments done by zooplanktologists to answer ecological questions could be applied profitably to geochemical and environmental engineering problems.

Literature Cited

1. Jørgensen, C. B. "Biology of Suspension Feeding," In "International Series of Monographs in Pure and Applied Biology"; Pergamon: Oxford, 1966; Vol. 27.
2. Madin, L. P. "Field Observations on the Feeding Behavior of Salps (Tunicata: Thaliacea)," *Mar. Biol.* 1974, 25(2), 143–147.
3. Harbison, G. R.; Gilmer, R. W. "The Feeding Rates of the Pelagic Tunicate *Pegea confederata* and Two Other Salps," *Limnol. Oceanogr.* 1976, 21(4), 517–528.
4. Gilmer, R. W. "Some Aspects of Feeding in Thecosomatous Pteropods," *J. Exp. Mar. Biol. Ecol.* 1974, 15(2), 127–144.
5. Alldredge, A. L. "Field Behavior and Adaptive Strategies of Appendicularians," *Mar. Biol.* 1976, 38(1), 29–39.
6. Paffenhöfer, G.–A. "On the Biology of Appendicularia of the Southwestern North Sea," *Tenth European Symp. Mar. Biol.* 1976, 2, 437–455.
7. Alldredge, A. L. "House Morphology and Mechanisms of Feeding in the Oikopleuridae (Tunicata, Appendicularia)," *J. Zool.* 1977, 181(2), 175–188.
8. Flood, P. R. "Filter Characteristics of Appendicularian Food Catching Nets," *Experientia* 1978, 34(2), 173–175.
9. Marshall, S. M. "Respiration and Feeding in Copepods," *Adv. Mar. Biol.* 1973, 11, 57–120.
10. Gauld, D. T. "The Swimming and Feeding of Planktonic Copepods," In "Some Contemporary Studies in Marine Science"; Barnes, H., Ed.; Allen and Unwin: London, 1966; pp. 313–334.
11. Roger, C. "Recherches sur la Situation Trophique d'un Groupe d'Organismes Pélagiques (Euphausiacea). I. Niveaux Trophiques des Espèces," In *Mar. Biol.* 1973, 18(4), 312–316.
12. Mackas, D.; Bohrer, R. "Fluorescence Analysis of Zooplankton Gut Contents and an Investigation of Diel Feeding Patterns," *J. Exp. Mar. Biol. Ecol.* 1976, 25(1), 77–85.
13. Judkins, D. C.; Fleminger, A. "Comparison of Foregut Contents of *Sergestes similis* Obtained from Net Collections and Albacore Stomachs," *Fish. Bull.* 1972, 70(1), 217–223.
14. Young, J. O.; Morris, I. G.; Reynoldson, T. B. "A Serological Study of *Asellus* in the Diet of Lake-Dwelling Triclads," *Archiv. Hydrobiol.* 1964, 60, 366–373.
15. Hargis, J. R. "Comparison of Techniques for the Measurement of Zooplankton Filtration Rates," *Limnol. Oceanogr.* 1977, 22(5), 942–945.
16. Lampert, W. "A Method for Determining Food Selection by Zooplankton," *Limnol. Oceanogr.* 1974, 19(6), 995–998.
17. Haney, J. F. "An *in situ* Method for the Measurement of Zooplankton Grazing Rates," *Limnol. Oceanogr.* 1971, 16(6), 970–977.
18. Pavlova, E. V. "On Grazing by *Penilia avirostris* Dana," *Tr. Sevastop. skoi Biol. Stn. Akad. Nauk Ukr. SSR* 1959, 11, 63–71.
19. Mullin, M. M.; Stewart, E. F.; Fuglister, F. J. "Ingestion by Planktonic Grazers as a Function of Concentration of Food," *Limnol. Oceanogr.* 1975, 20(2), 259–262.

20. Reeve, M. R.; Walter, M. A. "Observations on the Existence of Lower Threshold and Upper Critical Food Concentrations for the Copepod *Acartia tonsa* Dana," *J. Exp. Mar. Biol. Ecology* **1977**, *29*(3), 211–221.
21. Frost, B. W. "A Threshold Feeding Behavior in *Calanus pacificus*," *Limnol. Oceanogr.* **1975**, *20*(2), 263–266.
22. Mullin, M. M. "Some Factors Affecting the Feeding of Marine Copepods of the Genus *Calanus*," *Limnol. Oceanogr.* **1963**, *8*(2), 239–250.
23. Frost, B. W. "Feeding Behavior of *Calanus pacificus* in Mixtures of Food Particles," *Limnol. Oceanogr.* **1977**, *22*(3), 472–491.
24. Nival, P.; Nival, S. "Particle Retention Efficiencies of an Herbivorous Copepod, *Acartia clausi* (Adult and Copepodite Stages): Effects on Grazing," *Limnol. Oceanogr.* **1976**, *21*(1), 24–38.
25. Boyd, C. M. "Selection of Particle Sizes by Filter-Feeding Copepods: A Plea for Reason," *Limnol. Oceanogr.* **1976**, *21*(1), 175–180.
26. Petipa, T. S. "Feeding of the Copepod, *Acartia clausi*," *Tr. Sevastop. skoi Biol. Stn. Akad. Nauk Ukr. SSR* **1959**, *11*, 72–99.
27. Conover, R. J. "Feeding on Large Particles by *Calanus hyperboreus* (Kröyer)," In "Some Contemporary Studies in Marine Science"; Barnes, H., Ed.; Allen and Unwin: London, 1966; pp. 187–194.
28. Wilson, D. S. "Food Size Selection Among Copepods," *Ecology* **1973**, *54*(4), 909–914.
29. Poulet, S. A. "Grazing of *Pseudocalanus minutus* on Naturally Occurring Particles," *Limnol. Oceanogr.* **1973**, *18*(3), 564–573.
30. Richman, S.; Heinle, D. R.; Huff, R. "Grazing by Adult Estuarine Calanoid Copepods of the Chesapeake Bay," *Mar. Biol.* **1977**, *42*(1), 69–84.
31. Paffenhöfer, G.–A.; Strickland, J. D. H. "A Note on the Feeding of *Calanus helgolandicus* on Detritus," *Mar. Biol.* **1970**, *5*(1), 97–99.
32. Jones, D. A.; Jawed, T.; Tily, P. "The Acceptance of Artificial Particles by Planktonic Crustacea," *Chemosphere* **1972**, *1*(3), 133–136.
33. Jones, D. A.; Munford, J. G.; Gabbott, P. A. "Microcapsules as Artificial Food Particles for Aquatic Filter Feeders," *Nature* **1974**, *247*(5438), 233–235.
34. Poulet, S. A.; Marsot, P. "Chemosensory Grazing by Marine Calanoid Copepods (Arthropoda: Crustacea)," *Science* **1978**, *200*(4348), 1403–1405.
35. Conover, R. J. "Some Relations Between Zooplankton and Bunker C Oil in Chedabucto Bay Following the Wreck of the Tanker *Arrow*," *J. Fish. Res. Board Can.* **1971**, *28*, 1327–1330.
36. Paffenhöfer, G.–A. "The Effects of Suspended 'Red Mud' on Mortality, Body Weight and Growth of the Marine Planktonic Copepod *Calanus helgolandicus*," *Water, Air Soil Pollut.* **1972**, *1*, 314–321.
37. Corner, E. D. S. "On the Nutrition and Metabolism of Zooplankton. I. Preliminary Observations on the Feeding of the Marine Copepod, *Calanus helgolandicus* (Claus)," *J. Mar. Biol. Assoc. U.K.* **1961**, *41*(1), 5–16.
38. Friedman, M. M.; Strickler, J. R. "Chemoreception and Feeding in Calanoid Copepods (Arthropoda: Crustacea)," *Proc. Natl. Acad. Sci. U.S.A.* **1975**, *72*, 4185–4188.
39. Paffenhöfer, G.–A.; Knowles, S. C. "Ecological Implications of Fecal Pellet Size, Production, and Consumption by Copepods," *J. Mar. Res.* **1979**, *37*(1), 35–49.
40. Roman, M. "Feeding of the Copepod *Acartia tonsa* on the Diatom *Nitzschia closterium* and Brown Algae *(Fucus vesiculosus)* Detritus," *Mar. Biol.* **1977**, *42*(2), 149–155.
41. Heinle, D. R.; Harris, R. P.; Ustach, J. F.; Flemer, D. A. "Detritus as Food for Estuarine Copepods," *Mar. Biol.* **1977**, *40*(4), 341–353.
42. Poulet, S. A. "Feeding of *Pseudocalanus minutus* on Living and Non-Living Particles," *Mar. Biol.* **1976**, *34*(2), 117–125.

43. O'Conners, H. B.; Small, L. F.; Donaghay, P. L. "Particle-Size Modification by Two Size Classes of the Estuarine Copepod *Acartia clausi*," *Limnol. Oceanogr.* **1976,** *21* (2), 300–308.
44. Allen, J. D.; Richman, S.; Heinle, D. R.; Huff, R. "Grazing in Juvenile Stages of Some Estuarine Calanoid Copepods," *Mar. Biol.* **1977,** *43* (4), 317–331.
45. Smayda, T. J. "The Suspension and Sinking of Phytoplankton in the Sea," *Oceanogr. Mar. Biol. Ann. Rev.* **1970,** *8,* 353–414.
46. Fowler, S. W.; Small, L. F. "Sinking Rates of Euphausiid Fecal Pellets," *Limnol. Oceanogr.* **1972,** *17* (2), 293–296.
47. Small, L. F.; Fowler, S. W. "Turnover and Vertical Transport of Zinc by the Euphausiid *Meganyctiphanes norwegica* in the Ligurian Sea," *Mar. Biol.* **1973,** *18* (4), 284–290.
48. Schrader, H.–J. "Fecal Pellets: Role in Sedimentation of Pelagic Diatoms," *Science* **1971,** *174* (4004), 55–57.
49. Roth, P. H.; Mullin, M. M.; Berger, W. H. "Coccolith Sedimentation in Fecal Pellets: Laboratory Experiments and Field Observations," *Geol. Soc. Am. Bull.* **1975,** *86,* 1079–1084.
50. Honjo, S.; Roman, M. R. "Marine Copepod Fecal Pellets: Production, Preservation, and Sedimentation," *J. Mar. Res.* **1978,** *36* (1), 45–57.
51. Turner, J. T. "Sinking Rates of Fecal Pellets from the Marine Copepod *Pontella meadii*," *Mar. Biol.* **1977,** *40* (3), 249–259.
52. Ferrante, J. G.; Parker, J. L. "Transport of Diatom Frustules by Copepod Fecal Pellets to the Sediments of Lake Michigan," *Limnol. Oceanogr.* **1977,** *22* (1), 92–98.
53. Wiebe, P. H.; Boyd, S. H.; Winget, C. "Particulate Matter Sinking to the Deep-Sea Floor at 2000 m in the Tongue of the Ocean, Bahamas, with a Description of a New Sedimentation Trap," *J. Mar. Res.* **1976,** *34* (3), 341–354.
54. Knauer, G. A.; Martin, J. H.; Bruland, K. W. "Fluxes of Particulate Carbon, Nitrogen, and Phosphorus in the Upper Water Column of the Northeast Pacific," *Deep-Sea Res.* **1979,** *26* (1A), 97–108.
55. Honjo, S. "Sedimentation of Materials in the Sargasso Sea at a 5,367 m Deep Station," *J. Mar. Res.* **1978,** *36* (3), 469–492.
56. McCave, I. N. "Vertical Flux of Particles in the Ocean," *Deep-Sea Res.* **1975,** *22* (7), 491–502.
57. Cherry, R. D.; Higgo, J. J. W.; Fowler, S. W. "Zooplankton Fecal Pellets and Element Residence Times in the Ocean," *Nature* **1978,** *274* (5668), 246–248.
58. Bishop, J. K. B.; Edmond, J. M.; Ketten, D. R.; Bacon, M. P.; Silker, W. B. "The Chemistry, Biology, and Vertical Flux of Particulate Matter from the Upper 400 m of the Equatorial Atlantic Ocean," *Deep-Sea Res.* **1977,** *24* (6), 511–548.
59. Soutar, A.; Kling, S. A.; Crill, P. A.; Duffrin, E.; Bruland, K. W. "Monitoring the Marine Environment Through Sedimentation," *Nature* **1977,** *266* (5598), 136–139.
60. Lasker, R. "Molting Frequency of a Deep-Sea Crustacean, *Euphausia pacifica*," *Nature* **1964,** *203* (4940), 96.
61. Fowler, S. W.; Small, L. F. "Molting of *Euphausa pacifica* as a Possible Mechanism of Vertical Transport of Zinc-65 in the Sea," *Intern. J. Oceanogr. Limnol.* **1967,** *1* (4), 237–245.
62. Alldredge, A. L. "Discarded Appendicularian Houses as Sources of Food, Surface Habitats, and Particulate Organic Matter in Planktonic Environments," *Limnol. Oceanogr.* **1976,** *21* (1), 14–23.
63. Wheeler, E. H., Jr. "Copepod Detritus in the Deep Sea," *Limnol. Oceanogr.* **1967,** *12* (4), 697–702.

64. Weikert, H. "Copepod Carcasses in the Upwelling Region South of Cap Blanc, N.W. Africa," *Mar. Biol.* **1977**, *42*(4), 351–356.
65. Harding, G. C. H. "Decomposition of Marine Copepods," *Limnol. Oceanogr.* **1973**, *18*(4), 670–673.
66. Staresinic, N.; Rowe, G. T.; Shaughnessy, D.; Williams, A. J., III. "Measurement of Vertical Flux of Particulate Matter with a Free-Drifting Sediment Trap," *Limnol. Oceanogr.* **1978**, *23*(3), 559–563.
67. Steele, J. H.; Baird, I. E. "Sedimentation of Organic Matter in a Scottish Sea Loch," In "Detritus and Its Role in Aquatic Ecosystems"; Melchiorri–Santolini, V., Hopton, J. W., Eds.; Mem. Ist. Idrobiol. (1972) *29* (Suppl.), 73–88.
68. Silver, M. W.; Shanks, A. L.; Trent, J. P. "Marine Snow: Microplankton Habitat and Source of Small-Scale Patchiness in Pelagic Populations," *Science* **1978**, *201*(4353), 371–373.
69. Lal, D. "The Oceanic Microcosm of Particles," *Science* **1977**, *198*(4321), 997–1009.
70. Lal, D.; Somayajulu, B. L. K. "Particulate Transport of Radionuclides ^{14}C and [55]Fe to Deep Waters in the Pacific Ocean," *Limnol. Oceanogr.* **1977**, *22*(1), 55–59.

RECEIVED September 11, 1978.

11

Prediction of Oceanic Particle Size Distributions from Coagulation and Sedimentation Mechanisms

JAMES R. HUNT

Environmental Engineering Science, California Institute of Technology, Pasadena, CA 91125

An explanation is offered for observed oceanic particle size distributions considering only particle removal by coagulation and sedimentation. The analysis includes three coagulation mechanisms: Brownian, shear, and differential-sedimentation, for a continuous distribution of particle size. The size distribution is assumed to be in steady state with a constant flux of particle volume through the distribution. Predicted size distributions are power-law functions of the particle diameter with exponents −2.5 for Brownian coagulation, −4.0 for shear coagulation, −4.5 for differential-sedimentation coagulation, and −4.75 for gravitational settling. Observed size distributions for oceanic waters and digested sewage sludges are compared with the predictions. One consequence of the theory is the prediction of increased particle concentration at oceanic thermoclines in response to a decrease in fluid turbulence.

R emoval mechanisms of suspended particles control the transport of many environmental pollutants and govern the efficiency of water and wastewater treatment processes. Unfortunately, one of these removal mechanisms, coagulation, is not as yet quantified for the continuous particle size distributions encountered in oceanic waters and treatment plants.

In oceanic waters there is considerable interest in the residence time of particles and associated trace elements and in similarities of oceanic size distributions (Lal (*1*); Lerman et al. (*2*)). Various particle-removal

0-8412-0499-3/80/33-189-243$05.00/0

mechanisms have been considered, including dissolution (Lal and Lerman (3)), sedimentation of individual particles (McCave (4); Brun–Cottan (5)), and sedimentation of zooplankton fecal pellets (Smayda (6)). Acceleration of particle sedimentation by coagulation was discussed by Arrhenius (7), but more recently investigators have dismissed the importance of coagulation in the ocean (3, 5, 6).

In estuaries and water and wastewater treatment processes, Sholkovitz (8), Edzwald et al. (9), and O'Melia (10) have discussed the importance of coagulation in aiding the removal of colloidal and suspended matter. Coagulation in these systems has been examined experimentally either for a particle suspension uniform in size where Smoluchowski's theories can be tested, or for a continuous size distribution, where coagulation and sedimentation are inferred through changes in turbidity or suspended particle volume. These approaches have not adequately considered the dynamics of the particle size distribution during coagulation and sedimentation.

In this chapter, mechanisms of particle removal are limited to coagulation and sedimentation. Predictions of size distributions are obtained that are in reasonable agreement with measured size distributions from oceanic waters and digested sewage sludge. Sensitivity of the predictions to fluid turbulence and fluid density presents a plausible explanation for zones of higher particle concentration observed in the oceanic water column. The analysis does not include zooplankton fecal pellet production, particle breakup, or dissolution, nor does it directly incorporate biological productivity.

Mechanisms

In natural and polluted waters there exists a continuous distribution of particle sizes. The most appropriate function for describing the dynamics of a continuous size distribution is the particle size distribution $n(d_p)$ defined in the expression

$$dN = n(d_p) d(d_p)$$

where dN is the number of particles per fluid volume with diameters in the range d_p to $d_p + d(d_p)$. The particle size distribution has units of number per milliliter per micrometer, expressed as $[L^{-3}l^{-1}]$, where $[L]$ represents a fluid length unit and $[l]$ a particle length unit. Dimensional units are important in the analysis that follows.

Removal or loss of a particle of a given size from a volume of fluid is assumed to occur through only two physical processes: sedimentation and coagulation. Sedimentation removes particles from the volume of fluid, while coagulation transfers many smaller particles into fewer larger particles within the fluid volume.

The sedimentation flux of particles in the size range d_p to $d_p + d(d_p)$ is

$$\frac{g}{18\nu} \left(\frac{\rho_p - \rho_f}{\rho_f} \right) d_p^2\, n(d_p)\, d(d_p) \qquad [L^{-2}t^{-1}] \tag{1}$$

The flux is the Stokes' settling velocity of a spherical particle times the number of particles in that size interval. In the expression, g is the gravitational acceleration, ν the kinetic viscosity, and ρ_p and ρ_f the particle and fluid densities. The time unit is indicated by [t].

The loss of particles of sizes d_i and d_j by coagulation is

$$\beta(d_i, d_j)\, n(d_i)\, d(d_i)\, n(d_j)\, d(d_j) \qquad [L^{-3}t^{-1}] \tag{2}$$

where $\beta(d_i, d_j)$ is the collision function determined by the collision geometry and has dimensions of fluid volume per time $[L^3t^{-1}]$. There are three coagulation mechanisms, each with its own collision function for spherical particles. Collisions of the smallest particles are attributable primarily to thermal or Brownian motion with a collision function

$$\beta_b(d_i, d_j) = \frac{2}{3}\, \frac{kT}{\mu}\, \frac{(d_i + d_j)^2}{d_i d_j} \tag{3}$$

where k is the Boltzmann constant, T the absolute temperature, and μ the fluid viscosity. Fluid shear in laminar or turbulent flow will cause particles moving with the fluid to collide. The collision function for shear coagulation is

$$\beta_{sh}(d_i, d_j) = \frac{G}{6}\, (d_i + d_j)^3 \tag{4}$$

where G is the mean shear rate of the fluid. In laminar flow the shear rate is given by the velocity gradient, while in turbulent flow the shear rate is approximately (11)

$$G = \left(\frac{\epsilon}{\nu} \right)^{1/2} \tag{5}$$

where ϵ is the rate of turbulent energy dissipation.

The third coagulation mechanism, differential sedimentation, occurs when a particle falling at its terminal settling velocity collides with a slower settling particle. This is represented as a cross-sectional area of collision multiplied by the difference in Stokes' settling velocities of the colliding particles

$$\beta_{ds}(d_i, d_j) = \frac{\pi g}{72\nu} \left(\frac{\rho_p - \rho_f}{\rho_f} \right) (d_i + d_j)^2 |d_i^2 - d_j^2| \tag{6}$$

These expressions for the collision function assume purely geometrical collisions and do not include electrostatic, van der Waals, or viscous forces. Corrections for these surface and fluid forces are available for Brownian coagulation and have been verified experimentally by Lichtenbelt et al. (12) in the absense of electrostatic forces for particles uniform in size. For shear coagulation, corrections have been computed for collisions between spheres of equal size, and experimental agreement with theory has been obtained only when electrostatic forces are absent (van de Ven and Mason (13); Zeichner and Schowalter (14)). Differential-sedimentation coagulation of hydrosols has not been examined theoretically or experimentally.

A general equation for the dynamics of the particle size distribution could be written incorporating sedimentation and coagulation mechanisms, but solution of such an equation, analytically or numerically, is not possible at this time. The main hindrances to a direct solution are the complexity of the general dynamic equation (11) and the unknown corrections to the collision functions to account for the neglected forces. A partial solution of the general dynamic equation has been obtained by using a similarity or self-preserving transformation. The concept was introduced by Friedlander (11), and has been applied to aerosol and hydrosol coagulation. The self-preserving transformation will not be used in this analysis because a general self-preserving transformation has not been shown to exist for simultaneous Brownian, shear, and differential-sedimentation coagulation and gravitational settling. Also, quantitative application of the self-preserving transformation requires knowledge of the unknown corrections to the collision functions for shear and differential-sedimentation coagulation. Because these circumstances prevent direct solutions of the particle size distribution, it is necessary to make some major simplifications while still retaining the characteristics of the coagulation and sedimentation mechanisms.

Simplification

The first simplification is to assume only one coagulation or sedimentation mechanism is dominant in a subrange of particle size. Figure 1 is a comparison of the collision functions for collision of an arbitrary particle with a particle of 1 μm in diameter. Values of the collision functions were obtained directly from Equations 3, 4, and 6 for low fluid turbulence and low density particles. Because the collision functions plotted do not include the previously mentioned particle surface and fluid forces, this plot only approximates the dominance of a coagulation mechanism over a particle size interval. For particles less than 1 μm, Brownian motion is the dominant collision mechanism, while particles from 1 to about 100 μm collide because of fluid shearing. Collisions

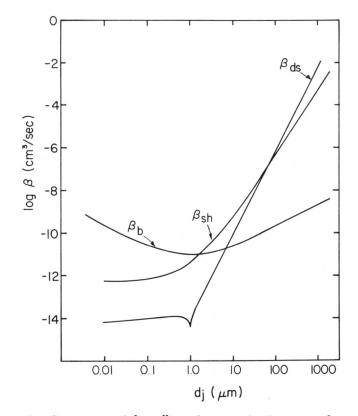

Figure 1.   Comparison of the collision functions for Brownian, shear, and differential-sedimentation coagulation ($d_i = 1.0 \ \mu m$; $T = 14°C$; $G = 3$ sec^{-1}; $(\rho_p - \rho_f)/\rho_f = 0.02$)

of a 1-μm particle with particles greater than 100 μm are by differential-sedimentation coagulation. Replotting Figure 1 for collisions other than those with 1-μm particles would shift the curves but still would retain the ordering of the dominant mechanisms: Brownian, shear, and differential sedimentation with increasing size of colliding particle. Particle removal by gravitational settling cannot be compared with the collision functions in Figure 1, but it is reasonable to assume that the largest particles would be removed from the fluid volume by settling.

Three further assumptions are necessary in the analysis. First, particles are assumed to have a low stability, independent of particle size. Destabilization can be achieved by compressing the electrical double layer in a high ionic strength solution or by neutralizing the surface charge through specific chemical interactions with hydrated metal ions or organic polymers. Second, the particle size distribution is in a dynamic steady state. For each small interval of particle size, the rate of particle volume transferred into the size interval by coagulation is balanced by either the volume coagulated from the interval or the particle volume lost from the fluid volume by sedimentation. This steady-state assumption implies the existence of a constant flux of particle volume through the distribution that is equal to the rate of formation of small particles and to the rate of large particle removal by sedimentation. The flux of particle volume is denoted by E with units of particle volume per fluid volume per time $[l^3L^{-3}t^{-1}]$.

The third assumption requires the particle size distribution to be a function of six variables: (1) particle diameter; (2) the flux of particle volume through the distribution; parameters representing: (3) Brownian coagulation; (4) shear coagulation; (5) differential-sedimentation coagulation; and (6) gravitational settling. Parameters for the coagulation and sedimentation mechanisms are obtained from fluid and particle constants appearing in Equations 1, 3, 4, and 6. The resulting parameters and the units obtained from the mechanism equations are:

Brownian

$$K_b = \frac{kT}{\mu} \qquad [L^3t^{-1}]$$

Shear

$$G \qquad [L^3]^{-3}t^{-1}]$$

Differential sedimentation

$$K_{ds} = \frac{g}{v}\left(\frac{\rho_p - \rho_f}{\rho_f}\right) \qquad [L^3]^{-4}t^{-1}]$$

Settling

$$S = \frac{g}{v}\left(\frac{\rho_p - \rho_f}{\rho_f}\right) \qquad [Ll^{-2}t^{-1}]$$

The parameters for differential-sedimentation coagulation and settling have the same grouping of constants but different units because differential sedimentation is second order in the particle size distribution from Equation 2, while gravitational settling is first order in the size distribution as seen in Equation 1. All parameters were chosen to be independent of particle size, assuming further that particle density is also independent of particle size.

Prediction

Based on these assumptions, the particle size distribution has the following functional form

$$n = n(d_p, E, K_b, G, K_{ds}, S) \tag{7}$$

The assumption of steady state removed the time dependence.

Predictions of the particle size distribution are obtained by considering subranges of particle size where only one coagulation mechanism or gravitational settling is dominant. For the smallest particle sizes Brownian coagulation is dominant and the functional relationship is

$$n = n(d_p, E, K_b) \tag{8}$$

which contains four variables (n, d_p, E, K_b) and three dimensions (L, l, t). Dimensional analysis is used to group the variables into a unique nondimensional expression (*15*)

$$A_b = n(d_p)\, d_p^{5/2} \left(\frac{K_b}{E}\right)^{1/2} \tag{9}$$

where A_b is a dimensionless constant. Solving for the particle size distribution gives

Brownian

$$n(d_p) = A_b \left(\frac{E}{K_b}\right)^{1/2} d_p^{-2.5} \tag{10}$$

Using the same technique for regions dominated by shear, differential sedimentation, and settling, the predicted size distributions are

Shear

$$n(d_p) = A_{sh} \left(\frac{E}{G}\right)^{1/2} d_p^{-4} \tag{11}$$

Differential sedimentation

$$n(d_p) = A_{ds} \left(\frac{E}{K_{ds}}\right)^{1/2} d_p^{-4.5} \tag{12}$$

Settling

$$n(d_p) = A_s \left(\frac{E}{S}\right)^{3/4} d_p^{-4.75} \tag{13}$$

Equations 10 and 13 were derived earlier by Friedlander (*16, 17*) for aerosols. The method of dimensional analysis arrives at predictions that contain dimensionless coefficients A_b, A_{sh}, A_{ds}, and A_s which must be determined experimentally.

Comparison of Predictions with Observations

Table I compares the predicted shape of the particle size distribution with observations from oceanic waters and sewage sludge digesters. The comparison is limited to the dependence of the particle size distribution on the particle diameter because fluid and particle parameters appearing in the predicted equations were not available. In plots of the logarithm of the size distribution vs. the logarithm of the particle diameter, Equations 10, 11, 12, and 13 become straight lines with slopes −2.5, −4, −4.5, and −4.75, respectively. Oceanic and digested sewage sludge size distributions are also observed to have one or more linear regions in such a plot, as summarized in Table I.

The particle size distribution of deep water from the Gulf of Mexico was measured by Harris (18) with an electron microscope over the size range 0.08 to 8.0 μm. For particles less than 2 μm in diameter the observed slope was −2.65, close to the Brownian prediction of −2.5. Particles in the size range 2 to 8 μm had a greater slope than expected for shear coagulation. Slopes close to the value predicted for shear coagulation were found in the two sets of North Atlantic data, both of which were obtained by electronic particle counters. Sheldon's data as reported by McCave (4) also contained some size distributions with a second linear region for the largest diameters which could be caused by either differential-sedimentation coagulation or gravitational settling. The size distributions reported by Lerman et al. (2) were limited to the size range from 1 to 6 μm and only one linear region was observed. Peterson (19) measured particle size distributions in natural and sewage-polluted California coastal waters with an electronic particle counter over the size range 1 to 20 μm. The observed slopes were close to the slope predicted for shear coagulation. Anaerobically digested sewage sludges were sized by Faisst (20) over the range of 1 to 70 μm. Both sludges had a region corresponding to shear coagulation and a second region with slopes exceeding differential-sedimentation and gravitational-settling predictions.

The observed distributions have particle diameter dependencies in reasonable agreement with the predictions, but these comparisons should not be taken as verification of the predictions. Verification requires coagulation experiments under controlled conditions where all the parameters appearing in the predicted equations are measured. A number of the assumptions required in the analysis appear reasonable. Both oceanic waters and digesters would likely be in a steady state because of the long residence times of the fluid and particles in suspension. The assumption of particle destabilization is reasonable in seawater because experiments by Sholkovitz (8) and Edzwald et al. (9) have shown that maximum destabilization for colloids and particles occurred at salinities less than that of seawater. Stability studies are not available for digested sewage sludge.

Table I. Predicted and Observed Slopes of Log $n(d_p)$ Vs. Log d_p

Dominant Mechanism Approximate Particle Size (μm)	Brownian < 2	Shear 2–40	Differential Sedimentation >40	Gravitational Settling >40
Predicted	−2.5	−4.0	−4.5	−4.75
Observed				
Gulf of Mexico (18)	−2.65	−4.4		
North Atlantic				
Sheldon in McCave (4)		−3.7 ± 0.2[a]	−4.65	
Lerman et al. (2)		−4.0 ± 0.3		
California coastal waters (19)		−3.8 ± 0.2		
Digested sewage sludge (20)				
Los Angeles City		−4.1	−5.2	
Los Angeles County		−4.1	−5.6	

[a] Standard deviation.

Application

In this section the sensitivity of the predicted size distributions to fluid turbulence and fluid density is examined above, in, and below an oceanic thermocline. This analysis shows that a coagulating size distribution at steady state has increased suspended particle volume in response to a decrease in fluid turbulence or an increase in fluid density. This may provide an explanation for the observed maximums in particle concentration and turbidity at oceanic thermoclines (Carder et al. (21)). Jerlov (22) has argued that higher thermocline concentrations result either from the minimum in eddy diffusivity at the thermocline which decreases turbulent transport through the thermocline, or from the increase in fluid density which traps low density particles.

The important feaures of a thermocline are the temperature gradient and resulting fluid density gradient. These gradients will have two effects on the predicted coagulating particle size distributions. First, turbulence is suppressed by the fluid stability corresponding to minimums in eddy diffusivity and shear rate G. Second, as the fluid density increases with depth, the settling velocity of particles decreases, which in turn decreases the values of the parameters for differential-sedimentation coagulation K_{ds} and gravitational settling S.

The effects of changes in shear rate and fluid density on the predicted particle size distribution are more easily illustrated graphically by transforming size distributions into volume distributions (11)

$$\frac{dV}{d(\log d_p)} = \frac{2.3\pi}{6} d_p^4 n(d_p) \tag{14}$$

The predicted volume distributions are

Brownian

$$\frac{dV}{d(\log d_p)} = \frac{2.3\pi}{6} A_b \left(\frac{E}{K_b}\right)^{1/2} d_p^{1.5} \tag{15}$$

Shear

$$\frac{dV}{d(\log d_p)} = \frac{2.3\pi}{6} A_{sh} \left(\frac{E}{G}\right)^{1/2} \tag{16}$$

Differential sedimentation

$$\frac{dV}{d(\log d_p)} = \frac{2.3\pi}{6} A_{ds} \left(\frac{E}{K_{ds}}\right)^{1/2} d_p^{-0.5} \tag{17}$$

Settling

$$\frac{dV}{d(\log d_p)} = \frac{2.3\pi}{6} A_s \left(\frac{E}{S}\right)^{3/4} d_p^{-0.75} \tag{18}$$

Because the dimensionless coefficients and parameters appearing in Equations 15–18 are not available, a surface water size distribution is assumed, and the sensitivity of the size distribution to changes in shear rate and fluid density is examined graphically.

Volume distributions for the surface water and for waters with a lower shear rate and a greater fluid density typical of a thermocline are plotted in Figure 2. For the surface water distribution, Brownian coagulation is assumed to be dominant for particles less than 1 μm in diameter,

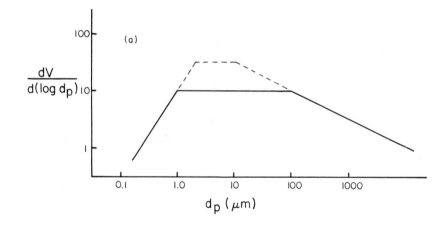

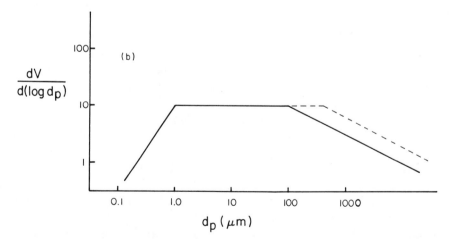

Figure 2.   Expected response of a coagulating volume distribution to changes in fluid properties at an oceanic thermocline: (a) order of magnitude decrease in shear rate ((————) surface layer; (– – –) decreased turbulence); (b) factor of 4 decrease in $(\rho_p - \rho_f)/\rho_f$ ((————) surface layer; (– – –) increased fluid density). Ordinate units are arbitrary.

shear coagulation dominates the range 1 to 100 μm, and differential sedimentation controls the particle size distribution for particles greater than
100 μm. For the size distributions representing conditions in the thermocline, only the shear rate G or the differential-sedimentation parameter
K_{ds} is modified; all other parameters, including E, are held constant.
Conclusions are similar if gravitational settling is assumed dominant for
particles greater than 100 μm instead of differential-sedimentation coagulation.

The volume distributions are plotted in Figure 2 on log–log scales
to display the power-law relationships of the predictions, Equations 15,
16, and 17. This representation distorts the volume distribution, which
should otherwise be plotted on semilog scales so that the total volume
can be obtained by visually integrating the distribution.

Figure 2a illustrates the change in volume distribution in response
to an order-of-magnitude decrease in the shear rate. Grant et al. (23)
measured the change in the average turbulent energy dissipation rate
with depth from the surface through the thermocline and found a decrease of two orders of magnitude, corresponding to a one order-of-magnitude decrease in the shear rate (Equation 5). With the lowered shear
rate, the volume distribution in the shear subrange increases in magnitude according to Equation 16, but the range of shear coagulation dominance decreases. Brownian coagulation is important for larger particle
sizes, and differential sedimentation dominates at diameters smaller than
those in the surface distribution.

Figure 2b demonstrates the effect of increasing the fluid density for
very light particles such that the factor $(\rho_p - \rho_f)/\rho_f$ decreases arbitrarily
by a factor of four. The increase in fluid density lowers the settling
velocity of particles and limits the importance of differential sedimentation to larger particles. This is represented by a decrease in parameter
K_{ds}, which increases the volume distribution in the differential-sedimentation subrange according to Equation 17. Shear coagulation has a larger
subrange of dominance.

The coagulating size distributions predict an increase in suspended
particle volume in response to lower turbulence or higher fluid density.
Both of these conditions are present at the thermocline, while at greater
depths turbulence is not suppressed and the fluid density remains high.
The zones of higher particle concentration occur in the thermocline which
suggests that the effect of diminished turbulence is more important than
the density increase. An alternate explanation is that particles become
neutrally buoyant in the thermocline and are trapped. Detailed size distribution measurements at varying depths in oceanic waters are necessary
to determine the role of coagulation in increasing the particle concentration at the thermocline.

Conclusion

This chapter has presented a theoretical derivation of continuous particle size distributions for a coagulating and settling hydrosol. The assumptions required in the analysis are not overly severe and appear to hold true in oceanic waters with low biological productivity and in digested sewage sludge. Further support of this approach is the prediction of increased particle concentration at oceanic thermoclines, as has been observed. This analysis has possible applications to particle dynamics in more complex systems; namely, estuaries and water and wastewater treatment processes. Experimental verification of the predicted size distribution is required, and the dimensionless coefficients must be evaluated before the theory can be applied quantitatively.

Acknowledgment

The author thanks James J. Morgan for his support during the development of this theory.

Glossary of Symbols

A_b, A_{sh}, A_{ds}, A_s = dimensionless coefficients for Brownian, shear, and differential-sedimentation coagulation, and gravitational settling

d_p = particle diameter [l]

dN = number of particles per fluid volume in a small interval of particle diameter [L^{-3}]

E = particle volume flux through the size distribution [$l^3 L^{-3} t^{-1}$]

g = gravitational acceleration

G = shear rate [$L^3 l^{-3} t^{-1}$]

k = Boltzmann constant

K_b, K_{ds} = Brownian and differential-sedimentation coagulation parameters [$L^3 t^{-1}$], [$L^3 l^{-4} t^{-1}$]

[l] = particle length unit

[L] = fluid length unit

$n(d_p)$ = particle size distribution [$L^{-3} l^{-1}$]

S = gravitational-settling parameter [$L l^{-2} t^{-1}$]

[t] = time unit

T = absolute temperature

$\beta_b, \beta_{sh}, \beta_{ds}$ = collision functions for Brownian, shear, and differential-sedimentation coagulation [$L^3 t^{-1}$]

ϵ = rate of turbulent energy dissipation

$$\mu = \text{dynamic viscosity}$$
$$\nu = \text{kinematic viscosity}$$
$$\rho_p, \rho_f = \text{particle and fluid density}$$

Addendum

Theoretical predictions for Brownian and shear coagulation have been verified experimentally for clay particles in artificial seawater (*24*).

Literature Cited

1. Lal, D. "The Oceanic Microcosm of Particles," *Science* **1977**, *198*(4321), 977–1009.
2. Lerman, A.; Carder, K. L.; Betzer, P. R. "Elimination of Fine Suspensoids in the Oceanic Water Column," *Earth Planet. Sci. Lett.* **1977**, *37*(1), 61–70.
3. Lal, D.; Lerman, A. "Dissolution and Behavior of Particulate Biogenic Matter in the Ocean: Some Theoretical Considerations," *J. Geophys. Res.* **1973**, *78*(30), 7100–7111.
4. McCave, I. N. "Vertical Flux of Particles in the Ocean," *Deep-Sea Res.* **1975**, *22*, 491–502.
5. Brun–Cottan, J. C. "Stokes Settling and Dissolution Rate Model for Marine Particles as a Function of Size Distribution," *J. Geophys. Res.* **1976**, *81*(9), 1601–1606.
6. Smayda, T. J. "The Suspension and Sinking of Phytoplankton in the Sea," *Oceanogr. Mar. Biol. Annu. Rev.* **1970**, *8*, 353–414.
7. Arrhenius, G. In "The Sea," Hill, M. N., Ed.; Wiley–Interscience: New York, 1963; Vol. 3, pp. 655–727.
8. Sholkovitz, E. R. "Flocculation of Dissolved Organic and Inorganic Matter During the Mixing of River Water and Seawater," *Geochim. Cosmochim. Acta* **1976**, *40*, 831–845.
9. Edzwald, J. K.; Upchurch, J. B.; O'Melia, C. R. "Coagulation in Estuaries," *Environ. Sci. Technol.* **1974**, *8*(1), 58–63.
10. O'Melia, C. R. In "Physicochemical Processes for Water Quality Control"; Weber, W. J., Ed.; Wiley–Interscience: New York, 1972; pp. 61–109.
11. Friedlander, S. K. "Smoke, Dust and Haze: Fundamentals of Aerosol Behavior"; Wiley–Interscience: New York, 1977.
12. Lichtenbelt, J. W. Th.; Pathmamanoharan, C.; Wiersema, P. H. "Rapid Coagulation of Polystyrene Latex in a Stopped-Flow Spectrophotometer," *J. Colloid Interface Sci.* **1974**, *49*, 281–285.
13. van de Ven, T. G. M.; Mason, S. G. "The Microrheology of Colloidal Dispersions VII. Orthokinetic Doublet Formation of Spheres," *Colloid Polym. Sci.* **1977**, *255*(5), 468–479.
14. Zeichner, G. R.; Schowalter, W. R. "Use of Trajectory Analysis to Study Stability of Colloidal Dispersions in Flow Fields," *AIChE J.* **1977**, *23*(3), 243–254.
15. Lin, C. C.; Segel, L. A. "Mathematics Applied to Deterministic Problems in the Natural Sciences"; Macmillan: New York, 1974.
16. Friedlander, S. K. "On the Particle-Size Spectrum of Atmospheric Aerosols," *J. Meteorol.* **1960**, *17*, 373–374.
17. Friedlander, S. K. "Similarity Considerations for the Particle-Size Spectrum of a Coagulating, Sedimenting Aerosol," *J. Meteorol.* **1960**, *17*, 479–483.

18. Harris, J. E. "Characterization of Suspended Matter in the Gulf of Mexico —II Particle Size Analysis of Suspended Matter from Deep Water," *Deep-Sea Res.* **1977,** *24,* 1055–1061.
19. Peterson, L. L. Ph.D. Thesis, California Institute of Technology, Pasadena, CA, 1974.
20. Faisst, W. K. Chapter 12 in this book.
21. Carder, K. L.; Beardsley, G. F., Jr.; Pak, H. "Particle Size Distributions in the Eastern Equatorial Pacific," *J. Geophys. Res.* **1971,** *76*(21), 5070–5077.
22. Jerlov, N. G. "Maxima in the Vertical Distribution of Particles in the Sea," *Deep-Sea Res.* **1958,** *5,* 173–184.
23. Grant, H. L.; Hughes, B. S.; Vogel, W. M.; Moilliet, A. "Some Observations of the Occurrence of Turbulence in and Above the Thermocline," *J. Fluid Mech.* **1968,** *34,* 443–448.
24. Hunt, J. R., Ph.D. Thesis, California Institute of Technology, Pasadena, 1980.

RECEIVED September 25, 1978.

Characterization of Particles in Digested Sewage Sludge

WILLIAM K. FAISST[1]

Keck Laboratories, California Institute of Technology, Pasadena, CA 91125

Some physical properties of several digested sludges were measured to better characterize the sludge particle system. The particle size distributions of two sludges were measured with a Coulter Counter. Number counts were in excess of $10^{12}/L$ and the majority of the particles had diameters less than 5 µm. Particle surface areas were at least 20 m^2/L. Particle sizing by filtration was attempted but proved unsuccessful because of membrane pore clogging. Sedimentation experiments for sludge in artificial seawater showed decreasing sedimentation velocities with increasing dilution (and hence lower particle number concentration). A comparison of sedimentation velocities for four sludges from different sources showed that sedimentation velocities increased with increasing solids content. Data presented suggest that flocculation of the sludges in seawater increased the sedimentation velocities.

Sewage sludge is the liquid–solids suspension resulting from the sedimentation phase of wastewater treatment. Quantities of sludge produced in the United States are increasing as the population grows and as wastewater treatment facilities are upgraded in an attempt to improve receiving water quality. Treatment and disposal are difficult and costly for sludge because of its contrary physicochemical nature and because it contains most of the trace metals and persistent synthetic organics from the influent sewage. Sludge treatment and disposal often represent one-quarter to one-half the total cost of wastewater treatment.

[1] Current address: Brown and Caldwell, Consulting Engineers, 1501 North Broadway, Walnut Creek, CA 94596.

0-8412-0499-3/80/33-189-259$06.00/0

Sludge suspensions are typically less than 5% solids by weight, but the solids are usually the most important part of the system. They may contain more than 99% of trace metals such as cadmium, zinc, and lead (1), 95% of organics such as dichlorodiphenyltrichloroethane (DDT) and polychlorinated biphenyls (PCB) (2), and substantial amounts of other organics. Pathogenic agents such as the eggs of the parasite *Ascaris lumbrociodes* are also found in sludge (3). The particle size distribution has been found by Karr and Keinath (4) to influence sludge dewaterability strongly. Sludge particulate matter has also been implicated in the alteration of the ocean bottom and overlying water column near the Los Angeles County Sanitation District's (LACSD) White Point effluent outfalls (5, 6, 7). The size distribution of sludge particles will directly affect its physical behavior on ocean discharge and its impact on many kinds of marine life.

In the presentation that follows, experiments to characterize the sludge particle system, including particle size and particle density, are described, and particle size distributions are reported. Quiescent sedimentation experiments for several sludges in seawater are also described. These experiments were carried out to better understand the general nature of the sludge particle system and to use as inputs for specific modeling studies for sludge discharges to the ocean. Since sludge behavior in the complicated ocean system is affected by both physical and chemical interactions not easily described mathematically nor easily simulated in the laboratory, this work attempted to bring together several types of information on the sludge particles to develop simple yet reasonable assumptions for modeling efforts (see Ref. 1 for further details). The detailed particle size information should also be useful when considering the particularly vexing problems of thickening or dewatering digested sludges.

Sludge Physical Properties

The most basic measure of the solids in sludge is total residue on evaporation (TROE), which includes both the suspended and dissolved matter remaining after evaporation at 105°C. The dissolved fraction— nonfilterable residue on evaporation (NROE)—is typically less than 2,000 mg/L, vs. 20,000–50,000 mg/L for digested sludges TROE. TROE and percent volatile residue for the nine sludge samples used in this work are given in Table I. The percent volatile is fairly constant for sludges from the three sources, reflecting good stabilization as a result of anaerobic digestion.

Particle Size and Particle Density. An introduction to the importance of particle size distribution for sludge has already been presented.

Table I. Digested Sludge Solids Analyses

	Solids	
Source	TROE[a]	Percent Volatile
Hyperion Digester 5C	.02374	57.1
LACSD[b] Primary[c]	.02429	51.9
LACSD[b] Primary[c]	.02528	52.3
LACSD[b] Primary[d]	.02880	54.8
Hyperion Digester 3C[d]	.02402	58.2
Hyperion Thermophilic[e]	.01831	56.6
Hyperion Mesophilic[e]	.02452	55.6
LACSD[b] Primary[e]	.02315	54.0
OCSD[e,f]	.02839	45.9

[a] TROE is the total residue on evaporation, expressed as g solids/g wet sludge.
[b] LACSD stands for County Sanitation Districts of Los Angeles County.
[c] Used for shallow columns sedimentation experiments.
[d] Used for particle sizing by Coulter Counter.
[e] Used for tall column sedimentation experiments.
[f] OCSD stands for Orange County Sanitation District.

Particle size together with particle density also affect the terminal settling velocity of particles released in the water column. For particles settling in environments where the Reynolds number:

$$R = \frac{\rho d v_s}{\mu} \tag{1}$$

is less than 0.5, Stokes Law holds:

$$v_s = \frac{g}{18} \frac{\rho_s - \rho}{\mu} d^2 \tag{2}$$

where v_s is the particle settling velocity; g the acceleration of gravity; ρ_s the mass density of particle; ρ the mass density of fluid; μ the absolute viscosity; and d the particle diameter.

Both ρ and μ vary with temperature. For natural materials, ρ_s varies from close to 1.0 g/cm^3 for some biological growths to 2.65 g/cm^3 for typical siliceous minerals to about 4.0 g/cm^3 for garnet sands. For Stokes Law v_s is proportional to d^2, where particle diameter is the more sensitive parameter. The importance of the density term ($\rho_s - \rho$) on sludge sedimentation is difficult to evaluate but is most critical where ρ_s approaches ρ, for example, with biological flocs.

The solids in sewage have been classified classically by size according to Rudolfs and Balmat (8) as:

1.	Settleable solids	>100 μm
2.	Supracolloidal solids	1 μm–100 μm
3.	Colloidal solids	1 mμm–1 μm
4.	Soluble solids	<1 mμm

These size ranges have general usefulness but fail to recognize the paramount biological and physical importance of subclasses, especially those between about 0.5 μm and 20 μm. Typical sedimentation processes in sewage treatment remove the settleable and some supracolloidal solids. Biological treatment and/or chemical coagulation can be used partially to remove colloidal and soluble solids. The fractions removed by sedimentation and/or biological treatment are called raw sludge.

Information about the particle size distribution of digested sludge is very limited. Raw and digested sludges from the City of Los Angeles Hyperion Treatment Plant have been wet-sieved, apparently using standard soil sieves. The data from this work appear in Table II (9) and on Figure 1 (10). The visible matter was listed as seeds, vegetable skins, fibers, bits of plastic sheeting or film, bits of bone, pieces of metal foil, fish scales, particles of soap, and bits of rubber and hair. With the possible exception of metal foil and bone, all of these materials should be volatilized by the standard volatile-solids testing (combustion at 550°C). Bits of eggshell, grains of sand, and glass chips are also common among the gritty matter found in sludge. Cigarette filters, coarse cellulose, and bits of plastic foam are representative of additional combustibles in the visible sludge fractions. Wet screening of sludge prior to ocean disposal effectively removes all visible matter.

Two important facts can be drawn from Table II and Figure 1. First, the process of sludge digestion at Hyperion clearly reduces the sludge mean particle size. Since one of the chief goals of sludge digestion is to stabilize the raw sludge, the particle size reduction, probably caused both by bacterial action on and degradation of components of the sludge and by mixing, is expected. While less than 50% of the raw sludge at Hyperion will pass a 53-μm hole, 70–80% of the digested sludge, depending on sludge digestion temperature, will pass the 53-μm screen.

Table II. Sieve Analysis of Hyperion Digested Sludge (9)

Sieve Mesh Size	Opening Size (mm)	Percent Retained (by Weight)
8	2.38	0.87
10	1.68	0.51
20	0.841	5.60
48	0.297	4.89
65	0.210	1.60
100	0.149	3.62
150	0.105	0.65
200	0.074	1.57
Passing 200 mesh	—	80.69

Journal of the Water Pollution Control Federation

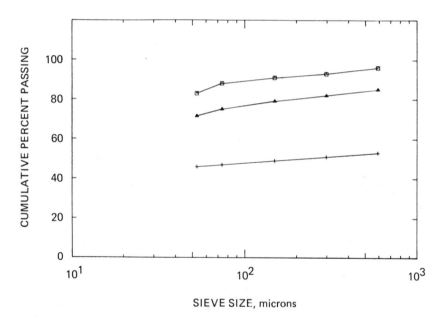

Journal of the Water Pollution Control Federation

Figure 1.   Sieve sizing of Hyperion sludges: raw (+), mesophilic (□), and thermophilic (▲) (10)

Second, the majority of digested sludge particles are less than 53 μm in diameter, and thus are slow to settle and difficult to remove by sedimentation processes. The particulate specific surface area (ssa), which increases as the particle diameter decreases (ssa $\propto 1/d$), should be high. These smaller particles should include bacterial cell bodies, fine cellulose, clay material, and mineral precipitates, such as metal sulfides, Fe_3O_4, etc.

Particle Sizing by Coulter Counter

The precise counting and sizing of sludge particles are possible using several techniques, including light and electron microscopy and an electronic particle-counting apparatus. However, microscopic methods are limited because of difficulties in sample preparation and handling. Electronic particle counting and sizing with a device, such as the Coulter Counter used for this work, allow for the rapid and accurate sizing of particles like sludge on a repetitive basis over a large particle-size spectrum. Sizing of the particles by Coulter Counter has been used for approximately 20 years on such diverse substances as blood cells, clays, pollens, plastic beads, water-treatment floc, and oil droplets. Extensive reviews of Coulter Counter applications and theory are available (Kubitschek (*11*), Wachtel and LaMer (*12*), Strickland and Parsons (*13*), and Treweek (*14*)).

The Coulter Counter depends on the conductivity difference between particles and the surrounding electrolyte for particle detection. The suspension is drawn through an orifice in the wall of a nonconductive tube by an advancing mercury column. A current is passed between electrodes inside and outside the tube; the passage of a nonconducting particle causes a voltage pulse which is proportional to the particle volume. This proportionality holds for particles with diameters from about 2 to 40% of the aperture diameter. For this work, the voltage pulses were amplified by the Coulter Counter electronics and scaled and sorted with a pulse height amplifier/multichannel analyzer (PHA/MCA). The 128-channel output from the MCA was then transferred automatically to paper punch tape with an attached teletype terminal.

Particle suspensions must be fairly dilute to avoid problems of coincident passage of several particles through the aperture. The volume sampled is set by electronic probes in the mercury column, which start and stop switches as the mercury passes. The instrument used for this work had settings for sample volumes of 50 μL, 500 μL, and 2,000 μL, and was operated with four apertures, 30 μm, 70 μm, 140 μm, and 280 μm. The Coulter Counter was calibrated with polystyrene latex (PSL) spheres. A computer program was used to convert channel counts and calibration information to particle diameters, surface areas, and volumes after editing spurious data from the paper tapes. Particles were assumed to be solid spheres, since output from the Coulter Counter for sludge is that for spheres of volume equivalent to the randomly shaped particles in the suspension.

The particle spectra were purposely taken with overlap, both to check the correspondence among apertures and in an attempt to provide continuous spectra over a large range of particle diameters. The spectra for the original plots represented the average results of three to five runs on the Coulter Counter. The data were also smoothed using a simple running average on sets of five adjacent points. Each point for plotting was obtained by accumulating seven channels of multichannel analyzer output.

The possibility of flocculation in the high ionic strength of the sodium chloride solution was countered by frequent agitation of particle solutions and sonification of samples before measurements were taken. Stirring the suspension between sampling runs was done for the 140-μm and 280-μm apertures to prevent losses by sedimentation. The teletype output for successive runs on given samples showed losses or variations of less than 10%.

Sizing Results. Figures 2 and 3 show particle size distributions, plotted as log (Δ number/Δ diameter) vs. log (diameter). Figures 4 and 5 show volume distributions for the same sludges, plotted as differential volume Δ(volume)/Δ(log diameter), vs. log (diameter). Six variations of aperture and instrumental settings were used to assemble a particle size distribution over a range from about 1 to 60 μm. The number distributions taken with different apertures match fairly well. However, the reader should note that the plots are log–log. The volume distributions appear neither as smooth nor as well fitted as the number distributions because of the semilog plots and the reasonably level distributions.

Number, surface area, and volumes for the entire sludge spectra are summarized in Tables III and IV. As expected, the number counts are very high, especially for small-diameter particles. The particles less than 5.4 μm in diameter constitute more than 98% of the total counts while contributing less than 40% of the measured volume. Measured surface areas are about 20 m^2/L, far greater than that typical for natural water systems.

Table III. Particle Size Distribution by Coulter Counter—Hyperion Mesophilic Sludge

Diameter Interval d(lower) to d(upper) (μm)	Number Sum (no./mL)[a]	Area Sum ((μm)^2/mL)[b]	Volume Sum ((μm)^3/mL)[c]
1.26 to 2.21	1.14×10^9	8.33×10^9	2.21×10^9
2.21 to 5.40	2.94×10^8	7.98×10^9	4.25×10^9
5.40 to 10.9	1.67×10^7	2.53×10^9	3.13×10^9
10.9 to 22.6	2.58×10^6	1.52×10^9	3.68×10^9
22.6 to 43.5	1.98×10^5	4.74×10^8	2.30×10^9
43.5 to 63.2	1.30×10^4	1.00×10^8	8.49×10^8
Totals	1.45×10^9	2.09×10^{10}	1.64×10^{10}

[a] Multiply by 1,000 for (number/L).
[b] Multiply by (1/10^9) for (m^2/L).
[c] Multiply by (1/10^{12}) for (cm^3/mL).

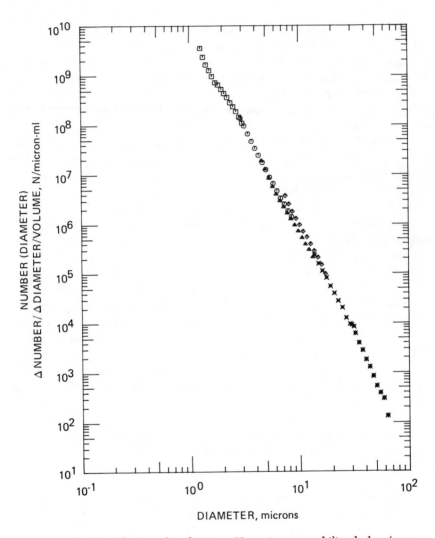

Figure 2. Particle size distribution—Hyperion mesophilic sludge (apertures: (□) 30 μm; (○) 70 μm, (△) 70 μm; (◇) 140 μm; (▯) 140 μm; (*) 280 μm)

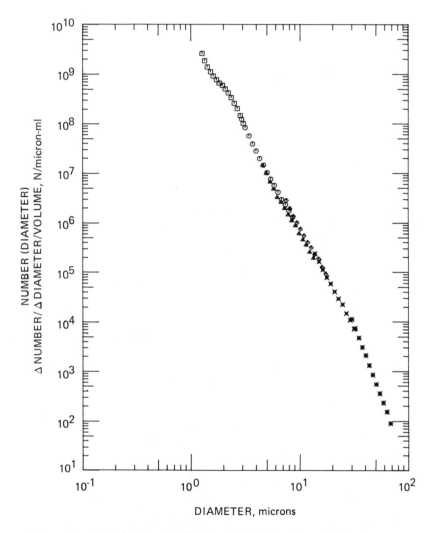

Figure 3.   Particle size distribution—LACSD primary sludge (apertures: (□) 30 μm; (○) 70 μm; (▲) 70 μm; (◇) 140 μm; (⊠) 140 μm; (*) 280 μm)

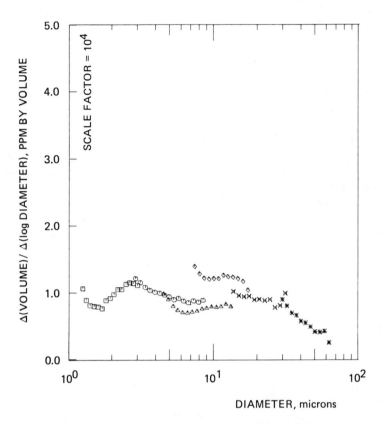

Figure 4. *Volume distribution—Hyperion mesophilic sludge (apertures:*
(□) 30 μm; (⊕) 70 μm; (△) 70 μm; (◈) 140 μm; (⬙) 140 μm; (*)
280 μm)

Table IV. Particle Size Distribution by Coulter Counter—LACSD Primary Sludge

Diameter Interval d(lower) to d(upper) (μm)	Number Sum (no./mL)[a]	Area Sum ((μm)²/mL)[b]	Volume Sum ((μm)³/mL)[c]
1.26 to 2.21	1.06×10^9	8.61×10^9	2.43×10^9
2.21 to 5.40	2.44×10^8	6.86×10^9	3.67×10^9
5.40 to 10.9	1.44×10^7	2.17×10^9	2.70×10^9
10.9 to 22.6	2.12×10^6	1.29×10^9	3.16×10^9
22.6 to 43.5	2.19×10^5	5.33×10^8	2.57×10^9
43.5 to 63.2	1.34×10^4	1.06×10^8	9.13×10^8
Totals	1.32×10^9	1.96×10^{10}	1.54×10^{10}

[a] Multiply by 1,000 for (number/L).
[b] Multiply by (1/10⁹) for (m²/L).
[c] Multiply by (1/10¹²) for (cm³/mL).

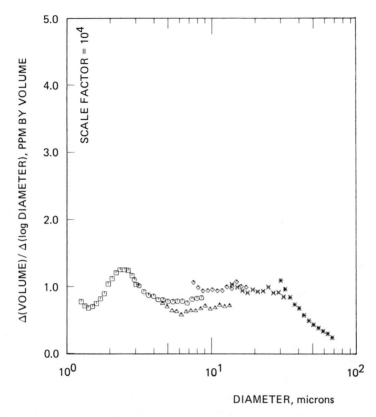

Figure 5. Volume distribution—LACSD primary sludge (apertures: (☐) 30 μm; (◯) 70 μm; (△) 70 μm; (◇) 140 μm; (⬭) 140 μm; (*) 280 μm)

The particle volume data from Tables III and IV can be combined with the weight information from Table I to estimate gross particle density. If NROE is assumed to be 2,000 mg/L and all of the sludge is assumed to fall in the size range from 1.26 μm to 63 μm, the resulting densities are 1.37 g/cm^3 for Hyperion mesophilic sludge and 1.69 g/cm^3 for LACSD sludge. These values are in the range of 1.0 to 2.65 g/cm^3 that would be expected for the heterogeneous sludge system.

Particle Sizing by Filtration

The classic technique for particle sizing in geology and soil mechanics is sieve analysis (either dry or wet). This method, generally applied to the larger sized particles greater than 400 mesh or 37 μm, is reasonably fast and gives good results for particles that are nearly spherical in shape.

An attempt has been made to use Millipore cellulose ester membranes to size sewage and sludge particles (see Chen et al. (15)). However, previous work on marine particle filtration (see Sheldon (16), Sheldon and Sutcliff (17), and Cranston and Buckley (18)) clearly shows that cellulose ester membranes are unsuited for such separation. Sheldon and Sutcliffe (17) point out that one manufacturer of such membranes, Millipore, actually recommends that their product not be used in such applications.

Sheldon and Sutcliffe (17) and Sheldon (16) used a Coulter Counter to examine the particle retention of nylon and stainless-steel meshes, Whatman glassfiber filters, and Nuclepore and Millipore membranes. With light particle loading and low vacuum (12–13 cm of mercury), the Millipore and glassfiber filters retained particles of much smaller diameter than the manufacturers' stated pore size. The meshes and Nuclepore membranes showed 50% retention of particles (by number) at their manufacturers' reported pore sizes.

Cranston and Buckley (18) examined the retention of latex beads, kaolinite clay, dissolved organics, and dissolved inorganics on five different membrane and fiber filters. Their interest was to select filters that gave the best particle retention and the most consistent gravimetric results. The Nuclepore membrane performed best except for retention of dissolved humic materials; Nuclepore membranes were particularly good for not retaining dissolved salts from seawater filtration.

Chen et al. (15) used centrifugation (up to 740g relative centrifugal force) to fractionate Hyperion digested sludge. With centrifugation times of at least 10 min, 82% of the sludge solids were removed. The centrate was then filtered through 8.0-μm, 3.0-μm, 0.8-μm, and 0.22-μm Millipore filters. Eighty percent of the centrate solids (by weight) passed the 8.0-μm Millipore filter, but all were retained by the 0.22-μm filter. The estimation of sludge particle size distributions from the data is crude at best. The initial centrifugation should remove some percentage of all particle-size fractions. The capture efficiency of the Millipore filters should also increase with filter loading.

Particle Sizing by Vacuum Membrane Filtration. In the original plan for this research, particle fractionation by sieving, that is, using microscreens (10, 20, or 30 μm) and Nuclepore membranes, was anticipated to be a viable means for particle sizing. Microscreens are available in both nylon and stainless steel. The Nuclepore membranes are thin polycarbonate sheets with very uniform round holes etched through them; available pore diameters decrease in size from 8.0 μm.

Microscopic examinations of dilute sludge suspensions revealed that the majority of particles, at least by number, were less than 10 μm in diameter. Since trace-metal analysis as well as particle sizing were planned, the use of stainless-steel meshes was ruled out. Nuclepore membranes were selected, but there was a possibility that solids loading per unit filter area large enough to allow for metal analysis by x-ray techniques might alter the sieve-like properties of the membrane.

Microscopic examination of 8.0-μm membranes loaded with both 1.1-μm and 15.8-μm PSL spheres showed some capture of 1.1-μm beads. (All filtration was done with vacuums of up to 74 cm of mercury.) Calculations were then made to determine the percent open area on the membranes and the theoretical loading of spheres necessary to clog all the pores. If a randomly shaped particle larger than a pore is lodged across a pore, fluid will continue to be drawn through the remaining opening, but the effective pore size is decreased significantly (see Figure 6).

The sample calculations of Table V, where it is assumed one particle per pore causes clogging, emphasize this point. Even if several particles per pore are necessary for clogging in vacuum filtration situations, less than 1 mg of material should be sufficient for complete clogging of a filter surface with a 3.5-cm diameter.

The actual performance of Nuclepore membranes was tested in this research using a 100:1 dilution of LACSD primary sludge with sludge filtrate. The sludge-filtrate dilution was applied in increasing volumes to sets of three membranes, mounted on a Millipore filter head, and the fluid drawn off by the vacuum. The membranes were then dried and the solids measured by weighing. The results, shown in Figure 7, demonstrate that the 8.0-μm membrane efficiency increases with increasing solids loading. The efficiency for the 1.0-μm membrane is constant (the correlation coefficient for the linear best fit is greater than 0.99). At the lowest loading, the capture on the 8.0-μm membrane is 46% of that on the 1.0-μm membrane. The total solids captured were 0.338 mg on a 8.0-μm membrane (weight = 14 mg) and 0.734 mg on a 1.0-μm membrane

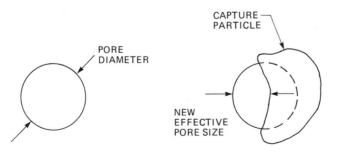

Figure 6.   Effects of filter clogging

Table V. Theoretical Clogging

Pore Diameter (μm)	Pore Density (number/cm²)[a]	Percent Open Area	Number of Pores per 3.5-cm Diameter Target
8	1×10^5	5	9.6×10^5
5	4×10^5	7.8	38.5×10^5
3	2×10^6	14.2	19.2×10^6
1	2×10^7	15.6	19.2×10^7

[a] As reported by the manufacturer.
[b] Volume of a spherical particle that would just clog the pore.

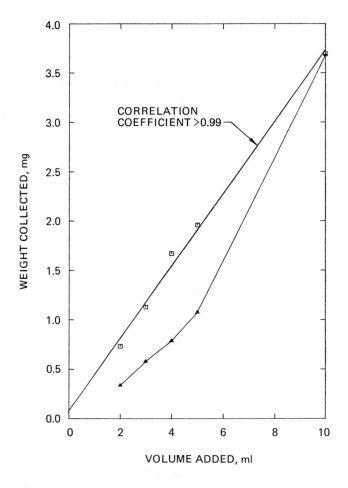

Figure 7. Solids retention by Nuclepore membranes loaded with a 100:1 dilution of sludge with sludge filtrate ((□) 1.0 μm; (▲) 8.0 μm)

of Nuclepore Membranes

Volume per Particle[b] *(cm³)*	*Volume for Clogging*[c] *(cm³)*	*Weight Range for Clogging*[d] *(mg)*	
2.68×10^{-10}	2.57×10^{-4}	.257	.684
6.54×10^{-11}	2.52×10^{-4}	.252	.668
1.41×10^{-11}	2.71×10^{-4}	.271	.718
5.24×10^{-13}	1.01×10^{-4}	.101	.267

[c] Assuming 1 particle/pore causes clogging.
[d] Assuming particle density extremes are 1.0 g/cm³–2.65 g/cm³.

(weight = 20 mg). At any lower total-solids loadings it is doubtful that accurate and reproducible weight determinations could be made. (The linear weight-capture per volume added for the 1.0-μm filter confirms the choice of these filters for use in solids capture during sedimentation experiments, where total loading per filter varied up to a factor of 10.)

Sedimentation Experiments

Extensive work on the theory and use of sedimentation columns with both flocculant and nonflocculant suspensions has been done by McLaughlin (*19, 20*). Single and multidepth samplings from columns up to 1.2 m deep were made on sand, clay, and sewage-particulate suspensions. Sweep flocculation was postulated to occur when faster particles overtake slow ones and the particles with different initial fall velocities become attached. As expected, the effect of flocculation was found to increase with depth in the column. Experiments made with several different sewage treatment plant effluents showed great variation in velocity distribution between samples. However, consistent results from samples from one source at one point in time were possible. McLaughlin noted the difficulty of extending laboratory results to real-world applications, since turbulence is important in natural systems but was not simulated in his quiescent experiments.

Previous work on the sedimentation of sludge or sewage effluent solids in seawater or seawater-like liquids has been done by Brooks (*21*), Myers (*5*), and Morel et al. (*22*). Brooks (*21*) conducted experiments with Hyperion digested sludge, diluted 20:1 with seawater. Solids were collected with Gooch crucibles. His data (see Figure 8) showed that 50% of the solids had a settling velocity v_s of less than 2.8×10^{-2} cm/sec. (These data have been adjusted from 31.5°C, the experimental temperature, to 12°C.) Rapid flocculation of the sludge was observed.

Myers (*5*) diluted a 24-hr composite of LACSD Whites Point effluent 1:1 with filtered seawater and captured the falling solids on Whatman type GF/C glassfiber filters. The composite sample contained

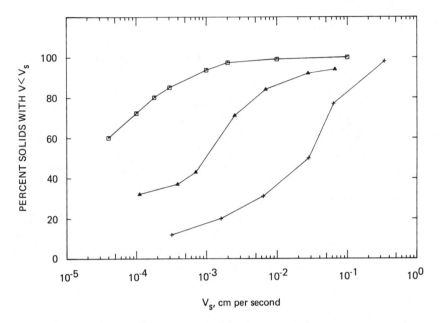

Figure 8. Sedimentation of wastewater solids in seawater-like media ((□) Morel (22), ~ 30 mg/L; (▲) Myers (5), ~ 150 mg/L; (+) Brooks (21), ~ 1600 mg/L)

primary effluent and centrate from digested primary sludge. The sludge centrifuges in the treatment plant captured 25–35% of the solids as sludge cake. The remaining fine solids were retained in the centrate. The 50% v_s reported by Myers was 1.0×10^{-3} cm/sec (see Figure 8).

Morel et al. (22) reported sedimentation experiments run on a 24-hr composite of sewage particulates and on the same composite diluted 10:1 with 0.5M NaCl. Solids were captured on membrane filters. Raw sewage showed a 50% v_s of 1×10^{-3} cm/sec. For the sewage/NaCl mixture, the 50% v_s was less than 1×10^{-5} cm/sec.

A direct comparison of these three experiments is difficult because the solids-capture techniques were not the same, different solids were used, and the sedimentation media and the temperature varied. However, a range of sedimentation velocities (1×10^{-5}–3×10^{-2} cm/sec) can readily be identified. The importance of experimental conditions and their control is evident for future work as are the choices of sedimentation fluid and the solids-separation technique. For example, in the experiments by Morel et al. (22), the use of NaCl as the sedimentation fluid may not stimulate flocculation as would the use of seawater which has a large concentration of divalent metal ions.

Sedimentation Experiments with Digested Sludges. Sedimentation experiments were made to observe the settling behavior after dilution in seawater of (a) sludge trace metals and (b) filterable solids. Such observations are necessary if predictive modeling is to be done for sludge disposal to the ocean. A properly designed experiment also provides insight into the flocculation of sludge in a high ionic strength medium (for example, seawater) and can be used to calculate sludge-particle diameters (or densities) if additional information on particle densities (or size distributions) is available. However, such experiments are limited because they are not hydrodynamically similar to ocean conditions. Turbulence is probably important for particle behavior in the ocean but was not simulated in these experiments.

Sedimentation experiments in this research were carried out at 10.5° ± 0.5°C in artificial seawater. The artificial seawater (pH 7.9), mixed according to the formula of Riley and Skirrow (*23*), was filtered through a 0.45-μm Millipore filter to remove the background solids. Two types of sedimentation apparatus were used—a shallow column with a standard 2-L graduated cylinder and a tall column with a special 10-L plexiglass column with side sampling ports (see Figure 9). Digested primary sludge (LACSD) was used for all runs in the shallow column, and samples were taken at a depth of 15 ± 0.3 cm with a volumetric pipette.

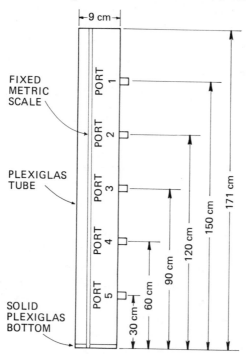

Figure 9. Multiport sedimentation column (nylon swaglok fittings adapted to take Neoprene septums were used as sampling ports)

In the tall column samples were taken at various depths. Solids were separated from the solution with a 1.0-μm Nuclepore filter. Various sample volumes were chosen to keep solids loading per filter as uniform as possible. All samples were dried at 105°C and weighed to $\pm$ 0.000005 g.

The experiments were designed to follow both the solids and selected trace metals upon sedimentation. Nuclepore membranes were chosen because of their favorable properties for metal analysis by x-ray fluorescence and because their particle-removal efficiency for the 1.0-μm size apparently increases linearly with solids loading, as described previously in the section on sizing by vacuum filtration. The shallow-column technique has several disadvantages which will be mentioned before experimental results are presented and discussed. Sampling was done at only one depth by inserting a clean pipette. This method may disrupt the surrounding fluid and fails to measure the effects of flocculation as a multiple-depth sampling technique might. Multiple dilutions (500:1, 100:1, and 50:1) were used in this work to test for flocculation effects. Repetition of experiments at 50:1 dilution indicated that the sampling technique was consistent.

Rapidly settling particles (large and/or very dense) were lost by sampling from the top only. These particles settled out before the initial samples were taken in a given run. Mass balances on the total solids showed that these particles did not constitute a large fraction of the total mass. However, these "fast" particles could be very important in the zone immediately surrounding any sludge discharge to the ocean. They are also removed easily by centrifugation of the sludge before ocean discharge.

Sample size was an additional limitation. At dilutions greater than 500:1, individual samples of 100 or 200 mL are necessary if significance is placed on the solids recovery. In a 2.0-L graduate, a 100-mL extraction lowers the water surface by 2.0 cm, or 5% of the total column depth, hardly a point sample. The tall column was thus developed to avoid these difficulties.

Shallow Column Results. The results of four sedimentation experiments with the shallow column (one each at 100:1 and 500:1 dilutions and two at 50:1 dilutions) are shown in Figure 10 with curves fitted as third-order polynomials. Experimental data are shown as points.

Comparison of the sedimentation curves in Figures 8 and 10 showed that the sludge at the 500:1 dilution settled more slowly than all samples except for those taken by Morel et al. (22). The filterable solids concentration at the 500:1 dilution was 50 mg/L. Morel's 10:1 dilution of effluent/NaCl should have had a solids concentration of 25–30 mg/L.

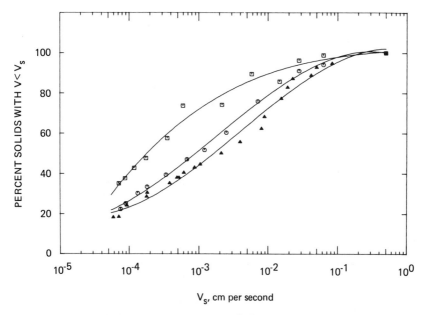

Figure 10. Sedimentation of LACSD sludge in seawater, shallow column (dilutions: (□) 500:1; (◐) 100:1; (▲) 50:1)

An experiment with a lower particle concentration would be expected to show less flocculation. (Possible effects from Morel's use of a NaCl solution have been mentioned already.) Myer's 1:1 dilution (final solids concentration of 150 mg/L) accounted for the relatively fast sedimentation in this system. In the case of Brooks' work (where the initial particulate concentration was 1,600 mg/L), visible flocculation was observed, and the rapid sedimentation measured was expected.

Increasing the dilution decreased the apparent sedimentation rate, indicating that the sludge flocculated at low dilutions. This result was not surprising; higher dilutions decreased the frequency of particle–particle collisions necessary for flocculation.

Tall Column Results. The results of four sedimentation experiments with the tall column are shown in Figures 11–14. All experiments were done at the same dilution (100:1), with samples taken at different depths and the water surface level recorded prior to taking each sample. Great differences in sedimentation velocity between ports were not observed, but the deeper port consistently showed faster sedimentation velocities, a strong indication of possible flocculation. A comparison of results for the shallow- and tall-column experiments with LACSD sludge shows essentially the same sedimentation-velocity curves at the 100:1 dilution.

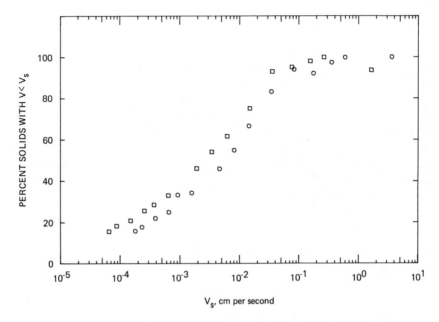

*Figure 11.   Sedimentation of OCSD sludge in seawater, tall column
(100:1 dilution; (□) port 3; (○) port 5)*

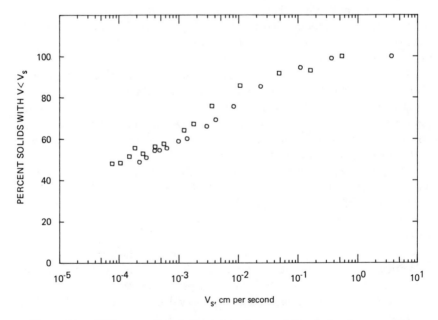

*Figure 12.   Sedimentation of Hyperion thermophilic sludge in seawater,
tall column (100:1 dilution; (□) port 3; (○) port 5)*

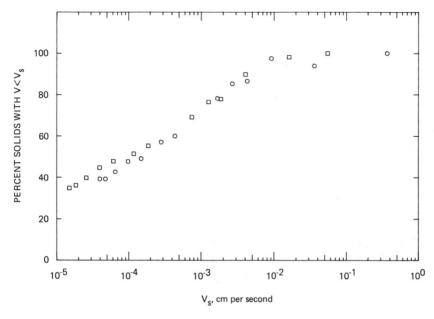

*Figure 13.   Sedimentation of Hyperion mesophilic sludge in seawater,
tall column (100:1 dilution; (□) port 3; (○) port 5)*

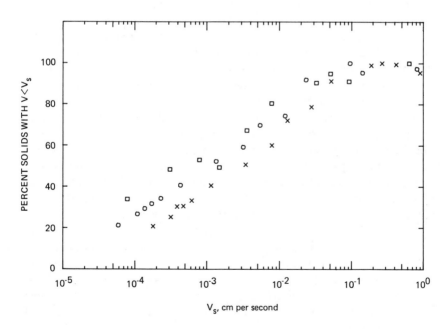

*Figure 14.   Sedimentation of LACSD primary sludge in seawater, tall
column (100:1 dilution; (□) port 2; (○) port 3; (×) port 5)*

The data for the four tall-column experiments show several interesting facts. Orange County Sanitation District's (OCSD) sludge settled most rapidly (Figure 11). This sample also had the highest total-solids concentration and the lowest percent volatiles. A higher solids concentration would allow more particle–particle interactions. The lower volatile percent implies a larger mineral content and, hence, perhaps a higher particle density. The Hyperion mesophilic sludge (Figure 13) and LACSD primary sludge (Figure 14), with almost identical solids concentrations, showed quite different sedimentation curves. This is likely attributable to their different origins—the Hyperion sludges include waste-activated as well as primary sludges. Both Hyperion sludges show similar sedimentation behavior in spite of a 33% difference in solids concentration and radically different digestion temperatures.

The increasing sedimentation velocities for samples taken at the lower ports on Figures 11–14 are good evidence that flocculation is occurring. This result is expected since differential settling velocities in the columns probably increased particle numbers in the lower region of the column, and since the flocculation rate increases with increasing particle number.

Summary and Conclusions

Some physical properties of several digested sewage sludges have been measured to better characterize the sludge-particle system. The particle size distribution for two sludges was measured with a Coulter Counter. Vacuum filtration techniques for particle sizing were also attempted. Sedimentation experiments were designed and carried out to determine the behavior of sludge in seawater, simulating some of the physical conditions of ocean disposal.

Particle size measurements for two sludges showed that the majority of particles by number count have diameters less than 5 μm. This size fraction, which is chosen selectively by many filter feeders in the ocean (24, 25), also disrupts the penetration of light in the water column if in suspension (6). Actual number counts were in excess of 10^{12} particles/L. Such small particles settle more slowly than particles of larger diameter and thus will be carried much greater distances if discharged to the ocean.

Particle sizing by filtration was unsuccessful because of pore clogging of the Nuclepore membrane. Results of filtration experiments confirm the results of work cited previously and lead to two major conclusions: (1) vacuum membrane filtration is not an acceptable means of particle sizing and (2) membranes with pore sizes of 1 μm or less should be used for suspended-solids analysis.

Sedimentation experiments at three dilutions for sludge in artificial seawater showed that sedimentation velocities decreased with increasing dilutions. Different sludges with varying solids content showed varying sedimentation velocities—velocities were higher with higher solids content. The data from both experiments strongly suggest that flocculation may be the cause of increased sedimentation velocity. The results of sedimentation experiments such as those described herein should be applied with care to the modeling of ocean discharges because of the lack of hydrodynamic similarity between ocean conditions and laboratory column experiments.

Literature Cited

1. Faisst, W. K. "Digested Sewage Sludge: Characterization of a Residual and Modeling for Its Disposal in the Ocean Off Southern California," EQL Report, Environmental Quality Laboratory, California Institute of Technology, Pasadena, CA, 1976, No. 13.
2. Young, D. R. Southern California Coastal Water Research Project, personal communication, 1976.
3. Brandon, J. R. "Parasites in Soil/Sludge Systems"; Sandia Laboratories: Albuquerque, New Mexico, 1978; Sand 77-1970.
4. Karr, P. R.; Keinath, T. M. *J. Water Pollut. Control Fed.* **1978**, *50*(8), 1911–1930.
5. Myers, E. P. "The Concentration and Isotopic Composition of Carbon in Marine Sediments Affected by a Sewage Discharge," Ph.D. Thesis, California Institute of Technology, Pasadena, CA, 1974.
6. Peterson, L. L. "The Propagation of Sunlight and the Size Distribution of Suspended Particles in a Municipally Polluted Ocean Water," Ph.D. Thesis, California Institute of Technology, Pasadena, CA, 1974.
7. Grigg, R. W.; Kiwala, R. S. *Calif. Fish Game* **1970**, *56*, 145–155.
8. Rudolfs, W.; Balmat, J. L. *Sewage Ind. Wastes* **1952**, *24*(3), 247–256.
9. Garber, W. F.; Ohara, G. T. *J. Water Pollut. Control Fed.* **1972**, *44*(8), 1518–1526.
10. Garber, W. F.; Ohara, G. T.; Colbaugh, J. E.; Rasket, S. K. *J. Water Pollut. Control Fed.* **1975**, *47*(5), 950–961.
11. Kubitschek, H. E. *Research (London)* **1960**, *13*, 128–135.
12. Wachtel, R. E.; LaMer, V. K. *J. Colloid Sci.* **1962**, *17*, 531–564.
13. Strickland, J. D. H.; Parson, T. R. "A Practical Handbook of Seawater Analysis"; Fisheries Research Board of Canada: Ottawa, 1968; Bulletin 167.
14. Treweek, G. P. "The Flocculation of *E. coli* with Polyethyleneimine," Ph.D. Thesis, California Institute of Technology, Pasadena, CA, 1975.
15. Chen, K. Y.; Young, C. S.; Jan, T. K.; Rohatgi, N. *J. Water Pollut. Control Fed.* **1974**, *46*(12), 2663–2675.
16. Sheldon, R. W. *Limnol. Ocean.* **1972**, *17*, 494–498.
17. Sheldon, R. W.; Sutcliffe, R. *Limnol. Ocean.* **1969**, *14*, 441–444.
18. Cranston, R. E.; Buckley, D. E., "The Application and Performance of Microfilters in Analyses of Suspended Particulate Matter," Bedford Institute of Oceanography, Dartmouth, Nova Scotia, Canada, BI-R-72-7, 1972, unpublished data.
19. McLaughlin, R. T. "On the Mechanics of Sedimentation in Artificial Basins," Ph.D. Thesis, California Institute of Technology, Pasadena, CA, 1958.

20. McLaughlin, R. T. *Proc. Am. Soc. Civ. Eng., J. Hydraulics Div.* **1959,** *85*(HY12), 9–41.
21. Brooks, N. H. "Settling Analyses of Sewage Effluents," unpublished report to Hyperion Engineers, 1956.
22. Morel, F. M. M.; Westall, J. C.; O'Melia, C. R.; Morgan, J. J. *Environ. Sci. Technol.* **1975,** *9*(8), 756–761.
23. Riley, J. P.; Skirrow, G., Eds. *Chem. Oceanogr.* **1975,** Vols. 1 and 2, 2nd ed.
24. Jorgensen, C. B.; Goldberg, E. D. *Biol. Bull. (Woods Hole, Mass)* **1953,** *105*, 477–489.
25. Haven, D. S.; Morales-Alamo, R. *Biol. Bull. (Woods Hole, Mass.)* **1970,** *139*, 248–264.

RECEIVED October 6, 1978.

13

Wastewater Particle Dispersion in the Southern California Offshore Region

JAMES R. HERRING [1]

Southern California Coastal Water Research Project, El Segundo, CA 92045

The nature and dispersion of effluent particles from municipal wastewater outfalls into the coastal ocean are little known, yet important, since these particles associate with several potentially toxic substances. This study of wastewater particle dispersion in the Southern California offshore region includes: (1) description of the nature and chemical and physical interactions of these particles; (2) determination of heavy metal and chlorinated hydrocarbon concentrations of those particles that deposit within a few kilometers of the outfall; (3) prediction, using measured particle settling velocities, of the dispersion and depositional field of the effluent particles; (4) determination of mass flux of the outfall particles to the bottom near the outfall site; and (5) testing the predicted dispersion field of effluent particles by releasing and detecting fluorescent-labeled particles from an outfall. The latter experiment shows that there is good agreement between the observed and predicted dispersion fields.

Each year the five major municipal outfalls in the Southern California area together release about 1.5×10^{12} L of effluent into marine waters. Figure 1 shows the release locations for the outfalls. The emission quantities and rates for each discharger are given in Table I. The effluent that is released to the oceans from these outfalls is mostly water; however, the seemingly small concentrations of particulates, usually between 100 and

---

[1] Present address: Branch of Regional Geochemistry, U.S. Geological Survey, Mailstop 925, Denver, CO 80225.

0-8412-0499-3/80/33-189-283$05.50/0
© 1980 American Chemical Society

Table I.   Emission Quantities and Flow Rates for the Various

		JWPCP	City of Los Angeles	
			Hyperion Effluent Line	Hyperion Sludge Line
Effluent Flow	10^9 L/yr	466	441	1.9
	% of total flow	33.5	31.7	0.1
Emitted Solids (metric tons/yr)		102,000	27,300	51,500[b]
Suspended Solids in Effluent (mg/L)		220	62	8,100[b]

[a] Data from (10). JWPCP refers to the plant operated by the County Sanitation District of Los Angeles County.
[b] Suspended and dissolved.

200 ppm, comprise a considerable quantity of material when integrated over a year. Annually there are approximately 3×10^5 metric tons of particulate emission or, on a per-particle basis, perhaps between 10^{20} and 10^{22} particles are emitted. It is this large number of particles and their largely unknown fate, nature, and behavior that are the focus of this chapter.

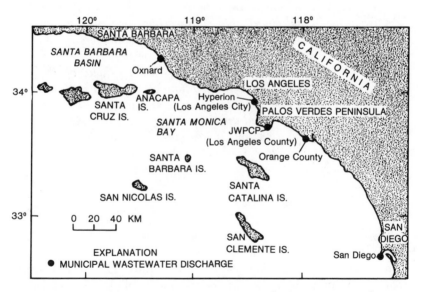

Figure 1.   Locations of municipal wastewater dischargers referred to in this chapter. Discharge amounts and rates are listed in Table I. The average depth of discharge is 60 m. The effluent is released to the ocean through perforated sections of the outfall pipe called diffusers.

Dischargers of Waste Effluent to Southern California Ocean Waters[a]

Oxnard	Orange County	City of San Diego	Other	Total
15.5	249	160	27.7	1,390
1.1	17.9	11.5	1.9	100.0
1,500	33,300	20,500	no data	236,100
98	132	128	no data	

Particles that are emitted from municipal ocean outfalls in the Southern California offshore zone are the subject of considerable concern. These particles associate with many of the 100 or more so-called pollutants that are being monitored in ocean outfall discharges (*12*). That is, the particles either contain or have sorbed onto their surfaces toxic species such as chlorinated hydrocarbons and certain trace metals. Furthermore, the particles, being mostly organic matter, can affect marine life other than simply acting as the transport vector for toxic species. This occurs, for example, if extensive quantities of effluent particulates concentrate over a relatively small area of the bottom. Higher than normal concentrations of organic debris can result in changes in both the amount and types of benthic infauna (*11*). Such considerations as pollutant carrying capacity or effects on benthic infauna make a knowledge of the dispersion of effluent particles important to those agencies that monitor effluent quality, establish emission standards, or measure resultant effects.

Sources and Particle Description

The mass of outfall particulates emitted annually into the ocean by the five major Southern California municipal waste dischargers is about 2.9×10^5 metric tons, dry weight. Since this mass may seem immense, certain perspectives aid its conceptualization. At a compaction density of 1 g cm^{-3} and no void volume, this dry mass has a volume equal to that of a cube 66 m on a side; if this volume of material were spread over the seabottom in a layer 1 mm thick it would constitute a blanket equal to a square 16 km on a side. However, the particles are neither dry nor without porosity. For example, if the particles constituting the above dry

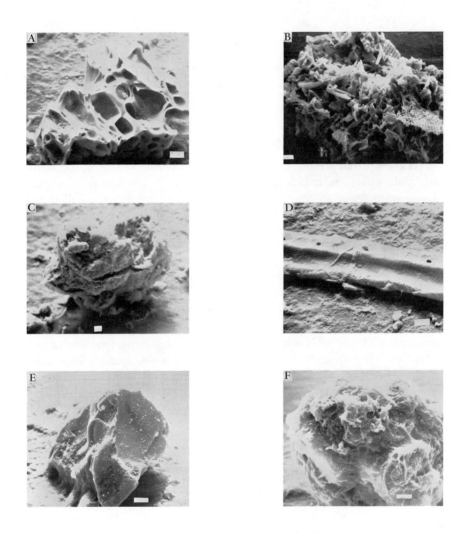

mass also contain an equal mass of water, and have a packing porosity or void volume of 75%, then the volume of the annually emitted particles is $2.3 \times 10^6 \, m^3$, which is equal to the volume of a cube 132 m on a side. Likewise, if this volume is spread into a layer 1 mm thick on the seafloor it would form a square with a side of 48 km.

The nature of emitted particles varies. In composition they range from inorganic minerals and debris to agglomerates of complex organic substances. Morphologically the particles range from materials with considerable structure, which permits or suggests their identification, to amorphous masses, with no discernible structure or inferable origin (Figure 2). Since no measurements of particle size are made by the dischargers, it is not possible to tabulate the total number of emitted particles except to suggest a range over which they might occur.

Ocean Discharge of Outfall Particulates

Upon discharge into marine waters, the movement and interactions of the particles are determined by external forces acting on the particles, internal forces arising from the chemical and physical nature of the particles, and forces acting on the fluid surrounding the particles.

The plume of discharged effluent, consisting of both liquid and solid fractions, is usually negatively buoyant compared to seawater at the depth of discharge. It rises from the typical discharge depth of about 60 m to an average height of 30 m from the bottom within about 3 min after dis-

Figure 2. *(facing page) Scanning electron micrographs of wastewater effluent particles taken from the discharge pipe of the City of Los Angeles, Hyperion Treatment Plant immediately prior to ocean discharge. Scale bars on the photographs are 10 μm in length.*

(A) *An organic particle taken from the 7-mile-long sludge discharge line. Dispersive energy x-ray analysis of the particle shows that it is enriched in K, P, and Si in addition to the organic constituents.*

(B) *A particle from the 5-mile-long effluent line. Note the aggregation of marine diatom fragments. Since this particle has not yet been in the ocean the diatom fragments probably have come from industrial sources, such as a commercial diatomaceous earth filter.*

(C) *A particle from the sludge line. X-ray analysis suggests an enrichment of elements lower than atomic number 10 in this particle. No elements of higher atomic number had recognizable peaks in the x-ray spectrum, hence the particle is an amorphous organic aggregate.*

(D) *A particle from the sludge line; possibly vegetable matter*

(E) *A particle from the sludge line again probably organic in composition as no peaks for elements with atomic numbers greater than 10 appear in the x-ray spectrum. Note the presence of bacterial colonies on the particle surface.*

(F) *Amorphous organic particle*

charge. By the time the plume attains this height it has been diluted with seawater about 100 times. The discharge depth is fairly uniform among major dischargers in Southern California, but the plume rise height at each discharge location can vary considerably depending on the time of day or year. The plume rise height is essentially a density realignment. The density structure of the seawater at the point of dicharge determines the rise height of the plume as the plume seeks to attain a level of equal density. The density structure of the seawater can be altered drastically over periods as short as a few hours by such effects as the heating of surface seawater by the sun, the passage of storms, or the incursion of upwelled cold water from deeper waters. Thus the plume rise height at any one time depends on many parameters and is subject to considerable change. Plumes have been known to erupt to the surface, although usually they are confined to depths below the thermocline in the water column. Thirty meters is taken to be only an average height of rise. Any detailed study or prediction of particle dispersion must also consider the wide variation in short- and long-term changes in the plume rise height at each location.

Discharge of effluent particles into the ocean results in certain chemical, physical, and biological changes in the particle. These changes may have important consequences to subsequent particle movement and dispersion. Examples of chemically induced changes include dissolution, flocculation, or particle growth. Dissolution reduces particle mass in proportion to the cube root, and sinking rate in proportion to the square root of the dissolution time. The effect of dissolution on changing particle population size and mass characteristics has been considered previously for the case of natural particles in the open ocean (6, 7), but not for waste effluent particles in coastal waters. Particle growth or agglomeration produces an opposite effect, that is, an increase in sinking rate resulting from the increase in mass. However, the increase in effective particle size and mass can be offset by the entrapment of seawater in the pore space of the agglomerated particle. Therefore, particle growth from flocculation can lead to two opposing changes in the sinking rate of the particle. These two opposite effects have not been studied for effluent particles in seawater, although I have observed flocculation for some wastewater sludges mixed with seawater.

Other physicochemical changes in the particles may be imposed by seawater salinity since the effluent is originally (prior to discharge) a freshwater mixture when compared with seawater. This salinity difference could be important to some organic particles, such as bacteria, where the osmotic pressure difference of 25 atm between the effluent and seawater could rupture the external membrane and seriously alter the particle nature.

The biological fate of the discharged particles while they are still in the water column must be important, although only scant knowledge exists about the biological processes that might occur. Many of the particles discharged from wastewater outfalls probably are scavenged by and incorporated into marine zooplankton. Subsequently these particles may be released in some altered state. Filtration of detritus from seawater has been observed in the case of fecal pellets from copepods in the open ocean and from laboratory experiments (9). These scavenging processes must also be important in coastal water for the case of effluent particle uptake, but the uptake rates have not been measured. A discussion of the interactions between marine zooplankton and suspended particles is provided elsewhere in this volume (8). Clearly, biological removal effects should be examined in terms of their importance to the dispersion and alteration of wastewater effluent particles.

Postdischarge Movement of Particles in the Ocean

After discharged effluent particles become entrained in the wastewater plume and rise to a new density equilibrium height, they begin to move in response to external forces. Two principal sets of forces act on the particle, one in a horizontal direction and one in a vertical direction. Horizontal movement of the particles is induced by the advective motion of the enclosing seawater. At the same time, the particles sink, float, or remain neutrally buoyant, depending on their density relative to that of the seawater. The particles in most of the effluent mass are denser than the surrounding seawater and begin to sink under the influence of gravity. The downward acceleration of the particles quickly becomes a constant fall velocity as the viscous drag forces of the seawater equal and oppose the gravitational force on the particle. The quotient of the fall height divided by the fall velocity yields the fall time of the particle, the time it takes the particle to reach the sea floor. During downward transit the particle continues to move horizontally in response to the prevailing ocean currents at that depth, location, and time. Thus the location of where each effluent particle finally settles to the bottom is given by the vector resulting from the two simultaneous velocities, downward and horizontal, divided by the fall time of the particle. Particles that float or are neutrally buoyant are dispersed by the surface or midwater ocean currents, respectively. The fate of these nonsinking particles is largely unknown, although a fraction of the floatable particles is scavenged by sea birds and other marine life in the vicinity of the outfall. These nonsinking particles, thought to constitute less than 10% of the discharged mass, will receive no further attention in this chapter.

Measurement of horizontal ocean currents near Southern California wastewater discharge sites has been the subject of considerable study using submerged current meters and drogues (1, 2, 3). Although subthermocline currents vary considerably in direction and speed, they have an average tendency to move upcoast (northward) at speeds of about 3 cm sec^{-1}. Predominant variation of this tendency is in a downcoast direction. Superimposed on this average tendency are the tidal oscillations and onshore–offshore motions. In the onshore–offshore direction there are longer and shorter period oscillations which, when combined, are approximately equal to the tidal oscillation, but less important to particle dispersion than alongshore currents (3). Other currents occur in addition to those that are subthermocline, notably surface currents and occasional near-bottom (within about 1 m of the bottom) currents. Surface currents are important to buoyant or entrained particles that rise above the thermocline. Generally these currents move downcoast at about 25 cm sec^{-1}; however, as in the subthermocline currents, there is considerable variation in their average directions and speeds. Near-bottom currents in the Southern California offshore region are largely uncharacterized. These currents are important to the horizontal dispersion of sinking particles during the final meter or two of their descent. Since near-bottom current speeds as high as 20 cm sec^{-1} have been observed (3), the effect of these currents on particle transport could be significant; they may also be significant in the resuspension of effluent particles off the seafloor. Such resuspended particles become susceptible to further horizontal current transport. The region-wide importance of this process in the Southern California offshore area is unknown.

The velocity of sinking particles is proportional to the difference in density between the particle and the surrounding seawater. The sinking velocity is also proportional to the square of the radius of the particle. These velocities can be calculated for spherical particles using Stokes' Law as follows:

$$U = \frac{2}{9} \frac{g(\rho_p - \rho_{sw}) \, r^2}{n}$$

where U is the settling velocity (cm sec^{-1}); g the acceleration of gravity (here 980 cm sec^{-2}); ρ_p the density of the particle (g cm^{-3}); ρ_{sw} the density of the seawater (g cm^{-3}); n the dynamic viscosity of the seawater (cgs units); and r the particle radius (cm).

Figure 3 shows the settling velocity as a function of particle density and diameter for a typical seawater density and viscosity value that may be encountered near a wastewater discharge site. Nonlinearity of the settling velocity contours on the semilogarithmic plot results as the

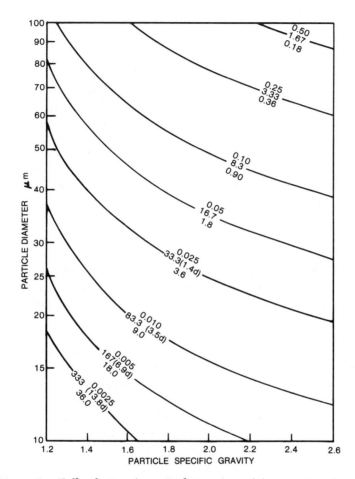

Figure 3.   Fall velocity of a particle in cm sec⁻¹ (upper of each set of 3 numbers) as a function of particle diameter and specific gravity as calculated from Stokes' Law.

Note that as the specific gravity of the particle approaches that of seawater (about 1.026), the settling velocity for a given diameter particle slows. Also shown is particle fall time in hours (middle of each set of 3 numbers), assuming a fall height of 30 m. The fall time decreases (that is, a faster settling velocity) approximately proportionally to the specific gravity increase and to the square of the diameter increase. This relationship breaks down as the specific gravity approaches that of seawater. In addition, the particle dispersion distance is shown in km (lower of each set of 3 numbers), assuming an average current velocity of 3 cm sec⁻¹ in a constant direction and a fall height of 30 m. Many effluent particles with low specific gravities can be moved at least several tens of kilometers from the outfall prior to deposition.

particle and seawater densities become close. The particle fall times over an average 30-m fall height are also shown on this figure as a function of the same range of particle density and diameter. Finally, dispersion distances for typical horizontal current speeds at the discharge localities, about 3 cm sec⁻¹, are also shown on the same figure. Only those particles with a fall time of about 4 days or less reach the seafloor within some immediate proximity (on the order of 10 km) of the discharge site. Since fall heights at a given location may vary considerably, the fall times will vary proportionally. The fall times shown in Figure 3 should be considered only as an average with extreme variations occurring over an approximate range of one-half to twice this value.

Measurement of Settling Velocity of Effluent Particulates

We have designed a series of experiments to determine the settling characteristics of wastewater effluent particles in seawater. Typically these experiments consist of obtaining a sample of effluent, composited over 24 hr, and then mixing the effluent sample with seawater to measure the settling characteristics of the effluent particles. Composite effluent samples are used to sample the different effluent character that may result over a 24-hr period. The variation between 24-hr composited samples from day to day, week to week, or year to year has not been studied. Typical dilutions of seawater to effluent for these experiments are on the order of 100:1 since this number is representative of the dilutions that wastewater plumes experience in the ocean. The experiments are performed in a series of 1-L graduated cylinders that are kept close to the temperature of receiving ocean waters (about 10°C) by using a water bath. Nonfiltered seawater is used in these experiments in a realistic attempt to provide natural particles that may interact with the effluent particles. At various times after homogenization of effluent and seawater, the cylinders are removed from the water bath and all but the bottom 50 mL are removed quickly through a vacuum siphon. The remaining slurry of effluent and natural particles in seawater is removed by pipette and filtered through a predried and tared Nuclepore filter of 0.2-μm pore size for gravimetric recovery of the particle mass. The filters are rinsed with particle-free deionized water, folded, and sealed with 0.1 μL of $CHCl_3$. The filters are dried at least overnight at 105°C and then weighed to ±3 μg on a microbalance. Simultaneous determination of the settling properties of the natural seawater particles in cylinders of seawater without effluent permits correction of the settled effluent particle fraction for contribution of the natural particles in the seawater. Settling velocity data for some of the Southern California wastewater dischargers are shown in Figure 4. From this experiment and a knowledge of the ocean currents in the vicinity of an outfall it is possible to calculate the depositional field of the emitted effluent particles.

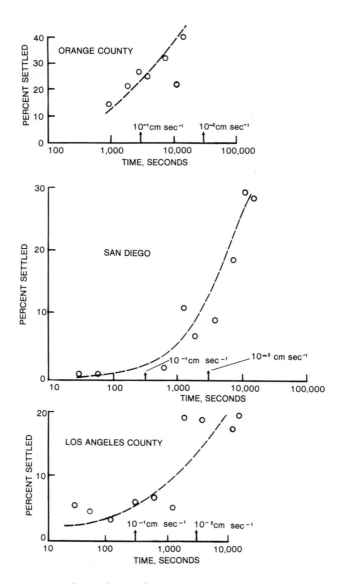

Figure 4. Particle settling velocity experiment curves for the effluents mentioned in Table IV.

Note the log scale for time. The abscissa lists the percent of the total effluent particles that have settled in the time given on the ordinate. Since all experiments were performed in 30-cm-high cylinders, the settling times can be converted to settling velocities by dividing the settling time into 30. Identifying marks shown on each curve correspond to settling velocities of $\geq 10^{-1}$ and $\geq 10^{-2}$ cm sec^{-1}. Curves through the data points are approximate fits to the data.

Calculated Depositional Field and Mass Flux of Emitted Particles

The following example details the calculation of the simplified dispersion field for settled effluent particles emitted from the San Diego wastewater treatment plant. This dispersion zone has been determined from extensive water current measurements which permit calculation of the horizontal extent of the dispersion zone and also, by measuring the current direction probability, that fraction of particulate material with a given settling velocity that will land within the zone. Other discharge sites have not had the same detailed study of current variance; hence only simplified dispersion zones have been calculated. Two fall velocities ($\geq 10^{-1}$ and $\geq 10^{-2}$ cm sec^{-1}) were used to calculate the extent of the dispersion zone. These two velocities represent, respectively, the material falling within a few km of the outfall and that falling within 10 to 20 km of the outfall. This latter velocity probably produces the maximum extent of the dispersion zone over which detectable effects from the effluent particles presently can be determined.

The simplified dispersion zone for the $\geq 10^{-2}$ cm sec^{-1} fraction is calculated as follows: the fall time from a 30-m height is 83 hr. In this time a mean upcoast flowing current of 3 cm sec^{-1} will produce an alongshore dimension for the dispersion zone of 9 km. The onshore–offshore dimension of the zone has been calculated by taking the length of the diffuser section, where the effluent is released through ports into the seawater, as the

Table II.   Dispersion Zone Characteristics for the
San Diego Outfall Area[a]

	Fall Velocity (cm sec^{-1})	
	$\geq 10^{-1}$	$\geq 10^{-2}$
Arrival time after discharge (hr)	8.3	72
Upcoast extent of fallout zone (km)	3	16
Downcoast extent of fallout zone (km)	1.5	5
Amount of material with given fall velocity that lands in dispersion zone (%)	70	50
Upcoast area (km^2)	1.4	6.7
Downcoast area (km^2)	0.7	2.1
Total area of dispersion zone (km^2)	2.1	8.8
Onshore–offshore width increase from diffusion of initial dilution zone (km)	0.08	0.42

[a] Upcoast and downcoast (north and south, respectively) extents of the dispersion zone for particles with a fall velocity of $\geq 10^{-1}$ cm sec^{-1} have been calculated based on the analysis of current meter data that assume a fall time for the particles at 7.5 hr. The actual fall time for a 30-m fall height at this velocity would average 8.3 hr.

initial width of the initial dilution zone. In the case of a "Y"-shape diffuser this measurement is the sum of the length of the two diffuser legs times the cosine of the complement of the included angle. The dispersion zone width increases over time because of diffusion and this increase has been calculated by assuming an increase of 0.4 cm sec^{-1} times the time. For a settling time of 83 hr this increase is 1.2 km. Therefore, for a settling velocity of $\geq 10^{-2}$ cm sec^{-1}, the area of the resulting bow-tie shape simplified dispersion zone for each of the wastewater dischargers is: Orange County—21.6 km^2; Hyperion—27 km^2; JWPCP (Los Angeles Co.)—19 km^2; and San Diego—8.8 km^2. For the case of JWPCP, the dispersion zone for the 120-in. diameter outfall is assumed to include the zone for the smaller 90-in. diameter outfall. No dispersion zone calculation has been made for the Hyperion sludge line since there is essentially little plume rise and the effluent is disgorged from the end of the outfall pipe with no consequent fall height.

For the San Diego dispersion zone calculation, the horizontal component of the water motion has been measured by current meter and has been resolved into two components: an upcoast and downcoast component (essentially along isobaths at the discharge site) and a perpendicular component (onshore–offshore). Net motion of the alongshore component is upcoast (north or west) at 3 cm sec^{-1} while that of the onshore–offshore component is virtually zero. Net motion and variance of the alongshore currents are better known than those of the perpendicular component. Hendricks (2, 3) has analyzed the alongshore motion of water at the discharge site. The onshore–offshore motion is conjectured to diffuse to twice the width of the onshore–offshore component of the initial dilution zone 72 hr after discharge. These horizontal motions and the resultant areas of particle fallout are tabulated in Table II for cases of fall velocities of 10^{-1} and 10^{-2} cm sec^{-1}.

The dispersion field dimensions and locations for particles settling at velocities greater than or equal to 10^{-2} cm sec^{-1} and 10^{-1} cm sec^{-1} are shown in Figures 5 and 6. The fallout zone is a narrow, northwest-trending strip approximately 1 km wide and 21 km long for settling velocities greater than or equal to 10^{-2} cm sec^{-1}. In attempting to correlate this fallout zone with observed field measurements (for example, volatile solids or infaunal index), it should be remembered that the fallout zone is predicted, not measured. It is intended to serve as a guide to where the particulate material should deposit, not as a description of where it has been found. Small deviations from the mean upcoast current direction will broaden the dispersion zone and proportionally reduce the flux of particles. In addition, changes in the mean current velocity or in the onshore–offshore component of the current will affect the shape of the dispersion zone.

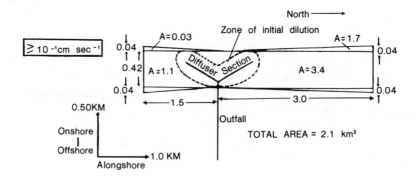

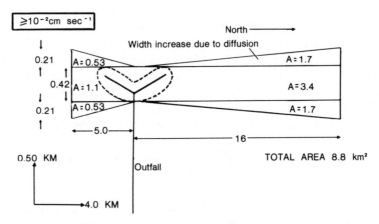

Figure 5. Dispersion field dimensions are given for San Diego effluent particulates with settling velocities of 10^{-2} and 10^{-1} cm sec^{-1}. The dimensions are in km and the areas in km^2. Note the exaggeration in the east–west direction; in reality the field is extremely elongated in the north–south direction.

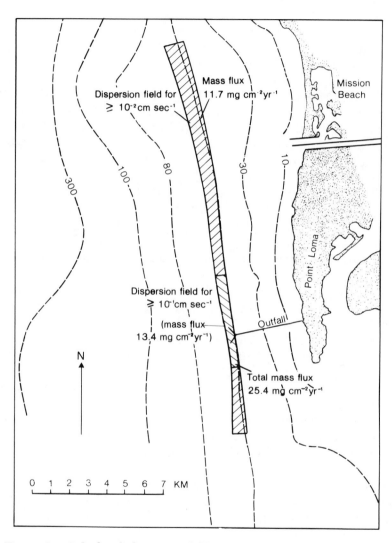

Figure 6.   Calculated dispersion field and mass flux for particles with settling velocities of $\geq 10^{-1}$ and $\geq 10^{-2}$ cm sec^{-1}

Mass Flux

For the San Diego discharge site, the water current probability model developed by Hendricks (4) predicts that 50% and 70% of the emitted particulates will deposit in the areas greater than or equal to the 10^{-2}-cm sec^{-1} and 10^{-1}-cm sec^{-1} dispersion zones, respectively. In 1977, 2.05×10^{10} g of solids were discharged from the San Diego effluent line. Settling velocity experiments show that approximately 2% and 10% of the effluent particulate material settles with velocities greater than or equal to 10^{-1} cm sec^{-1} and 10^{-2} cm sec^{-1}, respectively. Thus, 2,050 metric tons and 410 metric tons of particulates per year have fall velocities that equal or exceed 10^{-2} sec^{-1} and 10^{-1} cm sec^{-1}, respectively. If 50% of this material lands in the 8.8-km² zone shown in Figure 6 then the mass flux will be 0.103×10^{10} g/8.8 km² yr^{-1} or 11.7 mg cm^{-2} yr^{-1}. In the smaller area, 2.1 km², 70% of the material with a settling velocity equal to or exceeding 10^{-1} cm sec^{-1} will deposit. The resulting mass flux will be $0.70(4.10 \times 10^{8}$ g)/2.1 km² yr, or 13.7 mg cm^{-2} yr^{-1}. In this same area there also will be a contribution by deposition of particles settling at velocities greater than or equal to 10^{-2} cm sec^{-1}. If this material has an even distribution throughout its depositional area, then the mass flux in the smaller area will simply be the sum of the flux terms for the fall velocities greater than or equal to 10^{-2} and 10^{-1} cm sec^{-1}. The resultant sum of mass fluxes is 25.4 mg cm^{-2} yr^{-1} for the smaller 2.1-km² area. Table III lists these numbers.

Further modifications can occur to the effluent particle mass fluxes. For example, there are minor contributions from some of the effluent particles that fall slower than 10^{-2} cm sec^{-1}. Also, the dispersion of particulate material is assumed to be constant over the entire depositional area. In reality the depositional flux will be larger close to the outfall and smaller towards the flanks of the dispersion zone. Finally, enlargement of the dispersion zone by shifting currents or other processes will reciprocally reduce the mass flux to the dispersion zone.

The dispersion zone mass flux for San Diego also was calculated by Hendricks (4) on a more continuous basis than the previous simple example. The results of this calculation are shown in Figure 7. Smoother isoflux lines occur in this model because the calculation was continuously

Table III. Mass Fluxes of Effluent Particles to the Dispersion Area Listed in Table II

Settling Velocity (cm sec⁻¹)	Dispersion Area (km²)	Mass Flux (mg cm⁻² yr⁻¹)
$\geq 10^{-1}$	2.1	13.7
$\geq 10^{-2}$	8.8	11.7

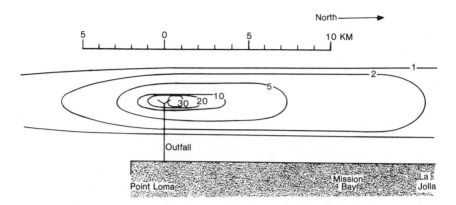

*Figure 7.   Mass flux distributions in mg cm^{-2} yr^{-1} for San Diego munici-
pal waste outfall particles, as calculated by T. Hendricks (4)*

stepped along 1.5-km–long cells using several different settling velocities.
The assumed onshore–offshore diffusion constant in this case was 0.4 cm
sec^{-1}. The onshore–offshore profile of the effluent here is assumed to be
top hat in shape rather than gaussian. The latter case produces higher
fluxes at the fringes of the onshore–offshore dispersion area.

One final comment about these mass flux calculations is necessary.
Deposition of effluent particles on the sea bottom is by no means perma-
nent. Several processes acting alone or in concert may result in the
removal of the effluent particle from its original site of deposition. These
processes include: (1) physical removal resulting from bottom-current
or storm-surge resuspension and (2) chemical removal resulting from
ingestion or disturbance (bioturbation). The simultaneous effect of several of these processes has been considered by Hendricks (4).

The PLOP Experiment

To investigate the toxic trace element and chlorinated hydrocarbon
concentrations on effluent particles that fall within about 20 km of the
outfall, an experiment called PLOP (Pollutant Loading Onto Particulates) was designed. In this experiment mixtures of unfiltered seawater
are mixed with effluent in large trash cans and allowed to settle for a
prescribed time. The analysis for trace metals is performed in a plastic
trash can using plastic stirring and syphon tubes. All plastic surfaces are
cleaned and soaked in 1N HNO$_3$ solution. By contrast, the analysis for
chlorinated hydrocarbon content of the particulates is performed in a
galvanized metal trash can using metallic stirring and syphon tubes. All
metal surfaces of the containers and apparatus used in this experiment

are cleaned and rinsed in analytical reagent-grade hexane to remove possible contaminants. Other technical aspects of the experiment are described elsewhere (5).

Typically the settling time for PLOP is calculated to require a particle settling rate greater than or equal to 10^{-2} cm sec^{-1}. All PLOP experiments were performed in mixtures of 30 parts unfiltered seawater and one part 24-hr flow-proportional effluent composite. One hour and forty-two minutes after stirring, the settled fraction is recovered by removing the overlying slurry through a syphon. The settled fraction is analyzed for the concentrations of trace species, which are compared to the initial concentration and, if necessary, corrected for a seawater blank. Next, the mass of each analyzed trace species that falls with the minimum velocity is determined. Table IV lists the total mass and flux of the various trace metals and hydrocarbons that were found in the rapid settling fractions. The composite effluent volume used in all experiments was 4 L. Thus the total annual mass of effluent-derived metals and hydrocarbons delivered to the sediments by the rapid-settling fraction can be found by multiplying the total pollutant mass in the settled PLOP fraction times one-fourth of the annual effluent emission, in liters. Combining this mass with measurements of the horizontal ocean current direction and speed at the discharge site permits calculation of the area over which this

Table IV. Annual Emission Amounts and Fluxes of Trace Metals Fraction of the Effluent Particulates

Discharger	Annual Flow (L) (10)	Dispersion Area (km²)	Chlorinated Hydrocarbons[b] Total DDT	Chlorinated Hydrocarbons[b] Total PCB
Orange County				
Emitted mass	2.5×10^{11}	21.6	0.39	4.2
Flux			0.0018	0.02
Hyperion Effluent Line (Los Angeles City)				
Emitted mass	5.0×10^{11}	27	1.6	11.5
Flux			0.0059	0.042
San Diego				
Emitted mass	1.6×10^{11}	8.8	4.6	6.5
Flux			0.052	0.073
JWPCP (Los Angeles County)				
Emitted mass	4.9×10^{11}	19	33.5	—[c]
Flux			0.18	—[c]

[a] Values have been corrected for seawater blank values but not for the concentration of the various species that are contributed by the same settling velocity fraction of the natural particles in the seawater used for the experiment. In the case of chlorinated hydrocarbons these natural values should be negligible; hence all of the reported concentrations can be taken as effluent-derived. The rapid settling fraction is isolated according to the techniques of the PLOP experiment, which is described in the text, and chemical analysis for the various trace constituents is performed as described in Ref. 10.

pollutant mass is dispersed. This number is the flux of that particular
metal or hydrocarbon to the bottom sediments near the outfall. If the
sedimentation rate is known, the flux of the pollutant species can be
calculated as a theoretical sediment concentration. The resultant fluxes
of trace metals and chlorinated hydrocarbons in the rapid-settling fraction of the effluent particulates are listed in Table IV.

Fluorescent Particle Tracer Experiment

The release into the effluent of identifiable particles, with settling
characteristics similar to those of typical effluent particles, permits the
monitoring of the temporal and spatial dispersion of the effluent particulates. There have been no previous attempts to label and trace particulates from the Southern California municipal waste outfalls. Bacteria
from the outfalls have been traced but their dispersion behavior is unknown since they probably attach to uncharacterized surfaces.

We released approximately 40 L of a fluorescent organic pigment
(Radiant Color Division of Hercules, Inc., Richmond, California; orange
red dispersion WD16) over a period of one week into the final effluent
of the San Diego Sanitation District. A series of Van Veen sediment
grabs was then taken to search for the pigment particles in surface sediments within the vicinity of the outfall. A sample of seawater and two

**and Chlorinated Hydrocarbons in the $\geq 10^{-2}$ cm sec^{-1} Settling
from Various Dischargers**[a]

Annual Metal Emissions (Emitted Mass (kg yr^{-1}); Flux (μg cm^{-2} yr^{-1}))								
Ag	Cd	Cr	Cu	Fe	Mn	Ni	Pb	Zn
20	110	460	770	16,000	340	260	310	1,100
0.093	0.051	2.1	3.6	74	1.6	1.2	1.4	5.1
539	535	4,820	6,068	64,015	1,201	1,472	2,650	8,339
2.0	2.0	18	22	237	4.5	5.5	9.8	31
100	100	930	1,606	52,000	690	440	900	3,700
1.1	1.1	11	18	590	7.8	5	10	42
350	800	15,000	5,900	270,000	1,800	3,700	5,000	23,000
1.8	4.2	79	31	1,400	9.5	19	26	121

[b] Chlorinated hydrocarbon mass emission and flux units are the same as those for
the metals. Total DDT includes op'-DDE, pp'-DDD, op'-DDT, and pp'-DDT; total
PCB includes the following arochlors: 1242 and 1254.
[c] PCB value contaminated.

or three sediment samples were analyzed from each of 37 grabs. After each sample was homogenized in a blender and allowed to settle for 1 min to remove coarse debris, 10–30-mL aliquots were extracted by syringe and filtered onto a glass fiber filter. The fluorescent pigment particles on the filter surface were detected visually using a binocular microscope at 26×. UV excitation was provided at 365 nm while the emitted fluorescence of the pigment particles occurred at 600 nm.

The pigment particles consist of a base polymer of triazine–aldehyde–amide. Specific gravity of the particles is 1.40 and the average particle size is about 5 μm with a range from 0.5 to 30 μm based on observations using scanning electron microscopy. The particles are insoluble in seawater. Prior to the release, an investigation of eight grab stations near the outfall site revealed no particles of the type that were to be discharged during the experiment. It is expected that this situation may change in the future as the particles are used increasingly in outdoor advertising.

Figure 8 shows the maximum number of fluorescent particles that occurred in sediment grab samples. Several important observations arise from the data. First, the maximum occurrence of labeled particles occurs quite close to the 60-m isobath, which is the discharge depth for the outfall and coincides approximately with the predicted dispersion zone. Second, there is considerably more onshore–offshore movement of the particles than is suggested by the theoretical dispersion zone. Although particle occurrence is low, the presence of particles at such extreme widths normal to the principal transport direction suggests that either significant sediment resuspension has occurred within a few days after discharge or that the principal current direction has meandered around a central isobath tendency. The implication of this latter possibility is important to considerations of the effluent particulate mass flux to the sediments. Based on the fluorescent particle data it seems that the dispersion zone during the time of release is perhaps six or seven times wider than that calculated theoretically and that fluxes should be reduced proportionally. However, deposition is still concentrated in the middle of the zone. Finally, the observed dispersion zone agrees well with measurements of subthermocline current direction and velocity made during the release of the labeled particles. These measurements show a constant upcoast direction with slight offshore tendency and velocities generally around 10 cm sec^{-1}, occasionally decreasing in intensity. Note that there are no particles south (downcoast) of the outfall or directly offshore from the diffuser. In summary, the agreement between the predicted and observed dispersion zones is quite good. Analogous future experiments with a more uniform size range of labeled particles will permit the calculation of actual particle flux to the sediment. In addition,

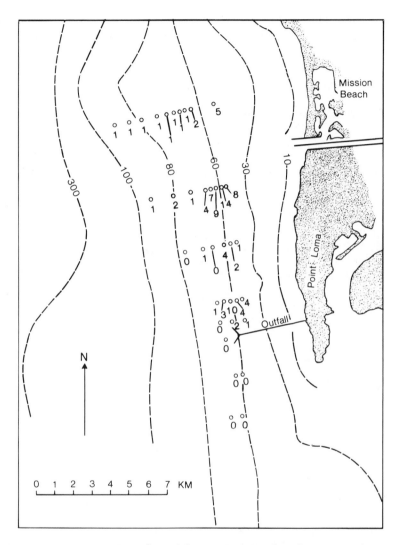

Figure 8. Maximum number of fluorescent particles that occurred in 3 replicate sediment samples from each station near the San Diego outfall.

Compare this inferred distribution with that predicted theoretically (Figure 6). Note the particle dispersion only in the northward direction and the confinement close to the 60-m isobath. Current measurement during the week-long release of the particles through the outfall revealed only northward-flowing currents ranging from extremely weak to 13 cm sec^{-1}. Thus the observed dispersion field of the tracer particles coincides well with the dispersion field that would have been predicted theoretically. Over longer periods of time the sub-thermocline currents are seldom this constant, so prediction is more difficult.

resampling of the sediments over time should show the extent to which resuspension and/or bioturbation processes are important in spreading the effluent particles throughout a larger area or to greater depth in the sediment.

Literature Cited

1. Hendricks, T. Annual Report, Coastal Water Research Project, El Segundo, CA, *1975*, 173.
2. Ibid., *1976*, 63.
3. Ibid., *1977*, 53.
4. Ibid., *1978*, 127.
5. Herring, J. R.; Abati, A. L. Technical Memo, Coastal Water Research Project, El Segundo, CA, unpublished.
6. Lerman, A.; Carder, K. L.; Betzer, P. R. *Earth Planet. Sci. Lett.* **1977**, *37*, 61.
7. Lerman, A.; Lal, D. *Am. J. Sci.* **1977**, *277*, 238.
8. Mullin, M. M. Chapter 10 in this book.
9. Roth, P. H.; Mullin, M. M.; Berger, W. H. *Geol. Soc. Am. Bull.* **1975**, *86*, 1079.
10. SCCWRP, Annual Report, Coastal Water Research Project, El Segundo, CA, 1978, 97.
11. Word, J. Q. Annual Report, Coastal Water Research Project, El Segundo, CA, 1978, 19.
12. Young, D. R. Annual Report, Coastal Water Research Project, El Segundo, CA, 1978, 103.

RECEIVED February 15, 1979.

14

Use of Particle Size Distribution Measurements for Selection and Control of Solid/Liquid Separation Processes

MICHAEL C. KAVANAUGH, CAROL H. TATE, ALBERT R. TRUSSELL, R. RHODES TRUSSELL, and GORDON TREWEEK

James M. Montgomery, Consulting Engineers, Inc., Walnut Creek, CA 94596, and Pasadena, CA 91101

Particle size distribution data for aqueous particulates larger than 1 μm in several freshwater and wastewater systems are shown to be modeled accurately with a two-parameter power-law function $dN/dl = Al^{-\beta}$, the exponent β providing an estimate of particulate contributions by size to the total particulate number, surface area, volume and mass concentration, and light-scattering extinction coefficient. The power-law exponent for several particulate systems computed from particle size distribution data determined by a variety of particle counters ranges from 1.8 to 4.5. A methodology for selection of solids/liquid separation processes in water treatment applications based on the use of size distribution data is demonstrated. Applications of data on total particle count, and various statistical parameters of the particle size distribution are shown for evaluation of pilot plant studies, and process control of particulate separation processes. When the power-law exponent is greater than three, submicron particles, which are not measured by particle counters, will control the magnitude of total count, surface area, and light-scattering extinction coefficient. In this case, measurements of both the particle size distribution and the particulate light-scattering characteristics are recommended.

The major objective of water treatment facilities is to provide a finished product at a reasonable cost, meeting specified standards irrespective of the source water quality. Achieving this objective requires engineered

0-8412-0499-3/80/33-189-305$06.00/0

facilities capable of removing materials ranging in size from those in true solution to coarse suspensions. The particulate fraction of these constituents is defined here to include both colloidal materials (1 nm– 1 μm) and coarse suspensions (greater than 1 μm), a range covering approximately seven size decades. Because many undesirable inorganic and organic constituents are associated with this particulate fraction, system design for particulate removal assumes a key role in facilities planning for potable water treatment, wastewater treatment, industrial water and wastewater treatment, and water reuse systems (30).

Process design of solids/liquid separation processes consists of four principal elements: the selection of technically feasible process alternatives, the selection of design criteria for each process, selection of the optimum process combinations, and selection of process control strategies. Knowledge of the physical, chemical, and biological properties of the particulates as a function of size would provide a rational basis for selection of process alternatives. However, such data are rarely available for fresh water systems (7), and determination of this information is likely to be prohibitively expensive as evidenced by the level of effort required to characterize particulates in the world's oceans (4, 10, 11). Thus until recently, process selection for solids/liquid separation problems has been based on easily measured characteristics of the particulate fraction including turbidity (principally 90° scattered light) and gravimetric determinations (total suspended solids).

Recent developments in the technology of aqueous particulate size distribution (PSD) measurements, however, provide the process engineer with rapid methods for determining the number densities and size distribution of particulates with at least one dimension greater than about 1 μm. For this size fraction, such data provide a direct measure of particulate counts based on a specified statistical length, and an estimate of the total mass or surface area in the particulate fraction above 1 μm. It also appears that particulate counting instruments may be more sensitive monitoring devices of particulate removal processes than the traditional turbidity instruments. Thus, particle size distribution measurements could provide a more accurate basis for decisions in the selection, design, and operation of solids/liquid separation processes. This chapter will present a number of examples to support this contention.

Measurement of Particle Size Distributions

Measurement of aqueous particle size distributions presents a number of challenges to the analyst because of the heterogeneous characteristics of particulates. Shape, density, refractive index, and other physical properties are usually nonuniform throughout the six- to seven-decade size range of the particulate fraction. No single measurement technique is able to measure the particle size distribution over this wide size range. These measurements are further complicated by changes in the distribution attributable to particulate aggregation or breakup during sampling, sample preparation, and in the sensing zone of the counting instrument. Many techniques suffer from precision and accuracy problems, and the statistical reliability of the data must be evaluated continually.

The principal techniques for particle size distribution measurements in natural waters are compared in Table I with respect to measurable size ranges and detection limits, sample handling, and ease of measurement. Additional details on these techniques are found in commercial literature or in texts on particle size measurements (for example, Ref. 1).

The electron microscope is the only instrument capable of measuring the size of particles in the colloidal range (1 nm–1 μm). Both the transmission and scanning electron microscopes require sample evaporation and coating of the particulates with carbon or gold. Thus, sample preparation can alter the original particle size distribution, unless freeze-drying techniques are used. In addition, size data must be determined manually, so these measurements may require up to 16 hr to complete. As a result, electron microscopy is used principally to measure the concentration of trace inorganic particulates in water, such as asbestos fibers (6).

Particle size distribution determination with the optical microscope is also a time-consuming technique, requiring up to 8 hr, depending on sample-handling requirements and the spread of the particulate size distribution. Automatic image analyzers for particle size distribution measurements of aqueous particulates have had limited success because of sample preparation problems, inaccurate counting resulting from multiple counts, and the high cost of equipment (19). Because of the small sample volumes scanned at any given magnification, detection limits are low. Despite these limitations, the optical microscope remains the essential technique for instrument calibration and determination of particulate shape factors.

For nonfibrous particulates with one dimension greater than 1 μm, the electrical sensing zone method (Coulter Principle) can reduce the total time required for particle size distribution determinations to less than 1 hr, depending on the size range of the distribution. In this technique, particulates suspended in a conducting solution (approximately 0.9% NaCl or its equivalent) are pumped through a small orifice through which a current is flowing. Each particle produces a pulse proportional to particle volume attributable to the change in electrical resistance in the orifice passage. The appropriate software provides a tabulated or graphical output of the particle size distribution. The particle size distribution of suspensions with a dynamic size ratio of less than 20 can be measured in less than 5 min.

Table I. Techniques for Measuring

| | | Size Limits | |
Measuring Principle	Equivalent Size Measured	Minimum (μm)	R^a
Electron microscopy	statistical length	0.001	50–200
Optical microscopy	statistical length	0.3	40
Electrical sensing zone method (Coulter principle)	volume diameter	1	20
Light-scattering low-angle, forward-scattering laser light source	cross section diameter	1	10–50
Light obscuration	cross section diameter	1	50–60

[a] R = ratio of maximum to minimum size for single sensing element, or single magnification.
[b] Speed depends on spread of distribution, number of size intervals required to characterize suspension; assumes PSD with range from 2 to 100 μm; includes sample preparation and instrument time.

This technique was developed originally for counting blood cells which are homogeneous in shape and of a narrow size range. Particle size distribution measurements of heterogeneous particulate suspensions with this technique are limited because of particle clogging of the sensor orifice and particle breakup. The recommended range for each orifice is approximately 2–40% of the orifice diameter (1). Because most natural particulate suspensions contain particles with sizes up to 50 μm or more, two to three orifices must be used with appropriate sample handling to avoid clogging and breakup in the smaller (30 and 70 μm) orifices. Fractionation of the distribution by serial filtration can be used, but the accuracy of this procedure is questionable (8).

In contrast to the previously mentioned techniques, some optical sizing methods offer the possibility of on-line process monitoring and control. Instruments designed on the basis of light scattering and light obscuration, using various light sources, analyze from 5 to 1000 mL of liquid sample, and thus exhibit improved detection limits compared to the electrical sensing zone method. Instruments using the light obscuration principle employ various sensors, each with a size measurement ratio (maximum to minimum size) of about 60. Thus the particle size distribution of a typical aqueous suspension with a size range from 2 to 100 μm could be determined using only one sensor. This reduces the time required for a single measurement to less than 10 min. When particulate densities exceed about 5000 mL^{-1}, samples must be diluted to

Aqueous Particle Size Distributions (PSD)

Typical Volume per Scan (mL)	*Sample Handling*	*Time Required for Typical PSD in Natural Waters*[b]
	evaporation, carbon coating	8–16 hr
1000X–10⁻⁵ 100X–10⁻²	concentration may be required	4–8 hr
0.05–2	sample must contain 0.9% salt solution; scalping may be necessary	45 min
10–1000		15 min
	dilution to reduce coincidence errors	
5–100		10 min

Typical Volume per Scan (mL): $1000X-10^{-5}$ $100X-10^{-2}$

[c] Depends on magnification, sample preparation, size range of interest.

avoid coincidence errors. Thus the principal application of these techniques will be monitoring the effluent quality from solids/liquid separation processes rather than influent water quality (*3, 32*).

Size Distributions in Natural Waters

Compared to the extensive data describing the ocean particulate (*10, 11*), size distribution data on particulates in fresh water systems and wastewaters are relatively scarce. Particle size distribution data for several low ionic strength solutions are shown in Figure 1 with the water source, particulate counting method, and references as noted. The size frequency distribution of the four heterogeneous suspensions shown can be modeled by a two-parameter power-law distribution function (*2*) given by the expression

$$\frac{dN}{dl} = A\ l^{-\beta} \tag{1}$$

where N is the particle number density, l the particle size, and A and β empirical constants. Least-squares fits of the data shown in Figure 1 with this distribution function gave correlation coefficients greater than 0.95.

The power-law size distribution function adequately characterizes a large number of particulate systems (*2, 22*). Data either determined by the authors or taken from the literature for particulates larger than 1 μm found in several low ionic strength environments are summarized in

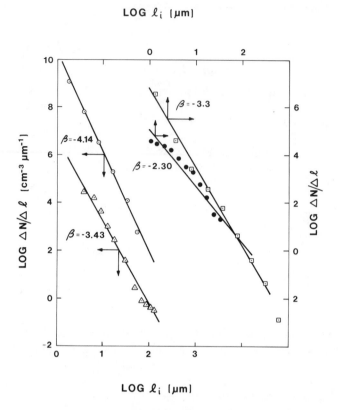

Figure 1.   Examples of particulate size frequency distributions in natural
waters and wastewaters. (△) Effluent, sedimentation basin, pilot-activated
sludge plant, light microscope (26); (●) Lake Zurich, 40 m, 6/2/76, Zeiss
Videomat (19); (□) Deer Creek Reservoir, Utah, 20 m, 3/76, HIAC-12
Channel Counter (16); (⊙) digested primary and secondary sludge, Coulter Counter (8).

Table II. Values of the power-law coefficient β for these suspensions
ranged from 1.8 to 4.5. Data on the ocean particulates indicate values of
β between 4 and 5 (10, 21, 22, 23).

Extrapolation of the power-law function into the submicron size
range is not yet justified, although some evidence (12, 28) suggests that
the power-law model can be extended to 0.01 μm. However, it is more
likely that a complete physical characterization of the seven decades of
aqueous particulates will reveal a size distribution more complex than
the simple power-law model, thus matching the developments in studies
of the atmospheric aerosol (see, for example, Ref. 9).

Table II. Summary of Particle Size Distributions in Natural Waters and Wastewaters: Power-Law Distribution Function

Source	Measuring Device	Coefficient of Power Law Distribution	Correlation Coefficient	Reference
Lakes, reservoirs				
Lake Zurich, Switzerland	Zeiss videomat	1.7–2.4	0.85	(19)
Lake Murten, Switzerland	Zeiss videomat	1.6–2.2	0.85	(18)
Deer Creek Reservoir, Utah	HIAC[a] (6)	3.4	0.98	(16)
Rivers				
American, CA	HIAC (6)	3.4	0.98	(16)
San Joaquin, CA	HIAC (12)	4.2	0.95	(16)
Sacramento, CA	HIAC (12)	3.8	0.96	(16)
Wastewaters				
Digested sludge	Coulter counter	4.2	0.99	(8)
Effluent, activated sludge	microscope	3.3	0.99	(26)
Effluent, lime-treated sewage (pH 10.5)	HIAC (12)	3.1	0.97	(17)

[a] HIAC Instruments Division, Pacific Scientific, California. Numbers in parentheses refer to number of size channels.

Applications of the Power-Law Distribution

Many of the chemical and microbiological contaminants found in natural waters and wastewaters are adsorbed on or incorporated within the particulates (7). In order to meet drinking water standards or effluent discharge criteria, removal of some fraction of the particulate-associated contaminants is required. The relative distribution of these contaminants between different size fractions of the particulates depends on the particulate surface properties, system chemistry, and the shapes and size distribution of the particulates.

Size fractionation of the particulates followed by appropriate analytical procedures can be used to estimate the relative contaminant apportionment within the size distribution (see, for example, Ref. 12). However, this procedure is time-consuming and of questionable accuracy when serial filtration is used (8).

In lieu of analytical data, estimates of the contaminant distribution can be made by computing the contribution of a given size fraction to the total particulate number concentration, area, and volume from knowledge of the particle size distribution. When the particle size distribution is described by the power-law model, a rapid estimate of contaminant distribution can be made from the magnitude of the model parameter β, which describes the slope of the frequency distribution.

Number, Area, and Volume Distributions. When the power-law model is applied, the number, area, and volume of particulates in the ith size fraction of interval size Δl are given by the expressions

$$\Delta N_i = A l_i^{*-\beta} \Delta l$$
$$\Delta A_i = C_2 l_i^{*2} \Delta N_i \tag{2}$$
$$\Delta V_i = C_3 l_i^{*3} \Delta N_i$$

where l_i^* is defined as the arithmetic mean size in the size interval of upper and lower class boundaries, l_{i+1} and l_i, respectively. C_2 and C_3 are the shape factors relating the size measured to the surface area and volume, respectively. For nonspherical particles, the shape factors must be determined microscopically.

To ensure statistical reliability, the interval size should be selected such that the ratio $\Delta l / l_i^*$ is nearly constant (1). A geometric progression, given as $l_{i+1} = \sqrt{2}\, l_i$, satisfies this constraint. It can then be shown that the relative contribution of the ith size class to the total number concentration N_T is given by

$$\frac{\Delta N_i}{N_T} = \frac{l_i^{*(1-\beta)}}{\sum_i l_i^{*(1-\beta)}} \tag{3}$$

When $\beta = 1$ the particulates are distributed equally in each of the class intervals. We have seen, however, that many aqueous particulates exhibit values of β greater than two. At this value of β, computations using Equation 3 show that the total number concentration is dominated by fine particulates. Data in Figure 1 confirm this analysis.

Using similar arguments, the contributions of the ith size class to the total area and volume concentration of the distribution are shown to be

$$\frac{\Delta A_i}{A_T} = \frac{l_i^{*(3-\beta)}}{\sum_i l_i^{*(3-\beta)}}$$

$$\tag{4}$$

$$\frac{\Delta V_i}{V_T} = \frac{l_i^{*(4-\beta)}}{\sum_i l_i^{*(4-\beta)}}$$

These relationships assume that the particle shape factors C_2 and C_3 are independent of particulate size. For homogeneous suspensions (for example, clays) and for many heterogeneous aqueous particulates, this appears to be a reasonable assumption (10, 21). When fibrous or unusually shaped particulates are present in large numbers, appropriate corrections to this assumption must be made.

For a hypothetical particulate suspension with sizes in the measurable range of 0.5 μm–100 μm and a $\beta = 3$, the relative contribution of given size classes to the total number, area, and volume concentration is shown in Figure 2, computed using Equations 3 and 4 with the number of class intervals $i = 12$. Particles smaller than 3 μm dominate the total number concentration for this distribution, while the volume concentration is controlled by particulates larger than 3 μm. Each size class contributes equally to the total surface area of the distribution, but for the size limits selected, particulates larger than 3 μm contribute approximately 75% of the total area.

The surface area of the particulates offers potential adsorption sites to numerous water pollutants. Assuming that the adsorption density is independent of size, the fraction of an adsorbed contaminant in the i size class ΔC_i is given by (17)

$$\frac{\Delta C_i}{C_T} = \frac{l_i^{*(3-\beta)}}{\sum_i l_i^{(3-\beta)}} \tag{5}$$

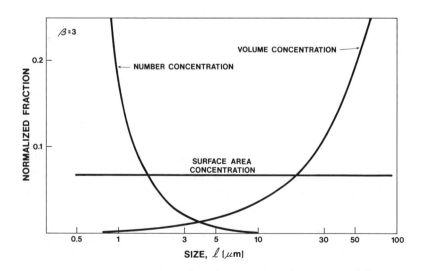

Figure 2. Relative contributions of particulate size classes to total number, surface area, and volume concentration for a hypothetical power-law frequency distribution function, $\beta = 3$

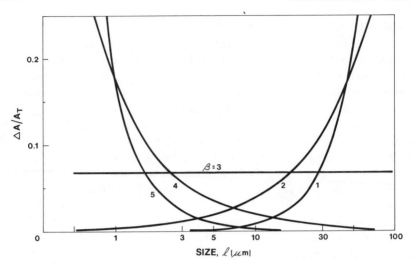

Figure 3.   Relative contribution of particulate size classes to total surface area concentration for a power-law frequency distribution with β = 1–5

which is identical to Equation 4 for the surface area contributions.

Using the same hypothetical particulate suspension as in Figure 2, the relative distribution of the total surface area concentration as a function of size for values of β ranging from 1 to 5 is shown in Figure 3. If $\beta > 3$, the total surface area concentration of the particulate fraction resides predominately in the fine-size fractions below 10 μm. Control of the concentration of an adsorbed contaminant would thus require removal of fine-sized particles by appropriate solid/liquid separation processes. If, however, $\beta < 3$, removal of coarser particulates ($1 > 10~\mu$m) may satisfy the effluent standards. The selection of solids/liquid separation processes will thus depend, in part, on the shape of the size distribution function, as reflected by the value of β.

Process Selection

When the required particulate removal efficiency has been determined for each size fraction in the distribution, the technical feasibility of available solids/liquid separation processes must be evaluated. Such a screening procedure will provide the framework for determining the extent of analytical modeling or pilot plant studies needed to develop process design criteria. The solids/liquid separation processes used in water and wastewater treatment exploit the physical–chemical properties of the particulates to achieve rapid and therefore economical separation from the treated water. Microscreening devices remove most particulates with at least one dimension larger than the minimum opening in the screen. The size and density of particulates determine whether gravity separation by sedimentation or flotation is an economical process alternative. The efficiency of granular-media filtration depends on the size, density, and surface-chemical properties (stability) of the particulates (25). When these alternatives are ineffective, modification of the surface properties and/or size distribution can be achieved using coagulation/flocculation.

A suggested screening method for evaluating alternative separation processes is illustrated in Figure 4. The three variables plotted are the total number concentration N_T, the mass concentration M, and the number volume mean size, l_{NV}, which are related by the expression

$$M = \rho C_3 \, l_{NV}{}^3 N_T \tag{6}$$

where ρ is the average particulate density, selected as 1020 kg m^{-3} for

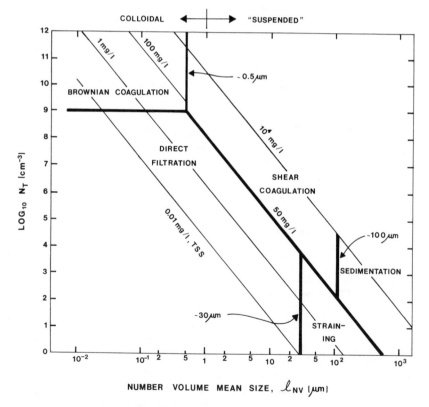

Figure 4. Solids/liquid separation process selection diagram. Mass concentration isopleths computed assuming spherical particles, specific gravity of 1020 kg m^{-3}. (Criteria for region boundaries discussed in text.)

this illustration. Spherical geometry is assumed $[C_3 = (\pi/6)]$. The parameter l_{NV} is a statistic of the particle size distribution, defined as $l_{NV} = (\sum \Delta N_i l_i^3 / \sum \Delta N_i)^{1/3}$, which is sensitive to the concentration of large-sized particles.

The boundaries delineated in Figure 4 were established from simple process kinetic models and empirical observations of process performance, using the criteria of 90% particulate removal within 1 hr. Particulate distributions with $l_{NV} \leq 0.5\,\mu m$ are predominantly colloidal and cannot be removed economically by gravity separation or screening. When the number concentration is greater than approximately 10^9 mL^{-1}, aggregation by Brownian coagulation satisfies the performance criteria at a temperature of $20\,°C$ computed from the Brownian coagulation model for monodispersed spherical colloids (33),

$$\frac{dN_T}{dt} = -\frac{4}{3}\,\alpha\,\frac{kT}{\mu}\,N^2 \tag{7}$$

In this model, α is the collision efficiency factor ($\alpha = 0.1$ for this example), k Boltzman's constant, and μ the dynamic viscosity.

For particulate distributions of relatively low mass concentration (< 50 mg/L) with $l_{NV} \geq 30\,\mu m$, microscreening devices may be a feasible separation process, provided the particulates have a rigid structure. Higher mass concentrations rapidly clog the screens, causing inefficient operation. More concentrated suspensions of large particulates are handled economically by sedimentation, which is efficient for removing particulate suspensions with $l_{NV} > 100\,\mu m$ (33).

In the domain below a mass concentration of about 50 mg/L and $l_{NV} < 30\,\mu m$, granular-media filtration without flocculation as a pretreatment step (direct filtration) easily satisfies the performance criteria for a 90% removal. Although removal efficiency appears to be a minimum in the size range 0.2 to 2 μm (17, 24), a proper selection of media size and depth will ensure satisfactory removals, provided that particulates are destabilized (17). Filters will tolerate mass concentrations higher than 50 mg/L if particulate density and shear strength are increased.

The boundaries of the process selection regions are guidelines only; they can be used for a preliminary assessment of technically feasible separation processes and pretreatment requirements (using particle size data). Lake waters typically contain low mass concentrations (< 10 mg/L) of particulates with number counts of 10^4 to 10^7 mL^{-1}, and l_{NV} between 5 and 20 μm (14, 16, 18). Thus direct filtration may satisfy particulate removal requirements. However, particulate concentrations in rivers may reach levels of up to 10,000 mg/L during heavy runoff, requiring the use of coagulation and sedimentation to provide an acceptable water quality. Other applications of this screening procedure include particulate removal from treated wastewater for spray irrigation, industrial water treatment, and particulate removal prior to groundwater recharge of reclaimed wastewater.

An application of the screening methodology suggested in Figure 4 is the design of water treatment facilities for a lake or reservoir supply. The engineering options are a deep intake structure or more extensive treatment facilities to meet the applicable drinking water standards for particulate removal. Regions of minimum particulate mass concentrations can be identified by turbidity depth profiles, but they provide no information on expected particulate physical properties.

During limnological investigations of Lake Zürich, Switzerland, monthly measurements were made of the total particle count and the particle size distribution as a function of depth over a 12-month period. A Kemmerer sampler was used to collect a 1-L sample which was conserved with formaldehyde, stored at 4°C in the dark, and counted within 48 hr. Sample handling, preparation, and details of the microscope counting technique using a Zeiss Videomat image analyzer and particle counter are described in detail elsewhere (19).

The comparative distributions for stratified and unstratified conditions in the lake are illustrated in Figure 5 (*top* and *bottom*, respectively). The total particle density, the number length mean size, l_{NL} ($l_{NL} = \sum l_i \Delta N_i / \sum \Delta N_i$), the number volume mean size l_{NV}, and the temperature are compared. For the stratified condition (Figure 5, *top*), N_T decreases from the surface to the top of the thermocline, but remains relatively constant below 30 m. The l_{NL} which is most sensitive to size fractions with large number concentrations exhibits few changes with depth, but statistically significant increases in l_{NV} occur at the lower limit of the thermocline, an indication of probable particulate aggregation resulting from various physicochemical and biological phenomena (14).

For the unstratified condition (Figure 5, *bottom*), N_T increased with depth. The value of l_{NL} again showed little variation with depth, but statistically significant variations in l_{NV} are observed, suggesting that aggregation occurred as particulates settled.

The data in Figure 5, as well as data not shown indicated that by locating the raw water intake structure of the treatment plant below the average depth of the thermocline, the direct filtration process could be used for particulate removal. Particulate counts of approximately 10^4 mL^{-1} ($l_{min} = 0.8 \ \mu m$) with l_{NV} 10 μm or less fall in the direct filtration region shown in Figure 4.

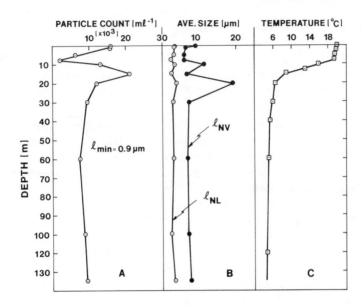

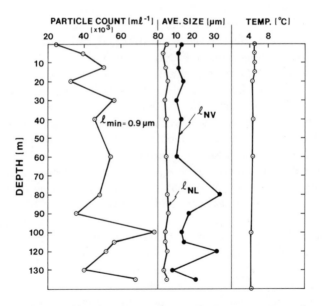

Figure 5. Depth profiles of particle number concentration ($l_{min} = 0.9$ μm), number volume and length mean sizes of the distribution, and temperature (Lake Zürich, Switzerland). Samples collected at Thalwil station, the deepest sampling point in the lake. (Top) Sample of 27 August 1975, stratified conditions; (bottom) sample of 28 January 1976, unstratified.

Evaluation of Pilot Plant Studies

In addition to providing a basis for process selection, particle size distribution measurements are also useful for evaluating predesign pilot plant studies. Turbidity or suspended solids measurements are traditionally used to assess process performance in such studies. However, for low-turbidity waters (< 5 turbidity units) particle counters may provide a more sensitive measure of particulate removal efficiency as well as data on the crucial question of removal efficiency as a function of size.

In a recent pilot study conducted for the Los Angeles Department of Water and Power (15), particle counting was used in conjunction with turbidity measurements to develop selected design criteria for particulate removal processes to treat the Los Angeles Aqueduct supply, a low-turbidity surface water from the eastern slopes of the Sierra Nevada Mountains in California. Several process trains were evaluated using combinations of coagulation, flocculation, sedimentation, and granular-media filtration. Two key design criteria were the duration and extent of mixing in the flocculation process preceding either sedimentation or granular-media filtration.

Flocculation experiments were conducted under steady-state conditions in a reactor with four mixed-flow tanks of equal volume in series. Mixing in each tank was provided by a motor-driven mixer with a horizontal paddle impeller. The intensity of mixing was characterized by the root mean square velocity gradient G.

Results of one test series are shown in Figure 6. The hydraulic residence in this test was 20 min, with G values tested ranging from 30 to 175 sec^{-1}. Turbidity and particle size distribution measurements (HIAC Model 320) were made in the influent and effluent reactor streams when steady-state conditions were reached (usually 1 hr after the start of an experiment). As shown in Figure 6, both the influent and effluent particulate suspensions exhibit a power-law distribution function. A G value of 30 sec^{-1} produced the maximum shift of the particulate surface area and volume from smaller to larger size fractions, as evidenced by the decrease in the slope of the distribution function from 3.6 to 2.1. No optimum G value was observed over the range tested, but maintaining G below 30 sec^{-1} should provide a satisfactory shift in the particle size distribution for the subsequent sedimentation step. This conclusion could not have been reached with turbidity measurements alone, which remained constant at 2.4 turbidity units (TU) both before and after flocculation.

PSD measurements were also used in the selection of design criteria for granular-media filters. One design problem evaluated was the duration and intensity of flocculation required prior to filtration. Results of

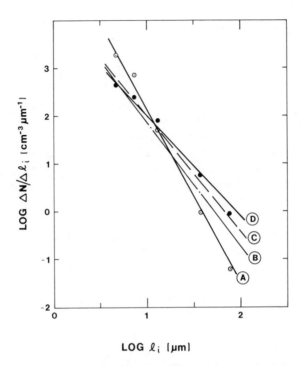

Figure 6.   Effect of increasing mixing intensity on size frequency distributions (15 mg/L alum as coagulant). Particle counting with HIAC six-channel counter and 2.5–150-μm sensor. (A) Influent sample (β = 3.6); (B) G = 175 sec⁻¹ (β = 3.6); (C) G = 70 sec⁻¹ (β = 2.5); (D) 30 sec⁻¹ (β = 2.1).

one test in this series are shown in Figure 7. Coagulants were 2 mg/L alum and 2 mg/L cationic polymer. A dual media filter was used with a flow rate of 15 m/hr (6 gpm/ft^2). Flocculation was provided at a mixing intensity of 70 sec^{-1} and a hydraulic residence time of 20 min. Samples were collected from the effluent of the filter at the point of maximum particulate removal, corresponding to minimum effluent turbidity and maximum rate of head loss development.

As seen in Figure 7, the anticipated shift in the shape of the influent size distribution curve occurs because of flocculation. However, the effluent particle size distributions of both filters are not statistically different. Influent and effluent turbidity measurements for both filters were also identical, thus indicating that flocculation prior to filtration could be eliminated. These results also suggest that the efficiency of particulate removal by granular-media filters with identical physical design variables (media type, size and depth, and flow rate) may depend primarily on particulate destabilization (17), with particle size being less important in determining removal efficiency than previously considered (33).

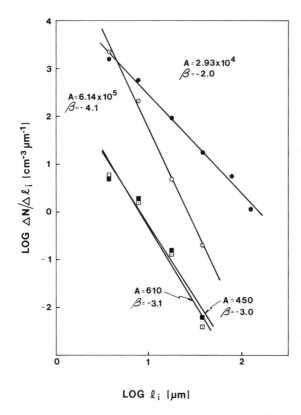

*Figure 7. Comparison of size frequency distributions; effect of floccula-
tion preceding granular-media filtration. (●) Filter influent, with flocculation; (○) filter influent without flocculation; (□, ■) filter effluents.*

Process Control

An additional potential application of particulate counting is process
control and monitoring. With the improvement in aqueous particle
counters, on-line measurement of number concentrations and size distributions for particulates larger than 1–2 μm is now feasible. Both feed-
forward and feedback process control applications can be envisioned.
Feed-forward control could be used to estimate the coagulant chemical
requirements needed for particle destabilization based on measurement
of particle count and estimation of particulate surface area (32). Feedback control possibilities include control of the particle size distribution
entering a filter, control of chemical dosing prior to granular-media filtration, and control of filter operation.

For process control or monitoring, the principal advantages of particulate counting compared to nephelometric (90°) light-scattering

measurements are provision of performance data as a function of particulate size (as was discussed), and under some conditions, superior sensitivity at low turbidity levels.

Power-Law Distribution and Light Extinction. It can be shown that the shape of a particulate size frequency distribution function influences the relative light-scattering properties of aqueous particulate suspensions. For independent scattering, the incremental attenuation of dI of an incident light beam I, passing through an aqueous suspension of path length dZ, is given by (20)

$$-dI = b \, dZ \tag{8}$$

where b is the extinction coefficient. The extinction coefficient is proportional to the cross-sectional area of the particulates and is given by the expression (9)

$$b = \int_0^\infty C_2' \, l^2 \, K_{\text{ext}}(l,\lambda,m) \, n(l) \, dl \tag{9}$$

C_2' is the cross-sectional surface area shape factor, $n(l)$ is the particulate size distribution function dN/dl, and K_{ext} is the extinction efficiency (a function of λ, the wavelength of the incident light, l, the particulate size, and m, the particulate refractive index relative to the surrounding medium). Thus, the contribution of any given size fraction of the distribution to the extinction coefficient is partly a function of the size frequency distribution as shown by the expression

$$\frac{db}{d \log l} = C_2 \, K_{\text{ext}} \, (l,\lambda,m) \, l^3 n(l) \tag{10}$$

For a hypothetical particulate distribution described by the power-law function, Equation 10 becomes

$$\frac{db}{d \log l} = C_2 \, K_{\text{ext}} \, A \, l^{(3-\beta)} \tag{11}$$

Particulates smaller than about 1/20 of the wavelength of the incident light exhibit Rayleigh scattering. The extinction coefficient for spherical, nonabsorbing particulates is given by (20)

$$K_{\text{ext}} = \frac{8\pi}{3} \left(\frac{1}{\lambda}\right)^4 \delta \tag{12}$$

where $\delta = (m^2 - 1/m^2 + 2)^2$. For particulates much larger than the wavelength of light $(1/\lambda > 10)$, the extinction efficiency is approximately

2 (20). Substituting these values of K_{ext} in Equation 11, expressions for the Rayleigh and large-sphere scattering regimes are obtained as follows:

Rayleigh

Large Sphere

$$\left. \begin{array}{c} \dfrac{db}{d \log l} = \dfrac{2}{3} \dfrac{\pi^5}{\lambda^4} \, \delta \, A \, l^{(7-\beta)} \\[2em] \dfrac{db}{d \log l} = \dfrac{\pi}{2} \, A \, l^{(3-\beta)} \end{array} \right\} \qquad (13)$$

Contributions to the extinction coefficient of size fractions in the Rayleigh and large-sphere scattering regions for values of β ranging from 1 to 5 are shown in Figure 8, for $\lambda = 0.6 \, \mu m$, and a suspension of silica spheres with $m = 1.16$. The quantity E is defined as $E \equiv (1/\pi A)[db/(d \log l)]$. Thus each curve corresponds to a hypothetical power-law distribution function with variable β's and equal values of A. Because the scattering between the Rayleigh and large-sphere regions cannot be described analytically (20), the relative light scattering, shown by the dashed lines, is schematic and based on detailed computations for light scattering by an aerosol having a power-law coefficient of $\beta = 4$ (9).

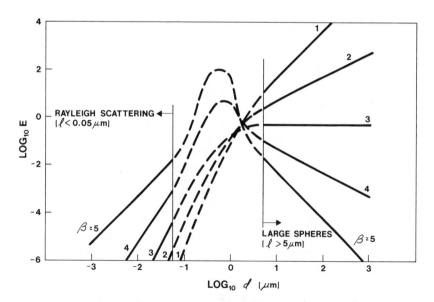

Figure 8. Estimated dependence of scattering extinction coefficient (no absorption) on particulate size frequency distribution assuming a power-law function, $\beta = 1$–5. Computations for Rayleigh and large-sphere scattering assume $\lambda = 0.6 \, \mu m$; silica spheres, m = 1.16.

Figure 8 indicates that for suspensions with $\beta > 3$, particulates in the colloidal range (10 nm–2 μm) control the magnitude of the extinction coefficient and are thus responsible for the decrease in the transmitted light. Because present counters cannot detect particles smaller than about 1–2 μm, light-scattering measurements would be a more sensitive indicator of changes in the particle size distribution when $\beta > 3$. However, when $\beta < 3$ the particle counters would detect the particulate fractions dominating the magnitude of the extinction coefficient.

A comparison of the sensitivity of a light-blockage particle counter (HIAC Model 320) and a nephelometric instrument was carried out during recent investigations conducted by the authors (16) on the applicability of granular-media filtration for treatment of the Columbia River at Kennewick, Washington. A variety of influent turbidities was used to simulate fluctuations in the turbidity levels in the river. In one test series, an influent turbidity of 40 TU was prepared from inorganic material found in the river sediments. Suspensions were fed to the dual-media granular filter at a flow rate of 15 m/hr with coagulants of 5 mg/L alum and 1 mg/L cationic polymer.

From the results shown in Figure 9, effluent particulate counts began to increase several hours before the effluent turbidity increased. The filter effluent history for five size classes between 2.5 and 100 μm is shown in Figure 10. As can be seen, particulates between 2.5 and 40 μm appeared in the filter effluent prior to the breakthrough of the larger (> 40 μm) particulates. Analysis of the effluent particle size distribution before and after the turbidity breakthrough shows a shift in the power-law exponent from approximately 2.5 to 3.5, with least-squares regression coefficients greater than 0.95 for the power-law fit.

As shown in Figure 8, with $\beta < 3$ particulates larger than 5 μm dominate the extinction coefficient and thus control the amount of transmitted light. For particles in this size range, light scattered at 90° to the incident beam is negligible compared to forward scattered light (20). Thus, the light-blockage instrument, which senses particulates by the change in transmitted light, would be expected to exhibit greater sensitivity than nephelometric measurements when $\beta < 3$ and large particles (> 5 μm) dominate the extinction coefficient.

These results have important implications for process control of solids/liquid separation processes. Because many suspensions in natural waters have particle size distributions with $\beta > 3$ (see Table II), presently available particle counters will not detect particulates that dominate the surface area concentration. Light-scattering devices may be needed to monitor the influent raw water supplies. However, if β values are less than 3 in effluent streams from solids/liquid separation processes, particle counters could be used for process monitoring and ultimately, process control.

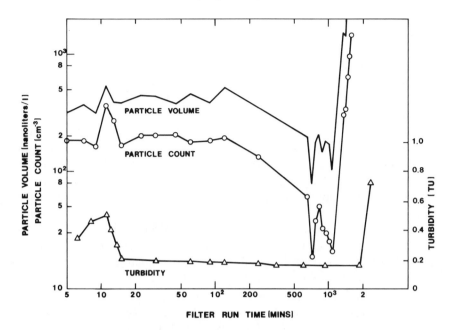

Figure 9.   *Effluent history with respect to particulate volume, number count, and turbidity.   Pilot studies of direct filtration, Columbia River water; influent turbidity $\cong$ 40 TU; influent particulate volume $= 6 \times 10^{-4}$ l/L; flow rate $= 15$ m/hr.*

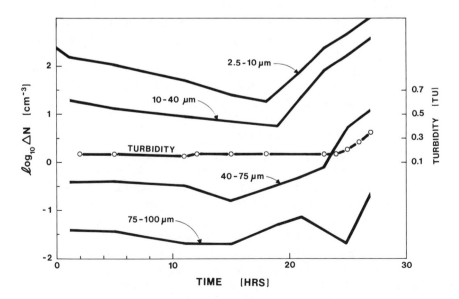

Figure 10.   *Effluent history of four particulate size classes compared to turbidity (experimental conditions as in Figure 9)*

Summary

Reduction of aqueous particulates is a major objective of water and wastewater treatment plants to meet aesthetic, health, and toxicity-based standards. Selection and design of particulate separation processes to meet these standards traditionally have been based on easily measurable collective parameters characterizing the particulate fraction, such as 90° scattered light (turbidity) or total suspended solids. Such measurements, though useful for some applications, correlate poorly with particle number counts or parameters of the particulate size distribution, data that could significantly improve the reliability of process selection and process monitoring. Previous methods for determining the particulate size distribution were time-consuming, and not suitable for on-line measurements. However, particle counters now available based on light-scattering or light-blockage principles may provide rapid and accurate measurements of the particle size distribution for many particulate systems in natural waters with at least one dimension greater than 1 μm.

Particulate size data reported for numerous particulate systems in natural waters can be modeled accurately with a two-parameter power law, given by $dN/dl = Al^{-\beta}$. The exponent of the power law, β, has been shown to be a useful estimator of the relative contribution of particulate size classes to the total number, surface area, mass and volume concentration, and extinction coefficient of the particulate fraction. Reported values of β range from 1.8 to 4.5 in low ionic-strength solutions.

For most particulates above 1 μm in natural waters and wastewaters, the power-law coefficient appears to be greater than 3. Therefore, adequate removal of the particulate fraction by sedimentation or flotation requires a reduction in β by, for example, coagulation/flocculation, which shifts the major portion of particulate surface area and mass into size classes above about 30 μm. If granular-media filtration is used as the particulate separation process, only particulate destabilization may be necessary to achieve desired removals.

The efficiency of solids/liquid separation processes for reduction of trace contaminants (such as heavy metals) and toxic organic compounds associated with the particulate fraction could be estimated if the chemical composition of the particulates as a function of size were known. However, such data are scarce and of questionable accuracy. As a first approximation, the distribution of an adsorbed constituent between various size classes in the particulate fraction can be estimated from a knowledge of the power-law coefficient. This combined with performance models of solids/liquid separation processes should provide an improved basis for process selection to meet increasingly stringent standards for water and wastewater treatment.

Particle counters also show promise for process control and performance evaluation, particularly in low-turbidity waters. Applications include more accurate evaluation of pilot plant studies and more sensitive control of particulate separation processes. However, when the power-law coefficient, β, is greater than 3, submicron particles, which escape detection with available counters, may control the magnitude of the total surface area concentration and the light-scattering properties of the particulate. It is likely that accurate process control and monitoring of solids/liquid separation processes will require both turbidimetric and particle size distribution measurements. Particle counting appears most promising as a feedback control sensor.

Literature Cited

1. Allen, T. "Particle Size Measurement," 2nd ed.; Chapman and Hall: London, 1975.
2. Bader, H. *J. Geophys. Res.* **1970**, *75*, 2822.
3. Beard, J. D.; Tanaka, T. S. *J. Am. Water Works Assoc.* **1977**, *69*, 533–538.
4. Bishop, J. K. B.; Ketten, D. R. *Deep-Sea Res.* **1977**, *24*, 511–548.
5. Black, A. P.; Hannah, S. A. *J. Am. Water Works Assoc.* **1965**, *57*, 901–916.
6. "Direct Filtration of Lake Superior Water for Asbestiform Fiber Removal," EPA Report, No. 670/2-75-050a, June 1975.
7. Committee on Safe Drinking Water of the National Research Council, "Drinking Water and Health"; National Academy of Sciences: Washington, D.C., 1977.
8. Faisst, W. K. Report No. 13, Environmental Quality Laboratory, California Institute of Technology, Pasadena, CA, June 1976.
9. Friedlander, S. K. "Smoke, Dust, and Haze"; Wiley–Interscience: New York, 1977.
10. Geochemical Ocean Sections Program—Collected Papers, *Earth Planet. Sci. Lett.* **1976**, *32*(2), 217–473.
11. "Suspended Solids in Water"; Gibbs, R. J., Ed.; Plenum: New York, 1974.
12. Harris, J. E. *Deep-Sea Res.* **1977**, *24*, 1055–1061.
13. Hulbert, H. M.; Katz, S. *Chem. Eng. Sci.* **1964**, *19*, 555–574.
14. Hutchinson, G. E. "A Treatise on Limnology"; Wiley–Interscience: New York, 1966; Vol. II.
15. James M. Montgomery, Consulting Engineers, Inc., Pasadena, CA, Water Quality Studies, Vol. IV, June 1977, report submitted to the Los Angeles Department of Water and Power.
16. James M. Montgomery, Consulting Engineers, Inc., 1978, unpublished data.
17. Kavanaugh, M. C.; Toregas, G.; Chung, M.; Pearson, E. A. *Prog. Water Technol.* **1978**, *10*(5/6), 197–215.
18. Kavanaugh, M. C.; Vagenknecht, A. *Gas, Wasser, Abwasser* **1975**, *55*, 554–559.
19. Kavanaugh, M.; Zimmermann, U.; Vagenknecht, A. *Schweiz. Z. Hydrol.* **1977**, *39*(1), 86–98.
20. Kerker, M. "The Scattering of Light and Other Electromagnetic Radiation"; Academic Press: New York, 1969.
21. Lal, D. *Science* **1977**, *198*(4321), 997–1009.
22. Lal, D.; Lerman, A. *J. Geophys. Res.* **1975**, *80*, 423–430.
23. Lerman, A; Carder, K. L.; Betzer, P. R. *Earth Planet. Sci. Lett.* **1977**, *37*, 61–70.

24. O'Melia, C. R.; Ali, W. *Prog. Water Technol.* **1978,** *10,* 123–137.
25. O'Melia, C. R.; Stumm, W. *J. Am. Water Works Assoc.* **1967,** *59,* 1393.
26. Parker, D. S., Ph.D. thesis, University of California, Berkeley, CA, 1970.
27. "Proceedings of Turbidity Workshop, National Oceanographic Instrumentation Center"; U.S. Department of Commerce: Washington, D.C., May 1974.
28. Sharp, J. H. *Limnol. Oceanogr.* **1973,** *18,* 441–447.
29. Sheldon, R. W.; Prakash, A.; Sutcliffe, W. H. *Limnol. Oceanogr.* **1972,** *17,* 327.
30. Stumm, W. *Environ. Sci. Technol.* **1977,** *11,* 1066–1070.
31. Swift, D. L.; Friedlander, S. K. *J. Colloid Sci.* **1964,** *19,* 621.
32. Tate, C. H.; Trussell, R. R. *J. Am. Water Works Assoc.* **1978,** *70,* 691–698.
33. Weber, W. "Physicochemical Treatment Processes for Water Quality Control"; Wiley: New York, 1972.

RECEIVED January 19, 1979.

15

Prediction of Suspension Turbidities from Aggregate Size Distribution

GORDON P. TREWEEK

James M. Montgomery, Consulting Engineers, Inc., 555 East Walnut Street, Pasadena, CA 91101

JAMES J. MORGAN

Environmental Engineering Science Department, California Institute of Technology, 1201 East California Boulevard, Pasadena, CA 91125

The turbidity of a coagulating, but noncoalescing, suspension cannot be determined directly because of interference between scattered waves and phase shifts in transmitted waves. This research developed a "coalesced-sphere" light scattering model which was applied to a noncoalescing, coagulating system to predict, within reasonable limits, the suspension turbidity. The application of the coalesced-sphere model for aggregates in the Mie scattering regime allows the calculation of the turbidity per size interval caused by particulates in the coagulated suspension. Knowledge of the turbidity fraction can lead to improved design of phase-separation devices, such as water and wastewater filters, so they remove the portion of the size distribution that contributes most significantly to the overall suspension turbidity.

In general, the removal of particulate matter in water and wastewater via coagulation/flocculation with cationic polymers occurs in three steps:

1. Destabilization—modification of the surface properties of the singlet particles so that interparticle repulsion is reduced and collision effectiveness is enhanced (coagulation).

2. Transport—movement of the particles into adhesive contacts via perikinetic (Brownian) or orthokinetic motion resulting in aggregate growth (flocculation).

3. Phase separation—removal of particle aggregates from the suspending medium.

More than 20 techniques have been used by prior investigators to evaluate flocculant effectiveness in accomplishing these steps, but aside from the efforts of TeKippe and Ham (4), little has been done to correlate the measurements taken by these different techniques. Many of the experimental techniques (settling velocity, residual scattering intensity, settled turbidity, refiltration rate) currently employed rely solely on measuring the final phase separation of particulate matter from the suspension. While these methods record the end result of successful coagulation and flocculation, the preceding steps are not followed quantitatively. In addition, measurement of phase separation alone tends to bias the investigator toward improvement of the physical parameters of the system at the expense of possible chemical alterations which would enhance earlier destabilization stages.

This research involves a quantitative study to evaluate the effectiveness of relative scattering intensity measurements ($I_{90^\circ}/I_{180^\circ}$) in recording the coagulation–flocculation of a suspended biocolloid, *E. coli*, via the addition of a cationic polymer, polyethyleneimine (PEI). To accomplish this goal, the changes in the scattering intensity must be correlated with changes in the particle size distribution. The assumption was made (and verified) that the aggregates of the singlet cells could be treated, for light scattering purposes, as coalesced spheres. Using the coalesced-sphere model for aggregates in the Mie scattering regime allows the calculation of the turbidity caused by the aggregated singlets in each size interval in the coagulated suspension.

Particle Size Distributions Via Electronic Particle Counters

A direct quantitative measure of change in aggregate diameter can be achieved via the electronic particle counting and sizing technique. Flocculation data recorded by this technique were used as the basis against which data taken by light scattering were compared. The intricacies of measuring particle size distributions in aggregating suspensions via electronic particle counters have been presented in the works of Birkner and Morgan (1, 2); Ham and Christman (3); TeKippe and Ham (4); Camp (5,6); Hannah, Cohen, and Robeck (7); and Treweek (8,14).

Light Scattering by Aggregated Particles

Light scattered from coagulating systems can be evaluated by one of three theories: Rayleigh, Rayleigh–Debye, or Mie, depending on the size regime of the aggregates. The Rayleigh theory for particles with d less than $\lambda/10$ treats both singlets and aggregates as Rayleigh scatterers in which an aggregate of k spheres, each of volume v, is equivalent to a single Rayleigh sphere of volume kv. For such an aggregate, the light scattered is k times greater than from k single spheres. The second theory, a Rayleigh–Debye-type treatment derived by Benoit, Ullman, DeVries, and Wippler (9), evaluates the scattering properties of aggregates whose constituent particles have: (1) refractive indices close to those of the suspending medium and (2) negligible "phase shift" at any point in the aggregate. The third interpretation, based on Mie theory, is used for larger particles without restriction on the index of refraction. Since bacteria cells are obviously larger than Rayleigh spheres ($d > \lambda/10$), only the latter two interpretations will be discussed further.

Rayeigh–Debye Scattering. In the analysis by Benoit et al. (9), primary particles within an aggregate scatter light which interferes immediately with that scattered by its aggregate neighbors. The intensity of light scattered at an angle Ψ by an aggregate of k monodisperse spheres randomly positioned to the incident light beam is expressed as:

$$I_k(\Psi) = kI_1(\Psi)[1 + (2/k)A_k(\Psi)] \qquad (1)$$

where $I_1(\Psi)$ is the intensity of light scattered at an angle Ψ by a single primary sphere (Rayleigh–Debye or Mie theory) and $A_k(\Psi)$ is a form factor that is a function of the geometry of the aggregate and of the scattering angle Ψ.

Benoit et al. (9) analogized the aggregate form factor $A_k(\Psi)$, which accounts for the interference between light waves scattered by particles within an aggregate, to the Rayleigh–Debye form factor $P(\Psi)$, which accounts for scattering interference between the various scattering centers within the same particle. The form factors and hence the scattering properties of aggregates depend on the geometrical arrangement of particles within the aggregates. Because an infinite number of arrangements of singlets is possible in a coagulating system, a precise evaluation of $A_k(\Psi)$ is not possible. Generally, $A_k(\Psi)$ can be expressed by:

$$A_k(\Psi) = \sum_{i=1}^{k-1} \sum_{j=i+1}^{k} \sin[(4h_{ij}/\lambda)\sin\Psi/2]/[4h_{ij}/\lambda\sin\Psi/2] \qquad (2)$$

where h_{ij} is the distance between centers of the ith and the jth particle in the aggregate, and λ is the wavelength of incident light.

Lipps et al. (10, 11) assumed a system mix of three types of aggregates (linear, planar, and close-packed three dimensional) in deriving form factors for singlet multiplicities up to 13. In coagulation experiments on polystyrene latex, Lips et al. (10) concluded that aggregates consisting of more than 13 singlets accounted for less than 0.5% of the scattered light and could therefore be ignored. Their results for smaller aggregates demonstrated the validity of the interference method in predicting light scattering intensity from the three types of aggregates.

Several major complications in this analysis arise when it is applied to suspensions of particles of the size and shape of *E. coli*. First, the scattering measurements for the primary particles must be corrected to account for the rod-like shape of the cells by applying the Rayleigh–Debye form factors (*12*). Second, an assumption must be made as to the system mix of aggregates, that is, the percentage of linear, planar, and three dimensional aggregates. For particles such as *E. coli*, with a relatively broad size distribution in the uncoagulated state, determination of the number of singlets per aggregate from the particle size distribution would be very difficult. Finally, form factors $A_k(\Psi)$ must be calculated from the assumed system mix and then substituted into Equation 1 to determine the scattered light intensity. For these reasons, and because the *E. coli* singlet does not meet the restrictions on the use of the Rayleigh–Debye theory $[2\pi d(m - 1)/\lambda << 1.0$, but for *E. coli* $2\pi d(m - 1)/\lambda$ equals approximately 1.2, where d (diameter) equals 1.3 μm, m (relative refractive index) equals 1.06, and λ (wavelength) equals 0.41 μm], the interference analysis of Benoit et al. (*9*) was not adopted.

Mie Scattering. For systems more complex than very small particles (Rayleigh) or small particles with low refractive indices (Rayleigh–Debye), the scattering from widely separated spherical particles requires solving Maxwell's equations. The solution of these boundary-value problems for a plane wave incident upon a particle of arbitrary size, shape, orientation, and index of refraction has not been achieved mathematically, except for spheres via the Mie theory (*12, 13*). Mie obtained a series expression in terms of spherical harmonics for the intensity of scattered light emergent from a sphere of arbitrary size and index of fraction. The coefficients of this series are functions of the relative refractive index m and the dimensionless size parameter $\alpha = \pi d/\lambda$.

The pattern given by the Mie series for spherical particles is complicated not only by interferences between the wavelets scattered by the volume elements of the same particle, but also by distortions in phase associated with the incident and emergent light. Nevertheless, the values of the Mie scattering efficiencies have been calculated accurately. For particles with radii comparable to or larger than the wavelength of the incident light, that is, $\alpha \geq 1$ and with m differing somewhat from unity ($m \leq 0.95$ or $m \geq 1.05$), calculations of the Mie series are tabulated in Kerker (*15*). If the assumption is made that the flocculated aggregates are equivalent, in a light scattering sense, to coalesced spheres, then Mie's solution for the scattering efficiency K_{sca} can be used along with the particle size distribution to determine the scattering coefficient.

The transmission (T) is defined as:

$$T = I_l/I_0 = e^{-NC_{ext}l} = e^{-\tau l} \qquad (3)$$

where I_l is the light intensity transmitted through a suspension of length l, I_0 the incident light intensity, N the number of particles in suspension, C_{ext} the extinction cross section, and τ the extinction coefficient.

The extinction cross section C_{ext} is composed of a scattering cross section C_{sca} and an absorption cross section C_{abs} (which is negligible for *E. coli*). Then the extinction coefficient or the scattering coefficient (turbidity) for a heterogeneous suspension of coagulated *E. coli* cells is given by:

$$\tau = \sum_i n_i (C_{sca})_i = \sum_i n_i \pi r_i^2 (K_{sca})_i \qquad (4)$$

Mie's solution for the scattering efficiency of spheres results in the expression:

$$(K_{sca})_i = (2/\alpha^2) \sum_{j=i}^{\infty} (|a_j|^2 + |b_j|^2)(2j+1) \qquad (5)$$

where a_j and b_j are the scattering components of the j multipole of the electric and magnetic field, respectively. They are complicated, complex functions of α and m. With increasing particle size, the light scattering diagram, consisting of numerous maxima and minima, becomes increasingly complicated.

In lieu of these expressions, Jobst (*16*) derived values for K_{sca}, the effective scattering efficiency of a particle, from the Mie equations for four special particle size–refractive index regimes. The values of K_{sca} and the restrictions on their application caused by approximations to the Mie efficiencies are listed in Table I. The Jobst approximations provide simple analytical solutions for particles with the dimensions and composition of bacteria, since α is quite large ($\alpha = 2\pi(0.65)/0.41 = 10$), yet

Table I. Jobst Approximations to Mie Scattering Efficiencies K_{sca} for Selected Size–Refractive Index Regimes[a]

Region	Restrictions	K_{sca}
Rayleigh	$\alpha(m-1) \ll \alpha \ll 1$	$32\alpha^4(m-1)^2/27$
Anomalous	$\alpha(m-1) \ll 1 \ll \alpha$	$2\alpha^2(m-1)^2$
Anomalous diffraction	$\alpha(m-1)$ arbitrary	$2 - (4 \sin \rho)/\rho + 4(1 - \cos \rho)/(\rho)^2$
Diffraction	$1 \ll \alpha \ (m-1) \ll \alpha$	2

[a] $\alpha = 2\pi r/\lambda$, $\rho = 2\alpha(m-1)$.

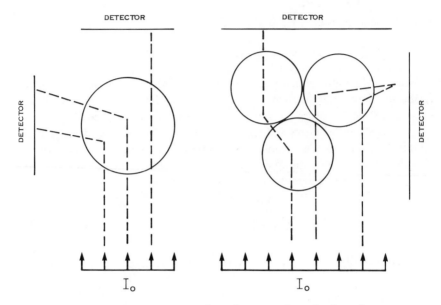

Figure 1.   An aggregate scatters less than a coalesced sphere because of rescattering in the forward direction and destructive interference between scattered waves

$\alpha(m-1)$ is small $[10(1.06-1)] = 0.6$. The error introduced by the use of the Jobst approximations, in lieu of the exact Mie calculations, is not significant in comparison to the potential error introduced via the coalescence assumption (*10, 11, 17, 19, 30*).

Research (*19, 30*) has shown that a real aggregate scatters less than an assumed coalesced aggregate of equivalent volume for two reasons: (1) a greater likelihood of light being scattered a second time back in the original direction and (2) a longer average difference in path length for waves scattered by separate particles than within coalesced particles. The stronger interference between scattered waves decreases the amount of light abstracted from the primary beam and increases the amount of transmitted light. These phenomena are illustrated in Figure 1. Because of these reasons, an aggregate does not behave precisely as a coalesced sphere, so the exact Mie scattering efficiencies are an unnecessary refinement for our model.

Light Scattering by Coagulating Systems

For a given weight (or volume) of material, initially composed of monodisperse singlets of very small size so that Rayleigh's law is obeyed ($d < \lambda/10$) undergoing coagulation, the suspension turbidity will increase as the third power of the ratio of the aggregate radius to the singlet radius, since the number of particles per unit volume in the system decreases inversely as the third power of the aggregate radius, while the scattering of each aggregate increases as the sixth power of this radius. The turbidity is the product of these quantities; that is,

$$\tau = \sum_i n_i \pi r_i^2 (K_{sca})_i \tag{6}$$

where K_{sca} equals $32\alpha^4(m-1)^2/27$.

When the initial singlets and their aggregates are comparable in size to the wavelength of the incident light $(\alpha(m-1) << 1 << \alpha)$, the scattering per aggregate is proportional to the fourth power of its radius, and the turbidity of a given weight of material will increase linearly with the ratio of the aggregate radius to the singlet radius:

$$\tau = \sum_i n_i \pi r_i^2 (K_{sca})_i \tag{7}$$

where K_{sca} equals $2\alpha^2(m-1)^2$.

For large aggregates with small refractive indices, and arbitrary values of $\alpha(m-1)$, the scattering per aggregate is approximately proportional to (1) the fourth power of the radius in the range $\alpha < 20$, (2) the third power of the radius for the range $20 < \alpha < 30$, and (3) the second power of the radius in the range $\alpha > 30$. Thus in one portion of the anomalous diffraction regime, the turbidity of a coalescing system is insensitive to changes in the aggregate diameter $(20 < \alpha < 30)$, and decreases for larger aggregate diameters $(\alpha > 30)$.

$$\tau = \sum_i n_i \pi r_i^2 (K_{sca})_i \tag{8}$$

where K_{sca} equals $2 - (4 \sin \rho)/\rho + 4(1 - \cos \rho)/\rho^2$ and ρ equals $2\alpha(m-1)$.

For still larger aggregates $(1 << \alpha(m-1) << \alpha)$, the scattering is proportional to the square of the aggregate radius so that the turbidity of the system will always decrease with the increasing diameter of the aggregate:

$$\tau = \sum_i n_i \pi r_i^2 (K_{sca})_i \tag{9}$$

where K_{sca} is 2.

If the aggregates of rod-shaped bacteria cells are assumed to coalesce into spheres, then the foregoing equations predict the variation of scattering efficiency with the size-refractive index parameter ρ. Experimentally, a variation in the size parameter can be achieved by varying the molecular weight of the cationic polymer used as the flocculant, with higher molecular weight species producing larger aggregates.

The resulting Jobst approximations to the Mie scattering efficiencies from the foregoing theoretical discussion are presented in Figure 2 for the anomalous region and the anomalous diffraction region. The application of this model to the actual *E. coli* system resulted in the scattering efficiency–spherical diameter relationship shown in Figure 3. Also shown is the approximate range of aggregate sizes that are obtainable by varying the flocculant molecular weight.

For a system involving the flocculation of singlet *E. coli* cells, the equation for the scattering efficiency in the anomalous diffraction regime can be used throughout the experimental calculations. The required scattering efficiency is composed of two parts: the initial "2" caused by diffraction and the terms $4 \sin \rho/\rho + 4(1 - \cos \rho)/\rho^2$ caused by refraction phenomena. The refraction causes oscillation of the scattering efficiency about the diffraction value of "2." Maxima in the values of the scattering efficiency with increasing ρ are caused by favorable interference of diffracted and refracted light; minima are caused by unfavorable interference. As ρ approaches zero, refraction is the most important phenomenon; as ρ approaches infinity, diffraction is the most important phenomenon. While this coalescing-sphere approach is valid strictly for emulsions and liquid aerosols, it has been used with aggregated systems of rigid singlets because of its simplicity and good correlation with experimental results (*17, 18*).

The turbidity of a suspension of aggregates coalesced from singlets is highly dependent on the particle size regime involved. Table II summarizes the turbidities for suspensions of coalesced aggregates as a function of the turbidity of a suspension of singlet particles, using the Jobst approximations for the Mie scattering efficiencies in the anomalous, the anomalous diffraction, and the diffraction regimes. For aggregates in the anomalous region, coalescence leads to increased turbidity whereas in the diffraction regime, coalescence leads to decreased turbidity. The anomalous diffraction regime serves as a transition between the anomalous and the diffraction regimes.

Lichtenbelt et al. (*19*) made a detailed comparison of extinction cross sections for real doublets and for hypothetical doublets coalesced into spheres. Using the Rayleigh–Debye theory for α up to three, Lichtenbelt et al. determined that the coalescence assumption leads to 10% larger values of the scattering cross section than would be found for real doublets created in coagulation. This is because a real doublet is less compact than a coalesced doublet of the same volume; therefore, the interference between light waves, scattered by different parts of the real

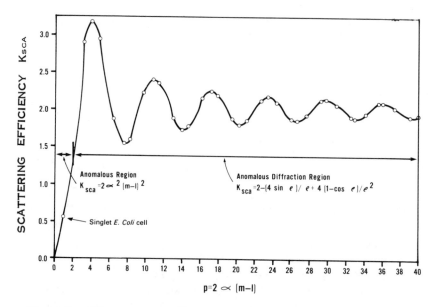

Figure 2.   The scattering efficiency oscillates about 2.0, the diffraction efficiency, with increasing values of the size–refractive index parameter p

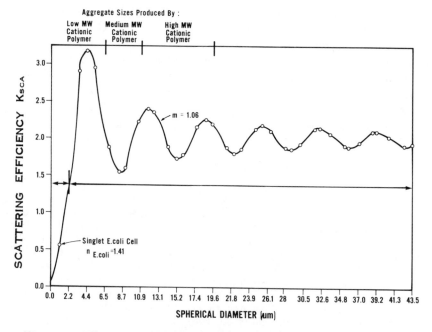

Figure 3.   Changes in scattering efficiency with aggregate diameter enable the suspension turbidity to increase, remain constant, or decrease depending on the size regime of the aggregate

Table II. Turbidity τ_c of a Suspension of Coalescing Aggregates as a Function of the Turbidity τ_0 of a Suspension of Uncoalesced Singlets

Region	Turbidity
Rayleigh	$\tau_c = (r_c{}^3/r_o{}^3)\tau_0$
Anomalous	$\tau_c = (r_c/r_o)\tau_0$
Anomalous diffraction	$\tau_c = (1.51/r_0 r_c)\tau_0$
Diffraction	$\tau_c = (r_o/r_c)\tau_0$

r_o = singlet radius
r_c = aggregate radius

doublet, is stronger because of the larger average difference in path length. Consequently, scattering is less for a real doublet than for a hypothetical coalesced doublet. Lower scattering means greater transmission and lower turbidity.

Although the coalescence assumption may lead to small errors in the scattering cross section, it does enable the rapid correlation of turbidity data (τ_i) with electronic particle counter data (n_i and r_i), provided the relative refractive index and size regime are known. Because the electronic particle counter measures the aggregate volume, and therefore determines only an equivalent spherical radius, the data from the electronic particle counter relate only to the coalesced-sphere approach.

In summary, the difficulties in determining aggregate form factors, particle form factors, phase shifts, and distribution functions combine to make the Rayleigh–Debye approach too complicated for practical application. On the other hand, the coalesced-sphere approach using the Jobst approximations to the Mie scattering efficiencies allows rapid correlation of turbidity with particle size distributions. Consequently, a coalesced-sphere approach was adopted for experimentation in the E. coli–PEI system.

Experimental Materials and Methods

A species of the coliform group, E. coli CR63, was selected for the experimental work as representative of biocolloids present in municipal wastewaters. Grown on a glucose-enriched Geneva M9 media (20), the monodispersed coliforms had a cylindrical shape: 0.8 μm in diameter, 2–3 μm in length, with a mean equivalent spherical diameter of 1.3 μm at the end of log-phase growth. This particular species has a smooth surface, free of fillae or fimbrae. In addition to a collidine–hydrochloric acid buffer (21), minor additions of 0.1M sodium hydroxide were made during the growth cycle to maintain the suspension pH constant at 7.0 ± 0.1. The water-jacketed chemostat was maintained at 25°C. With continuous stirring of the growth media, the seeded culture grew to a final cell concentration of 3.0 (±0.5) × 10^7 cells/mL. The E. coli cell size and low relative refractive index ($m = 1.06$) (27) permitted the use of Mie scattering theory with the Jobst approximations for scattering efficiencies (Table I).

A series of PEI polymers were selected as flocculants because of the range of molecular weights available, from 600–60,000, corresponding to 14–1400 monomer (C_2NH_5) units. PEI in solution functions as a cationic polyelectrolyte, strongly attracted to anionic colloids (*22*). By varying the molecular weight of PEI, the size of the resulting aggregates could be established (Figure 3).

In the experimental phase, 900 ML of stationary-phase *E. coli* cells were removed from the chemostat and placed in a stirrer–reactor assembly, which provided both rapid-mix and flocculation velocity gradients. A 100-mL aliquot of a specified PEI concentration was then added to the *E. coli* suspension during a 2-min rapid mix ($G = 190$ sec^{-1}). The suspension was flocculated for 60 min at $G = 20$ sec^{-1}, with turbidities and particle size distributions recorded before flocculant addition and after flocculation. The initial suspension had a pH $= 7.0$ (± 0.1), ionic strength $I = 0.06M$ NaCl, $T = 25°C$, and a cell concentration of 2.5 (± 0.5) $\times$ 10^7 cells/mL before flocculation. The flocculation velocity gradient ($G = 20$ sec^{-1}) was set by regulating the angular velocity of a two-pronged stirrer to yield the desired torque.

A Brice Phoenix Model 2000 spectrophotometer was used to measure the intensity of light scattered at 90° and transmitted at 180° through the stationary-phase culture of *E. coli* Strain CR63. From the scattering intensity ratio $I_{90°}/I_{180°}$, the turbidity may be calculated via Equation 10, which uses the ratio of galvanometer deflections recorded experimentally (*23*):

$$\tau = \frac{16\,IC}{3\,(1.049h)}\,u^2\,(R_w/R_c)\,aF\,(G_s/G_w) \tag{10}$$

where IC is the instrument correction factor; h the diaphragm width (1.2 cm); u the refractive index of solution (1.33); R_w/R_c the refraction correction factor (1.0); a the working standard constant; F the product of transmittances of neutral filters; G_s the galvanometer deflection for light scattered from solution at 90°; and G_w the galvanometer deflection for light transmitted through solution at 180°.

For an incident wavelength of 546 nm, the turbidity per centimeter of path length is $\tau = 0.048\,FG_s/G_w$. The turbidity calculated from the scattering ratio $I_{90°}/I_{180°}$ was linearly proportional to the turbidity determined from transmission by a Gilford Model 300 spectrophotometer. This indicated that the scattering intensity technique could be used to measure the suspension turbidity. In addition, the turbidity of the *E. coli* suspension was found to vary linearly over the concentration range 3×10^5 to 5×10^7 cells/mL. Both unflocculated and flocculated suspensions were removed directly from the reactor vessel, and their relative scattering intensity measured without making dilutions or multiple scattering corrections. At cell concentrations greater than 3.0×10^7 cells/mL, multiple scattering resulted in nonlinearity of relative scattering intensity (*24*).

For turbidity measurements, a 50-mL sample was pipetted from the stirrer reactor and placed in a Brice Phoenix scattering cell (Catalog No. T101). Care was taken to minimize disruption of the flocs during the transfer step. A sample was removed from the stirrer reactor, while determining the turbidity, and the particle size distribution was recorded in a Coulter Counter, Model B, used in conjunction with a Nuclear Data Model 555 pulse height analyzer (PHA) and multichannel analyzer (MCA). Modifications to the normal operating procedure to account for the effect on the pulse height of the distribution of singlets within an aggregate are discussed by Neis, Eppler, and Hahn (*25*), and Treweek and Morgan (14). Essentially, a conservation of volume condition between the unflocculated and flocculated state is used to correct the experimentally recorded pulse heights that result from the aggregate's passage through the detector.

Starting with an initial cell concentration of 2.5 (± 0.5) $\times$ 10^7 cells/mL, two serial dilutions of 10 to 1 resulted in 2.5 $\times$ 10^5 cells/mL, or 1.25 $\times$ 10^4 cells/50 μL. At this concentration a Coulter Counter 11-μm aperture will give the true count of cell concentration without necessitating coincidence corrections (26). As aggregation occurred with the addition of PEI, the particle number concentration decreased and fewer dilutions were required to achieve the same significance of count. Apertures of 30 and 70 μm were used to size the larger aggregates as they formed in the stirrer–reactor assembly.

Experimental Results

Particle size distributions and turbidity readings were recorded simultaneously from samples from the flocculating *E. coli*–PEI suspension. These data, together with the coalescent assumption and the Jobst scattering efficiency approximations, enable quantitative comparisons to be made between the two measurement techniques. The importance of using particle counts to supplement turbidity measurements can be seen readily in a comparison of Figures 4 and 5. Figure 4 represents the domain of percent reduction in the scattering ratio ($I_{90°}/I_{180°}$) after 60 min of flocculation at $G = 20$ sec^{-1}. This percent reduction is represented as a function of the flocculant concentration (mg/L) on the abscissa and

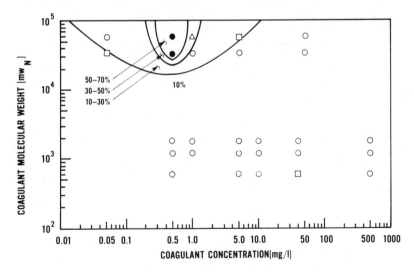

Figure 4.   Log mol wt$_N$–log [coagulant] domains of percent reduction in scattering ratio ($I_{90°}/I_{180°}$) after 60 min of flocculation at $G = 20$ sec^{-1} (percent reduction: ($\bigcirc$) < 10%; ($\square$) 10–30%; ($\triangle$) 30–50%; ($\bullet$) 50–70%; ($\blacksquare$) 70–90%; ($\blacktriangle$) > 90%)

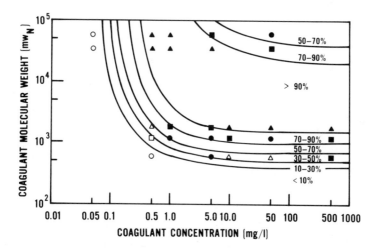

Figure 5. Log mol wt$_N$–log [coagulant] domains of percent reduction in number of particles after 60 min of flocculation at G = 20 sec^{-1} (percent reduction: (○) < 10%; (□) 10–30%; (△) 30–50%; (●) 50–70%; (■) 70–90%; (▲) > 90%)

the flocculant molecular weight on the ordinate. Large reductions in the suspension turbidity were achieved only for the high molecular weight flocculant over a limited concentration range between 0.2 and 1.0 mg/L. Over the full concentration range for molecular weights less than 10,000, the percent reduction in turbidity was less than 10%.

Figure 5 presents the corresponding domain of percent reduction in the number of particles after 60 min of flocculation at $G = 20$ sec^{-1} as a function of the flocculant concentration and molecular weight. For doses greater than 5.0 mg/L of PEI with molecular weight greater than 2000, at least a 90% reduction was achieved in the number of primary particles. The optimum dose for higher molecular weight species was between 0.5 and 5.0 mg/L; doses greater than 5.0 mg/L brought about restabilization caused by charge reversal on the bacteria cells (determined by measuring their electrophoretic mobility). As the flocculant molecular weight fell below 1000, the percent reduction in number of particles fell rapidly and approached less than 10%. Nevertheless, comparisons of Figures 4 and 5 indicate that substantial changes are brought about in the particle size distribution over a wide range of flocculant doses and molecular weights, but these changes in the particle size distribution are not recorded effectively by turbidity measurements.

As shown in Figure 3 and Equation 7, during the initial phases of coagulation the turbidity of the suspension theoretically can increase. Experimental verification of this fact is shown in Figure 6 where the bacteria are coagulated with a less than optimum dose of a low molecular

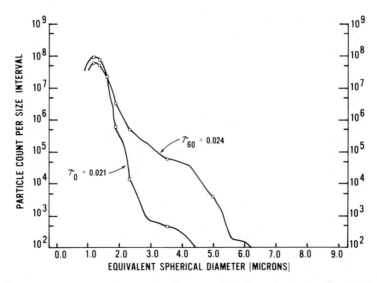

Figure 6. Differential size distribution—less than optimum dose of low molecular weight PEI produces small aggregates and increased suspension turbidity (mol wt—1200; polymer dose—0.50 mg/L; mixing intensity—20/sec; initial particle count—2.915E 07/ML; reaction time—(□) 0 min; (○) 60 min)

weight PEI. The differential size distribution indicates that a significant change is occurring over the course of the 60 min of flocculation, with substantial growth of aggregates in the 3 to 5 μm range. At the same time, the turbidity is increasing from an initial value of 0.021 to a turbidity at the end of 60 min of 0.024. In this case the increase in the scattering coefficient for the larger aggregates has more than compensated for the loss in the number of singlets.

Figures 7 and 8 illustrate that the optimum dose (5.0 mg/L) of both the low (1200) molecular weight and medium (1800) molecular weight PEI produces substantial changes in the particle size distribution, but the turbidity remains a conservative parameter in these systems. Both of these polymer doses produce substantial numbers of aggregates in the 3 to 8 μm range, but the increase in the scattering coefficient of these aggregates is balanced by the loss in numbers of singlets. Consequently, the suspension turbidity is insensitive to the changes in the aggregate diameter that are occurring in this flocculating system. In this portion of the anomalous diffraction regime, the turbidity is proportional to $n_i r_i^3$, a conservative parameter in a coagulating system.

Figures 9 and 10 illustrate the effectiveness of the optimum dose of two high molecular weight PEI species in producing flocculation of the singlet particles and in reducing the suspension turbidity. In these cases

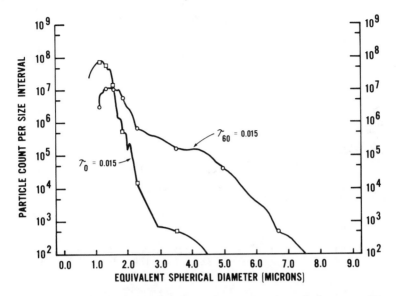

Figure 7.   Differential size distribution—suspension turbidity may be a conservative parameter in spite of significant changes in particle size distribution (mol wt—1200; polymer dose—5.00 mg/L; mixing intensity—20/sec; initial particle count—2.535E 07/ML; reaction time—(□) 0 min; (○) 60 min)

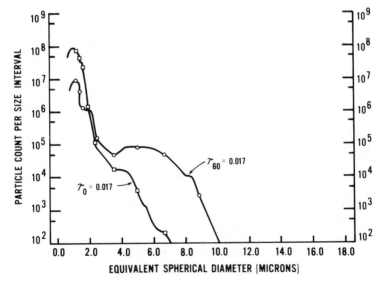

Figure 8.   Differential size distribution—optimum dose of medium molecular weight (1800) PEI produces larger aggregates but no change in suspension turbidity (mol wt—1800; polymer dose—5.00 mg/L; mixing intensity—20/sec; initial particle count—2.350E 07/ML; reaction time—(□) 0 min; (○) 60 min)

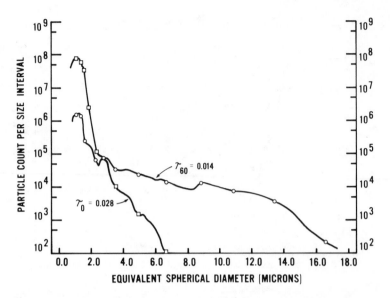

Figure 9.   Differential size distribution—optimum dose of high molecular weight (35,000) PEI produces large changes in both aggregate size distribution and suspension turbidity (mol wt—35000; polymer dose— 0.50 mg/L; mixing intensity—20/sec; initial particle count—3.011E 07/ ML; reaction time—(□) 0 min; (○) 60 min)

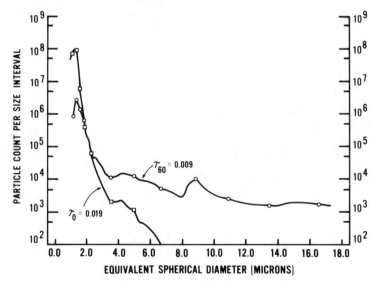

Figure 10.   Differential size distribution—growth of large aggregates reduces suspension turbidity by 50%, following flocculation with high molecular weight (60,000) PEI (mol wt—60000; polymer dose—0.50 mg/L; mixing intensity—20/sec; initial particle count—2.813E 07/ML; reaction time—(□) 0 min; (○) 60 min)

the PEI has reduced the number of singlet particles by approximately two orders of magnitude and has greatly increased the number of aggregates in the range 4 to 16 μm. These aggregates are entering the diffraction portion of the anomalous diffraction regime, and their scattering efficiencies approach 2.0. The turbidity of this suspension will always decrease with increasing radius of the aggregate as shown by Equation 9.

In each flocculation experiment, the initial turbidity and particle size distribution of the untreated *E. coli* suspension were measured. From this information the optical constant could be determined, that is:

$$\tau = A \sum_i n_i \pi r_i^2 (K_{\text{sca}})_i \tag{11}$$

where A is the optical constant reflecting geometry of the measuring system and K_i equals $2 - (4 \sin \rho)/\rho + 4(1 - \cos \rho)/\rho^2$.

Once the optical constant was known, the turbidity of the flocculated suspension calculated from Equation 11, using the known particle size distribution, could be compared with the experimentally measured turbidity. Correlations were made between particle size distributions and turbidity readings as the PEI molecular weight and dose were varied. The velocity gradient in the stirrer–reactor was held constant at $G = 20$ sec^{-1}. Other experiments indicate that the influence of varying the velocity gradient in the range $G = 20$ to 60 sec^{-1} on either turbidity or particle size distribution was minor.

The calculated turbidities for a wide variety of particle size distributions with aggregates ranging from 1 to 18 μm in diameter were within $\pm 20\%$ of the experimentally measured turbidities, that is:

$$\tau_{\text{pred}} = (1 \pm 0.2) \tau_{\text{act}} \tag{12}$$

where τ_{pred} is the turbidity predicted from particle size distribution and τ_{act} the turbidity measured experimentally.

When τ_{pred} was plotted against τ_{act}, a linear relationship was found between the two variables. For flocculation with the low molecular weight species, the calculated turbidity was consistently larger than the experimental value. This was attributed to the overestimation of the scattering cross section caused by the coalescence assumption and to interparticle interference of scattered light resulting in lower actual scattering intensities for aggregates (*19*). For flocculation with the high molecular weight species, calculated turbidity was consistently smaller than the experimental value. This was attributed to the difficulty in accurately sizing very large aggregates with the electronic particle counter (*14*). The correlation of experimental turbidities with those calculated from the anomalous diffraction–coalesced sphere model was felt to be good in light of the assumption made in deriving the model and the experimental difficulties in sizing the aggregates. The results were certainly acceptable for proceeding with the second phase of this research, that is, determining the turbidity distribution from the recorded particle size distribution established by the flocculant molecular weight and dose. This turbidity distribution is especially valuable to a sanitary engineer in determining the particle size range that must be removed in a water or wastewater filter to achieve desired turbidity reductions in the treatment process.

Figures 11, 12, and 13 are the differential turbidity distributions at optimum doses of PEI with molecular weight 1800, 35,000, and 60,000, respectively. These figures were prepared from the corresponding particle size distributions shown in Figures 8, 9, and 10, using the coalesced-sphere–Jobst approximations model. The total suspension turbidity before coagulation and flocculation is indicated on each figure by τ_0. The suspension turbidity after coagulation and flocculation for 60 min at a velocity gradient of 20 sec^{-1} is indicated by τ_{60}. These values were recorded experimentally, whereas the curves were prepared from the particle size distribution. Before coagulation and flocculation, more than 90% of the suspension turbidity can be found in particles smaller than 2 μm in diameter whereas, following coagulation and 60 min of floccula-

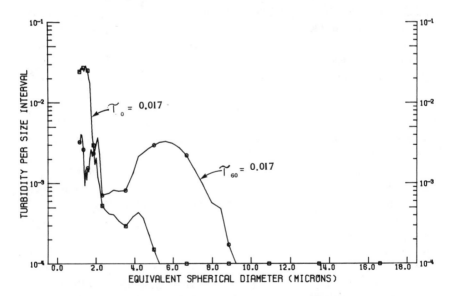

Figure 11.   Differential turbidity distribution—initially, 90% of turbidity is produced by particles smaller than 2.0 μm; following flocculation, 80% of turbidity is produced by aggregates larger than 2.0 μm (mol wt—1800; polymer dose—5.00 mg/L; velocity gradient—20/sec; initial particle count—2.350E 07/ML; reaction time—(□) 0 min; (○) 60 min)

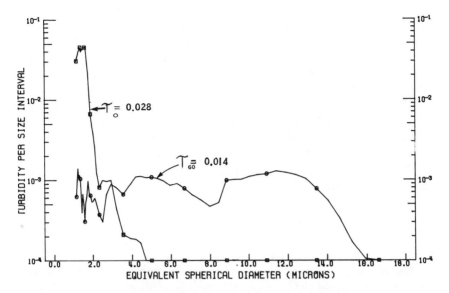

Figure 12.   Differential turbidity distribution—turbidity is reduced by 50% and shifted from small (< 2.0 μm) to large (> 2.0 μm) aggregates by optimum dose of high molecular weight PEI (mol wt—35000; polymer dose—0.50 mg/L; velocity gradient—20/sec; initial particle count—3.011E 07/ML; reaction time—(□) 0 min; (○) 60 min)

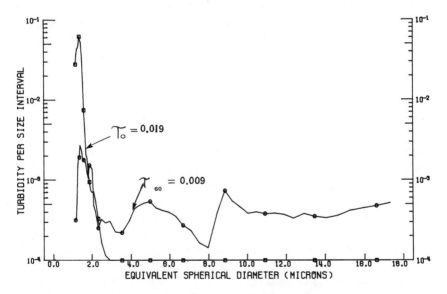

Figure 13.   Differential turbidity distribution—turbidity distribution indicates size range of aggregates that must be removed to achieve desired reduction in suspension turbidity (mol wt—60000; polymer dose—0.50 mg/L; velocity gradient—20/sec; initial particle count—2.813E 07/ML; reaction time—(□) 0 min; (○) 60 min)

tion, more than 80% of the residual turbidity is found in particles greater than 2 μm in diameter. In addition, the absolute turbidity following flocculation with the high molecular weight species is roughly 50% less than the absolute turbidity before flocculation. The differential turbidity distributions indicate the shift that occurs in the location of the turbidity-producing aggregates and can be used to determine the size range that must be removed by filters to meet specified reductions in suspension turbidity. If the design engineer knows the relationship between the media size and the removal efficiency of aggregates of a specified size by that media, then he can determine the media size required to achieve the desired turbidity reduction.

Summary

In these experiments an *E. coli* culture was grown to stationary phase, at which time the number and size distribution of the bacterial cells were determined via electronic particle counting. The turbidity of the bacteria cultures was determined with a Brice–Phoenix spectrophotometer. The bacterial culture was then aggregated with a strong cationic polymer PEI of varying molecular weight. The coagulated *E. coli* suspension was then flocculated at a constant velocity gradient $G = 20$ sec^{-1} for 60 min. After this stirring, a sample was removed and the resulting particle size distribution and turbidity of the suspension were recorded. The effect of varying the PEI molecular weight in the experiments was to create a broad particle size distribution ranging from 10 to 4,000 singlets clustered in aggregates.

Using the aggregate distribution determined by particle counting measurements, the suspension turbidity could be predicted using approximations for the Mie scattering efficiencies determined by Jobst (*16*). The Jobst approximations provide simple analytical solutions to the Mie scattering equations in the case of aggregates of the dimension and composition of bacteria, since α is quite large and $\alpha(m-1)$ is small. The exact Jobst expression used for the scattering efficiency depends on the size regime of the singlet bacteria cells and of the aggregates resulting from the flocculation. The turbidity of the flocculating suspension was found to increase, remain constant, or decrease as the flocculation proceeded, depending on the particle size regime involved. Because the electronic particle counter provides equivalent spherical diameters for the aggregated singlet cells, the aggregates were assumed to coalesce into spheres. Then the turbidity of the flocculated suspension could be represented by $\tau = \Sigma_i n_i \pi r_i^2 K_i$ where n_i is the number of aggregates of radius r_i, r_i the aggregate radius, and K_i the Jobst approximation for the Mie scattering efficiency. While this coalescing sphere approach is valid strictly for emulsions and liquid aerosols, it has been used with aggregated systems of rigid singlets because of its simplicity and good correlation with experimental results (*17, 18*).

In the size range of these experiments, from roughly 1 to 20 μm, substantial changes in the particle size distribution must occur before they are reflected in the turbidity. Figure 5 presents the percent reduction in the number of particles as a result of flocculation with five different molecular weight species of PEI at varying doses. As shown, polymers of molecular weight greater than 2×10^3 achieve more than a 90% reduction in the number of primary particles, with the exception of a region of restabilization. Figure 4 presents a corresponding domain of percent reduction in suspension turbidity after 60 min of flocculation at $G = 20$ $\sec^{-1}$. Obviously no simple, direct relationship exists between the number of aggregates in suspension and their turbidity. Turbidity measurements alone are inadequate to measure the effectiveness of coagulation–flocculation processes. Large and significant changes may occur in the particle size distribution without producing corresponding changes in the suspension turbidity.

A model using the coalesced-sphere approach, with Jobst approximations to the Mie scattering efficiencies, was developed that enabled the prediction of suspension turbidity from the aggregate size distribution. Using this model, turbidities were calculated within $\pm 20\%$ of the turbidities measured experimentally. The suspension turbidity can increase, remain constant, or decrease during the coagulation–flocculation process, depending on the number, size regime, and refractive indices of the constituent aggregates. These changes can be predicted by the coalesced-sphere model.

Although the coalesced-sphere model suffers from obvious theoretical and practical shortcomings, it is adequate to predict the turbidity fractions generated in each size interval. Knowledge of these fractions establishes the size range of aggregates that must be removed to achieve a specified reduction in suspension turbidity. Differential turbidity distributions indicate that before flocculation, more than 90% of suspension turbidity occurs in particles smaller than 2 μm in diameter, whereas after flocculation, more than 80% of the residual turbidity occurs in particles greater than 2 μm in diameter. This shift in the size range of turbidity-producing particles in the suspension can be used to establish the media sizes required to remove effectively the turbidity-producing aggregates in subsequent granular-bed filtration.

Literature Cited

1. Morgan, J. J.; Birkner, F. B. "Flocculation Behavior of Dilute Clay-Polymer Systems," Progress Report WP 00942-02, U.S. Public Health Service, Sept. 1966.

2. Birkner, F. B.; Morgan, J. J. "Polymer Flocculation Kinetics of Dilute Colloidal Suspensions," *J. Am. Water Works Assoc.* **1968**, *60*, 175–191.
3. Ham, R. K.; Christman, R. K. "Agglomerate Size Changes in Coagulation," *Proc. Am. Soc. Civ. Eng.* **1969**, *95*, 481–502.
4. TeKippe, R. J.; Ham, R. K. "Coagulation Testing: A Comparison of Techniques—Parts I and II," *J. Am. Water Works Assoc.* **1970**, *62*, 594–602, 620–628.
5. Camp, T. R. "Floc Volume Concentration," *J. Am. Water Works Assoc.* **1968**, *60*, 656–673.
6. Camp, T. R. "Discussion: Agglomerate Size Changes in Coagulation," *J. Sanit. Eng. Div., Am. Soc. Civ. Eng.* **1969**, *95*, 1210–1214.
7. Hannah, S. A.; Cohen, J. M.; Robeck, G. G. "Measurement of Floc Strength by Particle Counting," *J. Am. Water Works Assoc.* **1967**, *59*, 843–858.
8. Treweek, G. P. "The Flocculation of *E. Coli* with Polyethyleneimine," Ph.D. Thesis, California Institute of Technology, 1975.
9. Benoit, H.; Ullman, R.; DeVries, A.; Wippler, C. "Diffusion De La Lumiere Par Des Agregats En Suspension Dans Un Liquid," *J. Chim. Phys.* **1962**, *59*, 889–895.
10. Lips, A.; Smart, C.; Willis, E. "Light Scattering Studies on a Coagulating Polystyrene Latex," *Trans. Faraday Soc.* **1973**, *67*, 1226–1234.
11. Lips, A.; Willis, E. "Low Angle Light Scattering Technique for the Study of Coagulation," *J. Chem. Soc., Faraday Trans. 1*, **1973**, *69*, 1226–1234.
12. Debye, P. "Der Lichtdruck auf Kugeln von bebliebigem Material," *Ann. Phys. (Leipzig)* **1909**, *30*, 57–136.
13. Mie, G. "Beitrage zur Optik truber Medien," *Ann. Phys. (Leipzig)* **1908**, *25*, 377–445.
14. Treweek, G. P.; Morgan, J. J. "Size Distributions of Flocculated Particles: Application of Electronic Particle Counters," *Environ. Sci. Technol.* **1977**, *11*, 707–714.
15. Kerker, M. "The Scattering of Light and Other Electromagnetic Radiation"; Academic: New York, 1969.
16. Jobst, G. "Diffuse Straslung Dielektrisches Kugeln im Greszfalle, dasz Kugelmaterial and Umgebendes Medium Fast Gleich Brechungsindices Haben," *Ann. Phys. (Leipzig)* **1925**, *78*, 157–166.
17. Lips, A.; Levine, S. "Light Scattering by Two Spherical Rayleigh Particles Over All Orientations," *J. Colloid Interface Sci.* **1970**, *33*, 455–463.
18. Ottewill, R. H.; Shaw, J. N. "Stability of Monodisperse Polystyrene Latex Dispersions of Various Sizes," *Faraday Discuss. Chem. Soc.* **1966**, *42*, 154–163.
19. Lichtenbelt, J. W.; Ras, H. J.; Wiersema, P. H. "Turbidity of Coagulating Lyophobic Sols," *J. Colloid Interface Sci.* **1974**, *46*, 522–527.
20. Killenberger, E.; Sechaud, L. "Electron Microscope Studies of Phage Multiplication. II. Production of Phase-Related Structures During Multiplication of Phages T2 and T4," *Virology* **1957**, *3*, 256–274.
21. Gomori, G. "Buffers in the Range of pH 6.5 to 9.6," *Society Experimental Biology and Medicine* **1946**, *62*, 33–34.
22. *PEI Polymers*, DOW Chemical Company: Midland, Michigan, 1974.
23. *Instruction Manual*, Phoenix Precision Instrument Company: Philadelphia, 1963.
24. Bateman, J. B. "Osmotic Responses and Light Scattering of Bacteria," *J. Colloid Interface Sci.* **1968**, *27*, 458–474.
25. Neis, U.; Eppler, B.; Hahn, H. "Quantitative Analysis of Coagulation Processes in Aqueous Systems: An Application of the Coulter Counter Technique," presented at IWRA meeting, Chicago, Ill., June 1974.
26. Mattern, C. F.; Brackett, F. S.; Olson, B. J. "Determination of Number and Size of Particles by Electrical Grating: Blood Cells," *J. Appl. Physiol.* **1957**, *10*, 56–70.

27. Shie, P.; Rehberg, R. "Refractive Index of *E. Coli* in Solutions of Sucrose," *Biophys. J.* **1970**, *10*, 2112.
28. McCluney, W. R. "Radiometry of Water Turbidity Measurements," *J. Water Pollut. Control Fed.* **1975**, *47*, 252.
29. Pickering, R. J. "Measurement of 'Turbidity' and Related Characteristics of Natural Waters," *U.S. Geological Survey, Open-File Report* 76-153, January, 1976.
30. Koch, A. L. "Some Calculations on the Turbidity of Mitochondria and Bacteria," *Biochim. Biophys. Acta* **1961**, *51*, 429–441.

RECEIVED October 10, 1978.

Integral Water Treatment Plant Design

From Particle Size to Plant Performance

DESMOND F. LAWLER[1], CHARLES R. O'MELIA[2],
and JOHN E. TOBIASON[3]

Department of Environmental Sciences and Engineering,
University of North Carolina, Chapel Hill, NC 27514

Particle concentration and size distribution in raw water have extensive and complex effects on the performance of individual treatment units (flocculator, sedimentation tank, and filter) and on the overall performance of water treatment plants. Mathematical models of each treatment unit were developed to evaluate the effects of various raw water characteristics and design parameters on plant performance. The flocculation and sedimentation models allow wide particle size distributions to be considered. The filtration model is restricted to homogeneous suspensions but does permit evaluation of filter ripening. The flocculation model is formulated to include simultaneous flocculation by Brownian diffusion and fluid shear, and the sedimentation model is constructed to consider simultaneous contacts by Brownian diffusion and differential settling. The predictions of the model are consistent with results in water treatment practice.

Most pollutants of concern to human health and environmental quality are solid particles or are associated with solid particles. Examples include toxic metals and synthetic organics adsorbed on clays and detri-

[1] Current address: Department of Civil Engineering, University of Texas at Austin, Austin, TX 78712.
[2] Current address: Department of Geography and Environmental Engineering, The Johns Hopkins University, Baltimore, MD 21218.
[3] Current address: Wright–Pierce Architects/Engineers, 25 Vaughan Mall, Portsmouth, NH 03801.

0-8412-0499-3/80/33-189-353$09.00/0

tus, asbestos fibers, and pathogenic organisms. Such pollutants are transported in the natural environment and in water and wastewater treatment plants by processes that depend on the size, density, and concentration of the solid particles. Some imaginative investigations (1, 2) have been reported recently in which the effects of such physical factors on the transport of particulates in natural waters are examined.

The treatment of wastes and water supplies primarily involves the removal of particulates and, therefore, is accomplished by solid–liquid separation processes. However, present design procedures for such treatment systems do not utilize or even recognize the importance of the physical properties of particulates in solid–liquid separation. In fact, until very recently, efforts at measuring particle size in wastewaters and raw water supplies have been very limited. Among early efforts, the investigations at Rutgers University (3, 4, 5, 6) are especially notable. Very recent renewed interests in particle size determinations have used Coulter, Hiac, and Zeiss Videomat particle counters (7, 8, 9).

This chapter is written with three objectives in mind. First, the importance of the size and concentration of the particles to be treated in determining the effectiveness of some solid–liquid separation processes is evaluated. Second, past theories are used to examine how particle sizes and concentrations are altered by these treatments. Third, interrelationships among the individual unit processes that comprise a complete treatment system are investigated to provide a base for an integral treatment plant design. These aims are undertaken using a typical water treatment system as employed in practice to remove turbidity from surface water supplies. Before addressing these objectives, it is useful to review some mathematical expressions of particle size distributions, and to identify some important properties of these functions.

Particle Size Distributions

A suspension containing many particles of various sizes can be said to have a continuous size distribution. In fact, this particle size distribution may be the most important physical characteristic of the system. A cumulative size distribution is illustrated in Figure 1; cumulative number concentration (N) is plotted as a function of particle size expressed in volumetric units (v). Here N (number/cm^3) denotes the total concentration of particles with a size equal to or less than v (μm^3); the total concentration of all particles is given by N_∞. The slope of this curve, $\Delta N/\Delta v$ or dN/dv, is called a particle size distribution function and is represented as $n(v)$. In this case $n(v)$ has units of number/cm$^3\mu$m^3.

Particle volume is one of three common measures of particle size. Surface area(s) and diameter (d_p) are also used, so that three particle size distribution functions can be defined:

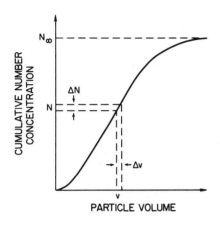

Figure 1. *Cumulative particle size distribution*

$$\frac{\Delta N}{\Delta v} = \frac{dN}{dv} = n(v) \qquad (\text{number/cm}^3 \mu\text{m}^3) \qquad (1)$$

$$\frac{\Delta N}{\Delta s} = \frac{dN}{ds} = n(s) \qquad (\text{number/cm}^3 \mu\text{m}^2) \qquad (2)$$

$$\frac{\Delta N}{\Delta d_p} = \frac{dN}{d(d_p)} = n(d_p) \qquad (\text{number/cm}^3 \mu\text{m}) \qquad (3)$$

These functions can be measured and used in both conceptual and empirical studies of coagulation and other particle transport processes. They are particularly useful in expressing data obtained with particle counters that directly determine the number of particles in each of many size intervals.

The volume concentration of all particles in the interval between Size 1 and Size 2 may be written as

$$V_{1\text{-}2} = \int_{1}^{2} dV$$

Dividing and multiplying the right hand side by $d(\log d_p)$ yields

$$V_{1\text{-}2} = \int_{1}^{2} \frac{dV}{d(\log d_p)} \, d(\log d_p)$$

A plot of $dV/d(\log d_p)$ vs. $\log d_p$ is illustrated in Figure 2. The total volume concentration provided by particles in the size interval from d_1 to d_2 is equal to the integrated (shaded) area in Figure 2. In preparing such plots from field measurements, it is frequently assumed that the particles are spherical. If this assumption is not true, the area is proportional (not equal) to the volume concentration.

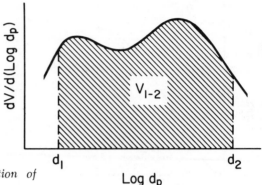

Figure 2. *Volume distribution of particles*

The definition of the particle size distribution function means that

$$dV = v \cdot n(v) \cdot dv$$

Dividing by $d(\log d_p)$, one obtains

$$\frac{dV}{d(\log d_p)} = \frac{v \cdot n(v) \cdot dv}{d(\log d_p)} = \frac{v \cdot n(v) \cdot dv}{d(d_p)} \cdot \frac{d(d_p)}{d(\log d_p)} \qquad (4)$$

Assuming spherical particles, $v = \pi d_p^3/6$ and $dv = (\pi d_p^2/2) \cdot d(d_p)$. Substituting these into Equation 4 and also noting that $d(\log d_p) = d(\ln d_p)/2.3 = d(d_p)/2.3d_p$, one obtains:

$$\frac{dV}{d(\log d_p)} = \frac{2.3\pi^2}{12} \cdot d_p^6 \cdot n(v) \qquad (5)$$

Equation 5 contains units of diameter (d_p) and volume $(n(v))$ and is often rearranged to include diameter as the measure of particle size. Recall that $n(v) = dN/dv$ and, for spherical particles, $dv = (\pi d_p^2/2) \cdot d(d_p)$. Making these substitutions in Equation 5 yields:

$$\frac{dV}{d(\log d_p)} = \frac{2.3\pi}{6} d_p^4 n(d_p) \qquad (6)$$

Similar relationships can be derived for the surface area (S) and number distributions. The results are as follows:

$$\frac{dS}{d(\log d_p)} = 2.3\pi d_p^3 n(d_p) \qquad (7)$$

$$\frac{dN}{d(\log d_p)} = 2.3 d_p n(d_p) \qquad (8)$$

For a plot of $dS/d(\log d_p)$ vs. $\log d_p$ that is similar to Figure 2, the integrated area under the resulting curve from d_1 to d_2 would represent the total concentration of surface area in a suspension provided by particles in the size interval from d_1 to d_2. A plot of $dN/d(\log d_p)$ vs. $\log d_p$ would provide similar information about the total number concentration.

The size distribution of atmospheric aerosols and aquatic suspensions is often found to follow a power law of the form

$$n(d_p) = A \cdot d_p^{-\beta} \tag{9}$$

in which A ($\mu m^{\beta-1}/cm^3$) is a coefficient related to the total concentration of particulate matter in the system. As will be noted later, the exponent β has been determined experimentally and also has been shown on theoretical grounds to result from the interactions of various physical processes such as coagulation and settling.

Measurements indicate that for particles above about 1 μm in size (that is, those detectable by present electronic or optical measurements) values of β typically range from 2 to 5. For example, Lerman et al. (2) report measurements of size distributions at four locations in the North Atlantic. Fifty-three size distributions derived from samples taken at depths ranging from 30 to 5100 m yielded a mean value of $\beta = 4.01 \pm 0.28$. Hunt (10) analyzed data obtained by Faisst (7) for two digested sewage sludges and determined values of β from 2.2 to 4.7.

Friedlander (11) has examined the effects of flocculation by Brownian diffusion and removal by sedimentation on the shape of the particle size distribution function as expressed by Equation 9. The examination is conceptual; the predictions are consistent with some observations of atmospheric aerosols. For small particles, where flocculation by Brownian diffusion is predominant, β is predicted to be 2.5. For larger particles, where removal by settling occurs, β is predicted to be 4.75. Hunt (10) has extended this analysis to include flocculation by fluid shear (velocity gradients) and by differential settling. For these processes, β is predicted to be 4 for flocculation by fluid shear and 4.5 when flocculation by differential settling predominates. These theoretical predictions are consistent with the range of values for β observed in aquatic systems.

The power-law size distribution (Equation 9) has some interesting characteristics. Substituting Equation 9 into Equations 6–8 yields

$$\frac{dV}{d(\log d_p)} = 2.3\frac{\pi}{6} A \cdot d_p^{(4-\beta)} \tag{10}$$

$$\frac{dS}{d(\log d_p)} = 2.3\pi A \cdot d_p^{(3-\beta)} \tag{11}$$

$$\frac{dN}{d(\log d_p)} = 2.3 A \cdot d_p^{(1-\beta)} \tag{12}$$

Solutions to Equations 10–12 for $\beta = 1, 3$, and 4 are presented in Figures 3A, B, and C. A suspension of particles ranging in size from 0.3 to 30 μm, with a density of 2.65 g/cm^3 and a mass concentration of 132 mg/L, has been assumed. This is equivalent to a volume concentration of 50 ppm. The resulting values of the coefficient A are 2.08×10^7 μm^3/cm^3, 3.22×10^6 μm^2/cm^3, and 1.09×10^4/cm^3 for $\beta = 4, 3$, and 1, respectively.

A particle size distribution that follows an empirical power law with $\beta = 1$ has an equal number of particles in each logarithmic size interval (Figure 3A). The surface area and volume concentration predominate in the larger sizes. Similarly, for $\beta = 3$, the concentration of surface area is uniformly distributed in each logarithmic size interval (Figure 3B). Here the volume or mass of solids is predominant in the larger sizes while the number concentration is dominant in the smaller sizes. Finally, for $\beta = 4$, the volume of solids is distributed equally in each logarithmic size interval, while the surface area and number concentrations are primarily in the smaller sizes (Figure 3C). This analysis is adapted from Kavanaugh et al. (12).

Some pollutants can be characterized in terms of mass or volume concentrations (examples include oil, suspended solids, and certain precipitates such as Cr(OH)$_3$). Other pollutants are concentrated at surfaces (such as DDT adsorbed on detritus and trace metals adsorbed on clays); for materials such as these, the surface concentration of the particulate phase is of interest. Still other pollutants (for example, pathogenic organisms) are best considered in terms of their number concentration.

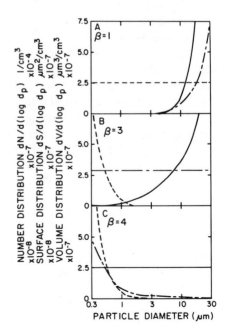

Figure 3. Effects of particle size distribution function on number (- - -), surface (— - - —), and volume (——) distributions (size interval 0.3 to 30 μm, particle concentration = 132 mg/L, and particle density = 2.65 g/cm^3)

Present routine methods of measuring particulates in water are based on their mass (suspended solids) or their light scattering properties (turbidity). Neither method provides enough useful information to characterize the transport of the particles in solid–liquid separation, nor the removal of the pollutants they carry. Measurements of particle size distributions, while more difficult, should be more rewarding. For example, if a particle size distribution exhibits a slope of 4 in the power-law function (Equation 9), then the area and number concentrations of the solid particles and the pollutants they carry can predominate in the submicron size range (Figure 3C). While these particles escape most routine measurements, those size distribution measurements that are available (for example, Lerman et al (2)) indicate that a value of $\beta = 4$ is a common occurrence.

The Water Treatment Plant

A schematic of a water treatment plant using coagulation, settling, and packed-bed filtration to remove particulates is presented in Figure 4. The raw water source containing particulates is: (1) mixed with chemicals in a rapid mix tank to provide chemical destabilization; (2) stirred slowly in a flocculation tank to provide contacts among the destabilized particles so that aggregation occurs; (3) settled in a sedimentation tank to remove particles by gravity forces; and (4) filtered by a bed of granular media to provide additional removal of solid particles.

The parameters used to characterize the raw water supply and the treatment system are presented in Table I. In this research, a "standard" water supply and treatment system was selected to represent a typical case. The performances of each unit process and of the overall treatment system were evaluated by methods described later. The effects of changes in the water source and the treatment system were determined by similar techniques and compared with the results of the standard case. The models used for the water source and the unit processes follow.

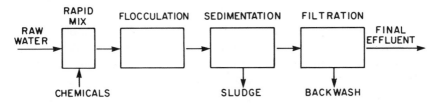

Figure 4.   Schematic of solid–liquid separation processes in typical water treatment plant

Table I.   Water Source and Treatment System[a]

System Component	Parameter	Standard Case	Variations Studied
Raw water	Volume concentration	50 ppm	5, 153 ppm
	Mass concentration	132 mg/L	13.2, 419 mg/L
	Particle density	2.65 g/cm^3	none
	Size range	0.3 to 30 μm	none
	β	4	3, and homo-geneous suspension
	Temperature	20°C	none
Rapid mix tank	Collision efficiency factor	1.0	none
	Coagulant mass	0	none
Flocculation tank	Flow type	plug flow	none
	Detention time	1 hr	none
	Velocity gradient	10 sec^{-1}	0, 50 sec^{-1}
Settling tank	Flow type	plug flow	none
	Detention time	2 hr	none
	Tank depth	4 m	2 m
	Overflow rate	2 m/hr	1 m/hr
Filter	Filtration rate	5 m/hr	none
	Media size	0.5 mm	none
	Media depth	60 cm	none
	Clean bed porosity	0.4	none

[a] The meaning and significance of each parameter are described in the text.

The Raw Water Supply. The concentration of solid material was assumed to be 50 ppm by volume for the standard case. Particle density was assumed to be 2.65, representative of inorganic materials; the corresponding mass concentration is about 132 mg/L. A particle size range from 0.3 to 30 μm was selected. There are few data to provide a base for this estimate. This range reflects a consideration that large particles will be removed effectively from the water source by natural processes, and that submicron particles can be present in surface water supplies even if not detected quantitatively by routine analysis for suspended solids. The particle size distribution throughout this size range was assumed to be described by the power-law distribution function (Equation 9) with $\beta = 4$. This is consistent with the available data.

The effects of particle concentration on the performance of a water treatment plant were evaluated using a concentration lower than the standard case by one log unit (13.2 mg/L) and also a concentration higher than the standard case by one-half log unit (419 mg/L).

The effects of particle size distribution have been evaluated by assuming $\beta = 3$ and comparing the results obtained with the standard case in which $\beta = 4$. In addition, the treatment of a homogeneous suspension has been investigated. Here a suspension having a volume concentration of 50 ppm but comprised of particles of uniform size is assumed in the water source. This corresponds to the uniform suspensions assumed in conventional applications of the Smoluchowski (13) flocculation equations.

The Rapid Mix Tank. The coagulation or aggregation of particles can be considered to involve two separate or distinct steps. In the first step, particles are destabilized or made "sticky" by the addition of chemicals. This is accomplished rapidly in the flash or rapid mix tank. In the second step, accomplished in the flocculation tank, the destabilized particles are brought into contact by physical forces so that aggregates are produced.

The effectiveness of the destabilization process cannot be predicted quantitatively for natural suspensions using present theories. Here we have assumed that destabilization is complete; that is, all contacts that subsequently occur are able to produce aggregates. This is sometimes described in terms of α, the collision efficiency factor. For all systems considered here, $\alpha = 1$.

The addition of destabilizing chemicals may or may not add significant quantities of solid particles to the water. For example, the destabilization of particles by addition of synthetic organic polymers occurs without addition of significant amounts of mass to the system. To a great extent, the destabilization of concentrated suspensions by polymeric hydroxometal complexes of Al(III) and Fe(III) can be considered to occur without significant addition of mass. However, the coagulation of dilute suspensions by metal salts requires the precipitation of metal hydroxide solids to enhance flocculation kinetics (14). For such cases the addition of coagulant significantly increases the concentration of solids in the system. In all of the systems considered, the addition of destabilizing chemicals was assumed to have no effect on the mass of solids in the water.

The Flocculation Tank. Smoluchowski (13) derived equations to describe the temporal change in the concentration of particles of a particular size k as they combine with other particles in the flocculation process. The formulation used here is based on the work of Friedlander (15). In discrete form, the basic equation is written as

$$\frac{dn_k}{dt} = \frac{1}{2} \sum_{i+j=k} \beta(i,j) \cdot n_j \cdot n_i - n_k \sum_{i=1}^{\infty} \beta(i,k) \cdot n_i \qquad (13)$$

Here n_i, n_j, and n_k are the number concentrations of particles with sizes i, j, and k respectively, t is time, and $\beta(i,j)$ is a collision frequency function that depends on the mode of interparticle contact.

The first term on the right-hand side of Equation 13 expresses the rate of formation of particles of size k (or volume v_k) from smaller particles having a total volume of v_k. The condition $i + j = k$ under the summation denotes the condition that $v_i + v_j = v_k$. The factor $1/2$ is needed since collisions are counted twice in the summation. The second term on the right-hand side of this equation describes the loss of particles of size k by growth to form larger aggregates; this occurs when a size k particle collides with (and attaches to) a particle of any size i.

For the flocculation tank, the significant transport mechanisms leading to interparticle collisions are assumed to be Brownian diffusion and fluid shear. Expressions for the collision frequency functions for these mechanisms were derived by Smoluchowski and are as follows:

Brownian diffusion:

$$\beta(i,j) = \frac{2\bar{k}T}{3\mu}\left[\left(\frac{1}{v_i}\right)^{1/3} + \left(\frac{1}{v_j}\right)^{1/3}\right][v_i^{1/3} + v_j^{1/3}] \tag{14}$$

Fluid shear:

$$\beta(i,j) = \frac{1}{\pi}[v_i^{1/3} + v_j^{1/3}]^3 \cdot G \tag{15}$$

Here $\bar{k}$ is Boltzmann's constant, T is the absolute temperature, μ is the fluid viscosity, v_i and v_j are the volumes of particles with sizes i and j, respectively, and G is the velocity gradient of the fluid. Equations 14 and 15 often are written in terms of the diameters of the particles rather than in terms of their volumes. However, when two particles collide, it is useful to consider that their total volume is conserved (coalesced-sphere assumption); therefore, the equations for flocculation kinetics are expressed in terms of particle volumes throughout this work.

For each particle size considered, an equation with the form of Equation 13 is needed. For the flocculation tank, 50 particle sizes ranging from 0.3 to 30 μm were considered. This span of two orders of magnitude in particle diameter corresponds to six orders of magnitude in particle volume. If these six orders of magnitude were divided into 50 equal arithmetic units, all particles in the smallest four orders of magnitude would reside in the first size unit. Hence, the size distribution was divided into equal logarithmic units so that each order of magnitude was described with the same detail.

Assumptions made in modeling the flocculation tank include: (1) ideal or plug flow exists throughout the tank; (2) the velocity gradient is identical everywhere in the tank; (3) flocculation occurs simultaneously by Brownian diffusion and fluid shear, and the collision frequency functions are simply additive; (4) the particles are distributed uniformly throughout the fluid as it enters the tank; and (5) aggregate breakup and viscous interactions during interparticle contacts are neglected. Assumptions 1 and 2 can be inconsistent. In this application it is assumed that plug flow exists at the large scale of the tank dimensions while a velocity gradient exists at the small scale of the particle dimensions.

A hydraulic detention time of 1 hr was assumed for all cases examined here; this is representative of actual treatment practice. The velocity gradient was varied, with values of 0, 10, and 50 sec^{-1} selected for examination. Ten sec^{-1} was used for the standard case; the range from 10 to 50 sec^{-1} represents most cases in practice. A zero velocity gradient does not occur in real systems, but this assumption allows only the effects of Brownian diffusion on flocculation to be determined. Some treatment plants have shorter flocculation times than 1 hr, and higher velocity gradients than 10 sec^{-1}; the reference or standard case chosen here has a Gt product of 36,000 which is consistent with water treatment practice.

Fifty equations (one for each particle size interval) in the form of Equation 13 were integrated numerically for 1 hr using an algorithm developed by Gear (16) to determine the particle size distribution in the effluent from the flocculation tank. This algorithm determines the step size to be taken based on the rate at which a function is changing at a particular point. During periods of very rapid change, small step sizes are taken and during periods of slow change, large step sizes are taken. The efficiency of this algorithm allowed the simulation of 1 hr of flocculation with approximately 1 min of computer time. The programmer must supply two subroutines to use this system, one to evaluate numerically each of the differential equations and one to evaluate the partial derivatives of each of the equations with respect to each of the dependent variables.

A flow diagram of the computer program is presented in Figure 5. Solution was accomplished using an IBM 370 computer. STIFDIF is a library subroutine based on Gear's algorithm for numerically integrating the differential equations. DIFFUN and PEDERV are the user-supplied subroutines for evaluating the differential equations and partial derivatives, respectively. BETA is a subroutine for evaluating the collision frequency factors (Equations 14 and 15) and summing them.

The use of the logarithmic division of the particle size range requires a subroutine termed FRAC. In using the discrete form of the flocculation equation (Equation 13), all particles must have one of the standard 50 sizes. Volume is conserved when two particles attach, that is, the volume of a new particle formed from two others is equal to the sum of the volumes of the two original particles. However, by using particle sizes that are equally spaced on the basis of the logarithm of their volumes, the arithmetic sum of any two standard volumes does not produce

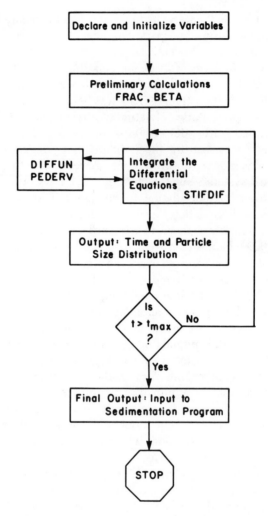

Figure 5.   Flow diagram of computer program for flocculation model

another standard volume. A method to account for this is illustrated in Figure 6. When particles of sizes i and j aggregate, one new particle of volume $(v_i + v_j)$ is formed, and this new particle has a volume between standard sizes k and $k + 1$. By assigning the fraction a/c of the new particle to standard size k and the fraction b/c to size $k + 1$, the three purposes of creating only one new particle from the collision of two, the conservation of volume, and the use of only standard sizes are all accomplished. It can be shown that the fraction to be assigned to k depends only on the logarithm of the step size of volume and on the difference between i and j. This system is valid for all combinations of particles except when the resultant particle volume is greater than the largest

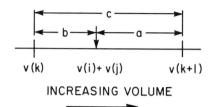

INCREASING VOLUME

Figure 6. Assignment of flocculated particles to standard sizes

standard size. In such cases it was decided to assign more than one particle to the largest (50th) size, the exact number being calculated as the ratio $(v_i + v_j)/v_{50}$. This method conserves volume but allows more than one (but less than or equal to two) particle to be created from the flocculation of two particles whose total volume is greater than the largest standard particle volume. The largest size thus serves as a sink for the flocculation process so that volume is not lost from the upper end of the size spectrum. Under conditions of extensive flocculation, where a significant portion of the original particles would grow to a size greater than the maximum size allowed in the model, this use of the largest size is obviously in error. However, it was deemed necessary to limit the storage space and time used on the computer.

The Sedimentation Tank. Mass is removed in the settling tank by gravity forces. In addition, contacts can occur between particles by such mechanisms as Brownian motion and differential settling. An unequal vertical distribution of particles occurs in the tank because of settling. The effects of heterogeneity in particle size and vertical variations in particle concentration were described by dividing the particle size range into 21 equally spaced logarithmic size units and by segmenting the tank depth into four equal units. The tank segmentation is illustrated in Figure 7. Each of the four vertical boxes is assumed to be mixed completely so that within each box the particles remain uniformly distributed. Within any time interval, particles can flocculate within each box and can settle from one box to another.

This simultaneous flocculation and settling is described as follows:

$$\frac{dn_k}{dt} = \frac{1}{2} \sum_{i+j=k} \beta(i,j) \cdot n_i \cdot n_j - n_k \sum_{i=21(m-1)+1}^{21m} \beta(i,k) \cdot n_i$$
$$+ \frac{w_{k-21}}{z} \cdot n_{k-21} - \frac{w_k}{z} \cdot n_k \qquad (16)$$

Here, the subscript k denotes particles with a particular volume and within a particular box, w is the Stokes settling velocity of the particle, z is the depth of each box, and m is the number of the box (Figure 7). The first two terms on the right-hand side of Equation 16, similar to those in Equation 13, describe the flocculation process within each box. The

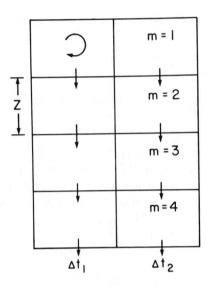

Figure 7. Partition of sedimentation tank for model calculations

summation in the second term is taken over all particle sizes in the same box as the kth particle. The third term describes the addition of particles into each of the lower three boxes by the settling of particles from the box above. The last term refers to the loss of particles from each box by settling to the box below. For the fourth or bottom box, this corresponds to the removal of particles from the system and into the sludge layer.

Flocculation by Brownian diffusion is characterized by the collision frequency function presented in Equation 14. For flocculation by differential settling, the corresponding function is:

$$\beta(i,j) = \left(\frac{6}{\pi}\right)^{1/3} \cdot \frac{g}{12\mu} \, (\rho_p - \rho_1) \, (v_i^{1/3} + v_j^{1/3})^3 \, | \, v_i^{1/3} - v_j^{1/3} \, | \quad (17)$$

Here g is the gravity acceleration and ρ_p and ρ_1 are the densities of the particles and the liquid, respectively.

With four vertical boxes and 21 particle sizes, there are 84 equations of the form of Equation 16 that must be integrated simultaneously. The format of the computer program is similar to that described previously for the flocculation model. The use of more particle sizes or vertical boxes does not significantly alter the results developed here, but does increase the computer time necessary to obtain a solution. With the 84 equations, the simulation of 2 hr of settling is accomplished with approximately 1 min of computer time.

The hydraulic detention time was assumed to be 2 hr in all cases. This is representative of treatment practice. Particle sizes ranged from 0.3 to 30 μm, and the particles were assumed to be distributed uniformly with depth at the tank inlet. Plug or ideal flow was also assumed. As stated earlier, simultaneous and additive flocculation by Brownian motion and differential settling were considered, and two cases of tank depth were examined. A depth of 4 m was used for the standard case, and a second example using a depth of 2 m was also investigated. These values correspond to overflow rates of 2 m/hr (approximately 1200 gal/day-ft²) and 1 m/hr (600 gal/day-ft²). Again these conditions are representative of actual practice.

The Packed-Bed Filter. Here the head loss development and the filter effluent quality throughout a filtration run are considered. Models for the performance of packed-bed filters are not as well developed as those presented previously for flocculation and settling. Early research by Kozeny (17) and Fair and Hatch (18) resulted in descriptions of the effects of media size, filtration rate, bed depth, bed porosity, and fluid temperature on the head loss through clean filters. More recently, the removal accomplished by clean filters has been described in terms of the physical characteristics of the suspended particles, the suspending fluid, and the filter bed by Friedlander (19), Yao et al. (20), Fitzpatrick and Spielman (21), and others with some success. However, these models for head loss and removal efficiency become incorrect as soon as a filter run begins. Quantitative concepts describing changes in head loss and filtrate quality as filtration proceeds are needed. Ives and co-workers (22, 23) have developed good mathematical characterizations of these changes, but these efforts use several empirical coefficients that are not yet related to properties of the suspension to be treated or the filter providing treatment. In the research reported here, a recent model by Ali (24) and O'Melia and Ali (25) was used.

O'Melia and Ali assume that particles removed from the flowing fluid and retained within a filter bed can act as collectors or filter media for particles applied subsequently to the bed. Stated another way, the actual filter media within a bed at any time after the start of a filtration run are comprised of the original sand, coal, or other media installed when the filter was constructed, together with those particles retained in the bed up to that time. The result is a model for filter ripening in which removal of suspended particles by packed-bed filters has an autocatalytic character. Saturation of the filter capacity and dislodging of retained particles are not included in this model. Hence, the model predicts that filter effluent quality will always improve. In addition, the model is developed for filters treating a monodisperse or homogeneous suspension. This requires that the effluent from the sedimentation tank, which is described in terms of the number concentration of each of 21 particle sizes, be described in terms of the concentration of a single size. The volume average diameter ($d_v = [(6/\pi)\Sigma nv/\Sigma n]^{1/3}$) was used.

In the filter model, the removal of suspended particles is based on the concept of the single collector efficiency (20) which is extended to account for the removal accomplished by previously retained particles. The single collector removal efficiency is considered to be the ratio of the rate at which particles are removed from suspension to the rate at which particles approach the collector. Removal involves the transport to and attachment on the collector surface.

This filter model considers that particles are transported from the flowing fluid to the filter media by Brownian diffusion, fluid flow (interception), and settling. The effects of each of these mechanisms are assumed to be additive. Happel's (26) equations for flow in a packed bed as used by Pfeffer (27) are assumed in calculating the diffusion and interception transport mechanisms.

The filtration equations describing contacts between particles and media and between particles in suspension and previously retained particles are not developed as well as the equations used to describe interparticle contacts in flocculation. This lack of knowledge about some aspects of the hydrodynamics and chemistry of the system is the principal reason for limiting filter models to monodisperse suspensions and also means that some empirical coefficients are necessary. In this research, values for these coefficients were taken directly from those reported by O'Melia and Ali (25).

Head loss development is also addressed in the model. A modification of the Kozeny equation for clean beds is used to describe the additional head loss caused by retained particles. The deposition of particles is assumed to increase the surface area exposed to and exerting drag on the flowing fluid while changes in bed porosity are assumed negligible. Not all of the retained particles are considered to contribute additional head loss because of the shadowing effects of the growth of dendrite-like particle deposits. In this study a factor based on the ratio of diffusional transport to total transport rate is generally used to describe the fraction of retained particles contributing additional exposed surface area.

The mathematical development results in a recursion equation describing the change in filtration efficiency with time over any bed depth. Calculations are started by determining the clean bed removal efficiency based on the transport mechanisms described previously. The bed is divided into a number of depth increments and a time series of removals is calculated for the first increment and subsequently applied to lower layers one after the other. Head loss is calculated as indicated previously and accumulated over time and depth. Thus a spatial and temporal distribution of filtration efficiency and head loss is produced.

The predictions of filtration performance obtained from the model are a function of the physical characteristics of the filter (bed depth, media size, porosity) and the suspension (flow rate, particle concentration, particle size, temperature) as well as chemical pretreatment (collision efficiency factor). The model is particularly sensitive to suspended particle size. Figure 8 shows the effects of particle size on head loss and filtration efficiency development over time. Parameter values assumed are shown in Table I and were used throughout the research. A concentration of 20 mg/L was assumed as the influent to the filter in developing Figure 8. Also, all retained particles were assumed to contribute additional exposed surface area.

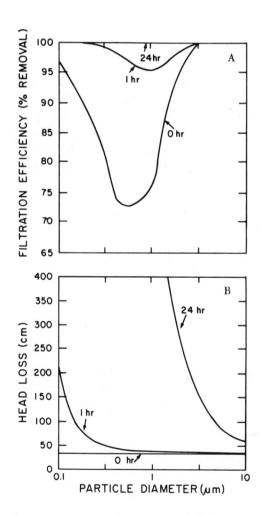

Figure 8. Filter performance: effects of suspended particle size and filtration time (influent concentration = 20 mg/L, particle density = 2.65 g/cm³; other filter parameters as indicated in Table I)

Figure 8B illustrates that head loss development is rapid for small particles but slow for large particles, since smaller particles contribute more surface area for the same mass concentration removed. Removal efficiency (Figure 8A) is a more complex function of particle size. Submicron particles are removed effectively by Brownian diffusion even on a clean bed ($t = 0$), and large particles are removed effectively by settling and interception. For particles of about $1\ \mu m$ in size, removal is least efficient (although still over 70% at $t = 0$ for the conditions considered), and a ripening period of about 1 hr is necessary to achieve 90% removal. These concepts are useful when considering the results and discussion that follow.

Results and Discussion

The results using this integral model for water treatment plants are presented in two ways throughout the remainder of this chapter. These are illustrated by Figures 9 and 10, prepared using calculated results for the typical or standard water treatment plant. In Figure 9, the performance of each treatment process is presented. The performance of flocculation and sedimentation facilities are expressed in terms of the change

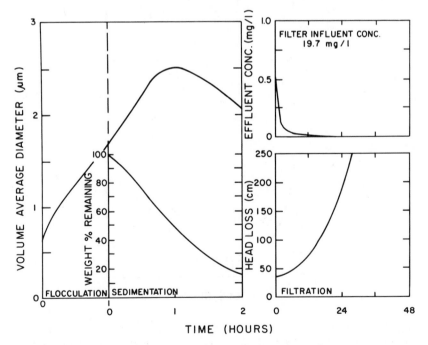

Figure 9.   Standard treatment system: predictions of plant performance

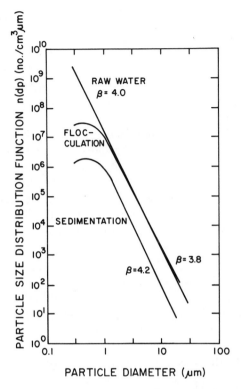

Figure 10. Standard treatment system: effects of flocculation and sedimentation on particle size distribution function

with time of the volume-average diameter of the particles passing through these treatment facilities. The performance of the settling tank is also described by the removal of solid mass (for example, suspended solids, expressed as weight percent remaining in suspension) as a function of time as the suspension flows through the tank. The performance of packed-bed filtration facilities is described by the changes with time in head loss and filter effluent concentration as predicted by the model.

The models for the flocculation and sedimentation tanks actually describe the heterogeneous distributions of particles in these tanks. The volume-average diameter is selected for use in Figure 9 as a means of summarizing information generated for 50 particle sizes in the flocculator and 21 sizes in the settling tank. A filter model for heterogeneous suspensions is not yet available; instead, a single particle size must be used. The volume-average diameter of the particles leaving the previous treatment unit, in this case the settling tank, is used for these calculations.

The results in Figure 9 reflect the effects of heterogeneity in suspended particle size on the performance of flocculation and settling facilities, but do not describe adequately changes in the particle size distribution that are produced by these processes. For this purpose, model calculations are presented in the form of Figure 10. Here, the particle size distribution function ($n(d_p)$) is plotted as a function of d_p on a log–log scale. Size distributions assumed in the raw water and calculated to result after coagulation and sedimentation in the standard treatment plant are presented.

The Standard Plant. The results presented in Figure 9 for the standard treatment plant indicate that effective removal (85%) of suspended solids is produced by the flocculation–sedimentation process. Filter run length, as indicated by the time it takes to reach a head loss of 250 cm, is 28 hr. Filter ripening is rapid, and essentially complete removal of suspended solids is accomplished by the filter throughout the run. The settling efficiency, head loss development, and filtrate quality predicted by the model are consistent with typical water treatment plant performance.

Figure 9 also indicates that the average particle size increases throughout flocculation as expected. In the settling tank, the average particle size increases for the first hour, and then decreases. In this tank, the processes of flocculation and settling occur simultaneously; the initial increase in the volume-average diameter followed by a decrease in this parameter reflects the balance of these two processes. Flocculation, a second-order reaction in particle number (see Equation 13), is a rapid reaction during the first hour of settling, since many particles are present. Such flocculation tends to increase the average particle size. As the particles become larger and removal by settling is accomplished more easily, the concentration of particles in suspension is reduced, the kinetics of the second-order flocculation reaction are slowed, and, since settling preferentially removes larger particles, the volume-average diameter of the particles remaining in suspension decreases.

The results for the particle size distributions for the standard plant are presented in Figure 10. The particle size distribution of the raw water is characterized by a value of $\beta = 4$ (slope $= -4$). Flocculation lowers β to 3.8 for most of the size range and reduces the number of submicron particles considerably. Sedimentation increases β to 4.2 for the main part of the distribution, and lowers particle concentrations throughout the size range, including substantial reduction of submicron particles. The reduction in the concentrations of the submicron particles by aggregation in the flocculation tank demonstrates that flocculation by fluid shear can be very effective for submicron particles if larger particles are present. This aggregation of submicron particles continues in the settling tank where contacts by differential settling occur. Particle size is not commonly measured in practice, so the model predictions for the changes in the particle size distribution during flocculation and settling cannot be compared quantitatively with field observations at present.

Raw Water Solids Concentration. Effects of the concentration of particulates in the raw water on the performance of the standard water treatment plant are presented in Figures 11 and 12. Results for a plant treating water containing 153 ppm of solids by volume (419 mg/L of solids having a specific gravity of 2.65) are shown in Figure 11, and results for 5 ppm by volume (13.2 mg/L) are shown in Figure 12. In both cases, the results for the standard system with an influent of 50 ppm by volume (132 mg/L) are included for comparison. High solids concentration in the raw water (Figure 11) produces rapid aggregation in the flocculation tank, extensive additional flocculation in the settling tank, effective solid–liquid separation by sedimentation (93%), long filter runs (47 hr to reach 250 cm head loss), and good effluent quality throughout the run. In contrast, low solids concentration in the raw water (Figure 12) produces only slight aggregation in the flocculation tank, a continuous but slow increase in average particle size by flocculation in the settling tank, and poor solid–liquid separation (21%) in the sedimentation tank. Filter runs are long if based on head loss development (58 hr

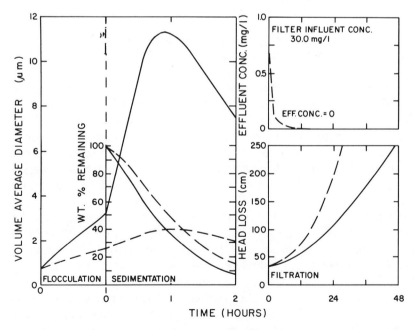

Figure 11.   High raw water particle concentration (——) (419 mg/L): predictions of plant performance—comparison with standard system (– – –) (132 mg/L)

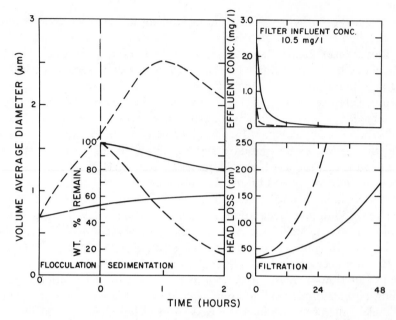

Figure 12. Low raw water particle concentration (———) (13.2 mg/L): predictions of plant performance—comparison with standard system (———) (132 mg/L)

to reach 250 cm head loss), but the ripening process is slow and the filter discharges more particles than in the standard case throughout the run.

The effects of raw water solids concentration on the particle size distribution functions for these cases are presented in Figures 13A and B. When water with a low solids concentration is treated (Figure 13B), there is little change in the particle size distribution, except for the submicron particles which are reduced considerably in both the flocculation and the settling tanks. Aggregation in the flocculation tank lowers β slightly below 4; solids removal in the settling tank increases β slightly above 4.

At high solids concentration, the treatment plant alters the size distribution function of the raw water substantially (Figure 13A). The reduction in the concentration of the submicron particles is dramatic. After flocculation, fewer submicron particles remain in the 419-mg/L suspension than in the 132-mg/L system (compare Figures 10 and 13A). Sedimentation continues this phenomenon so that the raw water containing 419 mg/L contains fewer submicron particles after settling than either the 132- or 13.2-mg/L suspensions (compare Figures 10, 13A, and 13B). This occurs because the rate of coagulation is second order in particle number and because the rate constants for contacts between

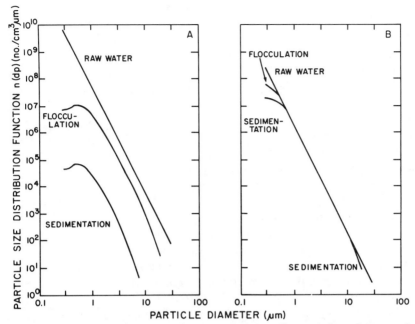

Figure 13. Raw water particle concentration: effects of flocculation and settling on particle size distribution function (A = 419 mg/L; B = 13.2 mg/L)

small and large particles are very high. Flocculation and sedimentation of the 419-mg/L suspensions yield particle size distributions that cannot be characterized by a single value of β over a broad range of sizes. Both flocculation and sedimentation increase β in the larger sizes to values greater than 4. The model does not calculate the upper end of the distribution correctly because of the maximum size limit allowed in the calculations. As stated earlier, the maximum size serves as a sink for particles that in actuality would grow to larger sizes. Conservation of mass requires that the graph of the particle size distribution function vs. particle size after coagulation must cross the plot of the influent distribution function. Model calculations do show that this occurs in the maximum size. This point is omitted in Figure 13A since the use of such an arbitrary maximum size does not reflect reality but is only necessitated by computer limitations. This limitation of the model is only significant in cases of extensive flocculation.

A comparison of Figures 9, 11, and 12 reveals that the performance of the water treatment plant is affected by the solids concentration in the raw water in complicated and unexpected ways. Influent water with the poorest water quality produces long filter runs and the best filter effluent quality. The standard influent produces good filtered water quality but the shortest filter runs. Influent water with the best quality (lowest solids concentration) produces long filter runs but the poorest filtrate quality. These results occur because of the combined effects of solids concentration on the kinetics of aggregation in the flocculation and settling tanks, of particle size distribution on flocculation kinetics and sedimentation efficiency, and of solids concentration and particle size on filtration efficiency, head loss development, and filter ripening. High solids concentrations permit rapid particle aggregation and, hence, effective settling. The large particle size applied to the filter in this case permits effective removal and low head loss development (Figure 8). For raw water containing low solids concentrations, flocculation is slow and sedimentation is not effective because the particles are small.

Effective treatment of waters containing low solids concentrations requires some modifications to the standard plant presented in Figure 4. One modification used extensively in the past for treating such water supplies is to add additional solids to the water by precipitating large quantities of aluminum or ferric hydroxide. This effectively transforms a water with a low solids concentration (Figure 12) to one with a higher solids concentration (Figure 9 or 11). Alternatively, direct filtration can be practiced. This involves deletion of the settling tank (and in some cases the flocculation tank) from the treatment system. However, becaused packed bed filters are less efficient when clean than when dirty or ripened, and because the low solids concentrations treated by direct filtration cause filter ripening to be slow, it is important to use deep filters in direct filtration processes of this type.

Particle Size Distribution Function. The effects of the size distribution of the particles in the raw water on plant performance are examined by comparing the results for a suspension having $\beta = 3$ (Figure 14) and the results for a homogeneous suspension (Figure 15) with the standard suspension ($\beta = 4$). Changes in the size distribution functions of these suspensions that are produced by flocculation and settling are presented in Figures 16A and B. The results indicate that the particle size distribution in the suspension to be treated has extensive effects on the performance of a water or wastewater treatment system using coagulation, settling, and filtration facilities.

An influent containing 132 mg/L of particles (50 ppm) distributed over the size interval from 0.3 to 30 μm and characterized by a size distribution function with $\beta = 3$ contains substantially more particles in the larger sizes than water with the same solids concentration and size range but with $\beta = 4$. It also has substantially fewer small particles and hence a higher volume-average diameter. A treatment plant receiving such an influent is predicted (Figure 14) to produce larger particles in the flocculation tank, more efficient removal of suspended solids by the settling tank, and longer filter runs than a plant treating a suspension with an

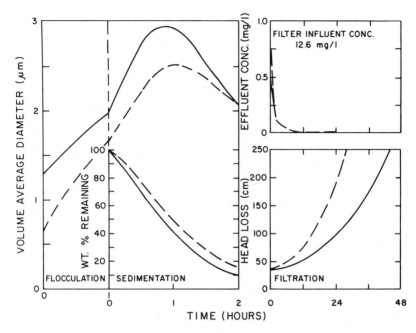

Figure 14. Raw water particle size distribution (——) (β = 3): predictions of plant performance—comparison with standard system (---)
(β = 4)

equal mass concentration but with β = 4. Filtration efficiency is excellent in both cases.

Effects of flocculation and settling on the size distributions of the suspension with β = 3 are presented in Figure 16A. Substantial reductions in the concentrations of submicron particles occur in both the flocculation and sedimentation tanks. The slope of the size distribution function for particles larger than 1 μm changes from −3 to about −3.4 in the flocculation tank, indicating that the volume distribution in this size range becomes more uniform for the suspension characteristics and flocculation conditions considered. Sedimentation removes the larger particles and further increases the negative slope of the size distribution function.

Most analyses of flocculation kinetics assume a homogeneous suspension (that is, a suspension containing particles of only one size) at the onset of flocculation. To compare such suspensions with the heterogeneous suspensions evaluated here, an influent containing 132 mg/L (50 ppm) of particles having the same volume-average diameter (0.688 μm) as the standard case was assumed. Such a suspension also has the

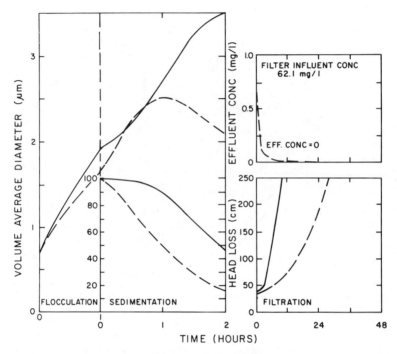

Figure 15.   Raw water particle size distribution (homogeneous suspen-sion (——)): predictions of plant performance—comparison with standard system (- - -) (β = 4)

same total particle concentration (N_∞) as the standard influent. Calcu-lated results of the performance of the water treatment plant when receiving this homogeneous suspension are presented in Figure 15. Re-moval efficiency by sedimentation is poor (53%) and filter runs are short; filter effluent quality is good. Extensive flocculation occurs throughout the settling tank.

Changes in the size distribution function of the homogeneous sus-pension are presented in Figure 16B. Both flocculation and sedimenta-tion broaden the distribution. Reduction in the concentration of the primary particles ($d_p = 0.688\ \mu$m) is substantial in both tanks; the con-centrations of the larger particles (10–30 μm) actually increase in the settling tank because of the flocculation reactions that occur in it.

Comparison of the predicted performances of the standard treatment plant receiving suspensions with $\beta = 4$, $\beta = 3$, and a homogeneous sus-pension (Figures 9, 14, and 15) demonstrates that the particle size distri-bution of the raw water substantially affects each unit of the treatment system, and the performance of the entire plant. Plant performance is worst for the homogeneous suspension and best for the suspension with $\beta = 3$.

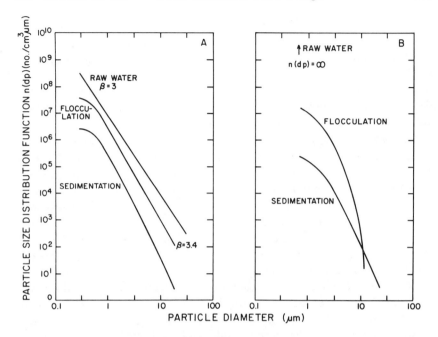

Figure 16. Raw water particle size distribution: effects of flocculation and settling on particle size distribution function ((A) β = 3; (B) homogeneous)

Removal efficiency of solid–liquid separation processes is usually based on removal of suspended solids or mass. However, suspended solids might not be an adequate measure, for two reasons. First, some pollutants are characterized better on a particle number or surface area basis. Second, a comparison of Figures 9, 14, and 15 shows that the effectiveness of the sedimentation and filtration processes is quantitatively different when the same mass, distributed in different ways, is treated. The performance of the individual treatment units and the entire system depends on the particle size distribution of the solids to be treated.

Particle size is probably the most important factor influencing the transport and removal of particles in the flocculation, sedimentation, and filtration of suspensions. The rates of particle contact in flocculation and settling tanks by Brownian diffusion, fluid shear, and gravity forces depend strongly on particle size (see Equations 14, 15, and 17). Removal of solids in settling tanks is accomplished by gravity forces and depends on particle size. Particles are removed from suspension during packed-bed filtration by Brownian diffusion, interception (fluid flow), and settling, all of which depend on suspended particle size. Despite the importance of particle size in flocculation, sedimentation, and filtration, the standard design criteria for each of these processes do not consider it.

This research indicates the importance of using available measurement techniques to describe the particle size distributions in raw water supplies and wastewaters, and in water and wastewater treatment plants. This information is needed to describe the suspensions, to evaluate treatment performance, and to provide a base for design improvement. It is also important to develop techniques to measure submicron particles, since these are of concern in some systems.

Velocity Gradient. Effects of the velocity gradient provided in the flocculation tank on plant performance are presented and compared with the standard case in Figures 17 and 18. Increasing the velocity gradient from 10 to 50 sec^{-1} (Figure 17) improves the removal in the settling tank from 85 to 92%, lengthens the filter run from 28 to 58 hr, and provides good filtrate quality throughout the run. As expected, particle aggregation is more rapid in the flocculation tank at the higher velocity gradient and continues to be rapid in the settling tank, after velocity gradient flocculation is terminated.

Plant performance in the absence of flocculation by fluid shear ($G = 0$ sec^{-1}) is examined in Figure 18. Some particle growth occurs by Brownian diffusion in the flocculator and by differential settling and Brownian diffusion in the settling tank. Removal efficiency by sedimentation is smaller than in the standard case (74% compared with 85%), and filter runs are reduced from 28 to 16 hr. Filtrate quality is good throughout the run.

Effects of the velocity gradient on the particle size distribution functions produced by flocculation and settling are presented in Figures 19A and B. At the high velocity gradient (50 sec^{-1}, Figure 19A), the size distribution function of the raw water is changed extensively throughout the size range from 0.3 to 30 μm by both flocculation and settling. Reduction in the concentration of submicron particles is substantial. A portion of the solids in the raw water is transformed to sizes larger than 30 μm, and so is not evaluated correctly by the flocculation and sedimentation programs.

When aggregation of the raw water is assumed to occur by Brownian diffusion alone in the flocculation tank ($G = 0$ sec^{-1}, Figure 19B), the concentrations of submicron particles are reduced, the concentrations of particles in the 1- to 3-μm size range are increased, and the distribution function of the larger sizes is not altered significantly. Subsequent sedimentation produces additional aggregation and substantial solids removal. The slope of the size distribution function in the upper region is changed so that $\beta = 4.2$.

Increasing the velocity gradient in the flocculation tank improves the performance of subsequent settling and filtration facilities. This effect is as expected. The integral analysis presented here describes the causes of these improvements and relates improved solid–liquid separation to

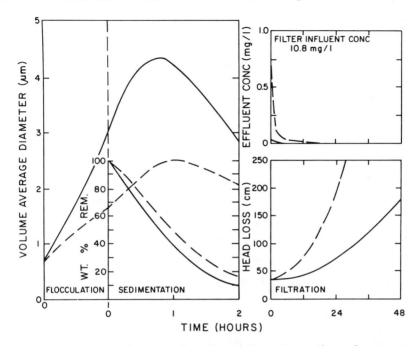

Figure 17. High velocity gradient (——) (G = 50 sec⁻¹): predictions of plant performance—comparison with standard system (— — —) (G = 10 sec⁻¹)

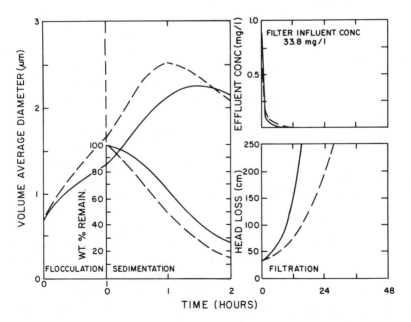

Figure 18. Zero velocity gradient (——): predictions of plant performance—comparison with standard system (— — —) (G = 10 sec⁻¹)

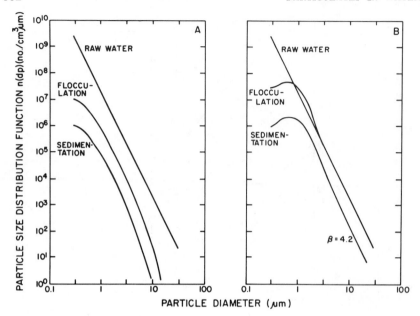

Figure 19.   Velocity gradient: effects of flocculation and settling on particle size distribution function ((A) G = 50 sec⁻¹; (B) G = 0 sec⁻¹)

flocculator design in a quantitative way. As the velocity gradient is increased, aggregation is more rapid in the flocculation tank, sedimentation is more efficient in removing mass because larger particles are formed, and the head loss development in the filters is slower because the concentration applied to the filters is decreased and the sizes of the particles applied to the filters are increased. The practical upper limit of increasing the velocity gradient in the flocculation tank is established by floc breakup, and is not addressed here.

Overflow Rate. Effects on plant performance that result from lowering the overflow rate of the sedimentation tank from 2 m/hr to 1 m/hr (1200 gal/day-ft² to 600 gal/day-ft²) by halving the tank depth while maintaining the detention time at 2 hr are presented in Figure 20. Results of the standard plant are included for comparison. Use of the shallow tank and low overflow rate produces slightly higher suspended solids removal in the settling tank (87% compared with 85%) and slightly longer filtration runs (31 hr compared with 28 hr). However, aggregation in the settling tank is reduced significantly, so that the particles applied to the filter are smaller than in the standard case. (The volume-average diameter of the particles in the filter influent is 1.47 μm.) These particles are somewhat difficult to remove by packed-bed filters (see Figure 8). The result is that a long ripening period occurs and solids are predicted to pass the filter continuously during the run.

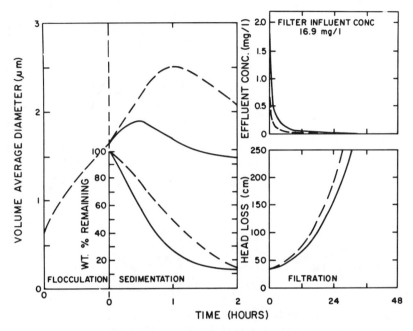

Figure 20.   Overflow rate in sedimentation tank (———) (1 m/hr): predictions of plant performance—comparison with standard system (– – –) (2 m/hr)

Effects of this reduction in the overflow rate on the size distribution of the particles leaving the settling tank are presented in Figure 21 and compared with the standard case. The standard settling tank with a depth of 4 m (overflow rate = 2 m/hr) produces greater reductions in the concentrations of particles in the size range from 0.3 to 30 μm than the shallow tank. The extensive aggregation that occurs in the 4-m settling tank produces particles larger than the 30-μm upper size limit, so these particles are not evaluated correctly in the settling model.

In this analysis, the overflow rate of the standard sedimentation tank was halved from 2 m/hr to 1 m/hr by halving the tank depth from 4 m to 2 m. This resulted in slightly improved sedimentation efficiency (mass basis), considerably less particle aggregation in the settling tanks attributable to a reduction in interparticle contacts by differential settling, and some reduction in filter performance. If the overflow rate had been halved by doubling the hydraulic detention time while maintaining the 4-m depth, additional aggregation would have occurred in the settling tank, and both settling and filter performance would have improved. While not presented here, these results can be anticipated by extrapolating the results for the standard plant (Figure 9) to a detention time of 4 hr in the settling tank. It is apparent that overflow rate and detention

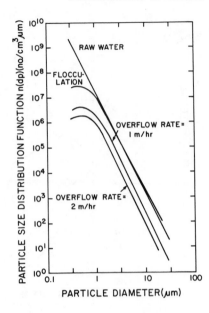

*Figure 21. Overflow rate in sedi-
mentation tank: effects on particle
size distribution function*

time are each inadequate to describe the performance of settling tanks; specification of both parameters is required. Furthermore, this research provides a quantitative assessment of these effects on sedimentation and subsequent filtration facilities.

Conclusions

The overall model presented for integral water treatment plant designs contains submodels for flocculation and sedimentation that are well established theoretically and experimentally. The approach selected for describing the characteristics of the water to be treated has been used extensively for over 20 years for aerosols and is coming into use for describing aquatic particulates. The submodel selected to describe filter performance is newer, more empirical, and less developed than the other components of the integral design model. The possible contributions of this research are the application of the integral approach, the feasibility of considering size heterogeneity through most of the treatment system, and the insights gained about the phenomena that occur and the performance that results in the plant.

Portions of the preceding discussion are summarized and supplemented in the following conclusions:

1. Water treatment plant design can and should be approached with an integrative perspective. Each unit affects the performance of subsequent units in predictable ways. The model presented here is a conceptual demonstration of the utility of this approach.

2. The predictions of the basic model for the typical or standard raw water and treatment system appear to be reasonable in comparison with experience in water treatment practice. This agreement indicates that other model predictions can be given serious consideration.

3. A better testing of model predictions requires determination of particle size distributions in raw waters and throughout treatment plants. These data will also provide needed understanding of treatment processes and more useful measures of plant performance. Existing instrumentation can be used for particle sizes larger than 1 μm. New techniques are needed for submicron particles.

4. Particle concentration and size distribution have extensive, complex, yet predictable effects on the performance of individual treatment units and on overall plant performance. In addition, depending on the pollutant of concern, appropriate measures of treatment efficiency can depend on these factors.

5. Solid–liquid separation processes and coagulation facilities should be designed to reflect important characteristics of the solids to be treated, including particle concentration and size distribution. This objective is feasible conceptually at present and can become feasible in practice within a few years if proper efforts are exerted.

6. Extensive flocculation can occur in settling tanks, indicating that overflow rate and detention time must be considered together as design criteria.

Acknowledgments

The authors would like to express their appreciation to Harvey Jeffries, Associate Professor, and Michael Kuhlman, graduate student, in the Department of Environmental Sciences and Engineering, University of North Carolina at Chapel Hill, for their assistance in developing the computer programs for the flocculation and sedimentation models, and to Brian Ramaley, a graduate student in the same department, for his aid in calculating the effects of particle size distribution.

Glossary of Symbols

A = coefficient in empirical power-law size distribution function (number $\cdot$ μm$^{(\beta-1)}$/cm^3)

d_p = particle size (diameter basis) (μm)

d_1 = diameter of Particle 1 (μm)

d_2 = diameter of Particle 2 (μm)

G = velocity gradient (sec^{-1})

g = gravity constant (cm/sec^2)

k = Boltzmann's constant (J/$^\circ$K)

N = number concentration of particles with a size equal to or less than a stated value (number/cm³)

N_∞ = total number concentration of particles (number/cm³)

$n(d_p)$ = particle size distribution function, diameter basis (number/cm³μm)

$n(s)$ = particle size distribution function, surface area basis (number/cm³μm²)

$n(v)$ = particle size distribution function, volume basis (number/cm³μm³)

n_k = number concentration of particles of size k (number/cm³)

S = surface concentration of particles (μm²/cm³)

s = particle size (surface area basis) (μm²)

T = temperature (°K)

V = volume concentration of particles (μm³/cm³, also ppm)

$V_{1\text{-}2}$ = volume concentration of particles having sizes between d_1 and d_2 (μm³/cm³, also ppm)

v = particle size (volume basis) (μm³)

v_k = volume of a particle of size k (μm³)

w_k = Stokes' settling velocity of a particle of size k (cm/sec)

z = depth of vertical compartment or box in settling tank (cm)

α = collision efficiency factor (dimensionless)

β = exponent in empirical power-law size distribution function (dimensionless)

$\beta(i,j)$ = rate coefficient for collision between particles of size i and j by some physical transport process (cm³/sec)

μ = absolute viscosity (g/cm sec)

ρ_1 = density of water (g/cm³)

ρ_p = density of particle (g/cm³)

Literature Cited

1. Lal, D.; Lerman, A. "Size Spectra of Biogenic Particles in Ocean Waters and Sediments," *J. Geophys. Res.* 1975, 80, 423–430.
2. Lerman, A.; Carder, K. L.; Betzer, P. R. "Elimination of Fine Suspensions in the Oceanic Water Column," *Earth Planet. Sci. Lett.* 1977, 37, 61–70.
3. Ruldolfs, W.; Gehm, H. W. "Colloids in Sewage and Sewage Treatment. I. Occurrence and Role. A Critical Review," *Sewage Works J.* 1939, 11, 727–737.
4. Rudolfs, W.; Balmat, J. L. "Colloids in Sewage. I. Separation of Sewage Colloids with the Aid of the Electron Microscope," *Sewage Ind. Wastes* 1952, 24, 247–256.
5. Hunter, J. V.; Heukelekian, H. "The Composition of Domestic Sewage Fractions," *J. Water Pollut. Control Fed.* 1965, 37, 1142–1163.
6. Rickert, D. A.; Hunter, J. V. "Rapid Fractionation and Materials Balance of Solids Fractions in Wastewater and Wastewater Effluent," *J. Water Pollut. Control Fed.* 1967, 39, 1475–1506.
7. Faisst, W. K. Chapter 12 in this book.
8. Beard, J. D., II; Tanaka, T. S. "A Comparison of Particle Counting and Nephelometry," *J. Am. Water Works Assoc.* 1977, 69(10), 533–538.

9. Kavanaugh, M. C.; Zimmermann, U.; Vagenknecht, A. "Determinations of Particle Size Distributions in Natural Waters; Use of Zeiss Micro-Videomat Image Analyzer," *Schweiz. Z. Hydrol.* **1977**, *39*, 86–98.
10. Hunt, J. R. "Prediction of Oceanic Particle Size Distributions from Coagulation and Sedimentation Mechanisms." Chapter 11 in this book.
11. Friedlander, S. K. "Similarity Considerations for the Particle-Size Spectrum of a Coagulating, Sedimenting Aerosol," *J. Meteorol.* **1960**, *17*, 479–483.
12. Kavanaugh, M. C.; Toregas, G.; Chung, M.; Pearson, E. A. "Particulates and Trace Pollutant Removal by Depth Filtration," *Progress Water Technol.* **1978**, *10*(5, 6), 197–215.
13. Smoluchowski, M. "Versuch Einer Mathematischen Theorie der Koagulations-Kinetic Kolloider Losungen," *Z. Phys. Chem.* **1917**, *92*, 129–168.
14. O'Melia, C. R. "Coagulation and Flocculation," In "Physicochemical Processes for Water Quality Control"; Weber, W. J., Jr., Ed.; Wiley: New York, 1972; pp. 61–109.
15. Friedlander, S. K. "Smoke, Dust, and Haze"; Wiley: New York, 1977.
16. Gear, C. W. "Numerical Initial Value Problems in Ordinary Differential Equations"; Prentice-Hall: Englewood Cliffs, New Jersey, 1971.
17. Kozeny, J. *Sitzungber. Akad. Wiss. Wien, Math. Naturwiss. Kl. Abt. 2A.* **1927**, *136*, 271–306.
18. Fair, G. M.; Hatch, L. P. "Fundamental Factors Governing the Streamline Flow of Water through Sand," *J. Am. Water Works Assoc.* **1933**, *25*, 1551–1565.
19. Friedlander, S. K. "Theory of Aerosol Filtration," *Ind. Eng. Chem.* **1958**, *50*, 1161–1164.
20. Yao, K. M.; Habibian, M. T.; O'Melia, C. R. "Water and Wastewater Filtration: Concepts and Applications," *Environ. Sci. Technol.* **1971**, *5*, 1105–1112.
21. Fitzpatrick, J.; Spielman, L. A. "Filtration of Aqueous Latex Suspensions through Beds of Glass Spheres," *J. Colloid Interface Sci.* **1973**, *43*, 350–369.
22. Ives, K. J.; Gregory, J. "Basic Concepts of Filtration," *Proc. Soc. Water Treat. Exam.* **1967**, *16*, 147–169.
23. Ison, C. F.; Ives, K. J. "Removal Mechanisms in Deep Bed Filtration," *Chem. Eng. Sci.* **1969**, *24*, 717–729.
24. Ali, W. "The Role of Retained Particles in a Filter," unpublished master's report, University of North Carolina, Chapel Hill, 1977.
25. O'Melia, C. R.; Ali, W. "The Role of Retained Particles in Deep Bed Filtration," *Prog. Water Technol.* **1978**, *10*(5, 6), 123–137.
26. Happel, J. "Viscous Flow in Multiparticle System," *Am. Inst. Chem. Eng.* **1958**, *4*(2), 197–201.
27. Pfeffer, R., "Heat and Mass Transport in Multiparticle Systems," *Ind. Eng. Chem. Fundam.* **1964**, *3*(4), 380–383.

RECEIVED December 4, 1978.

Errata

Just prior to publication, a numerical error was discovered in the work reported herein. The factor $(6/\pi)^{1/3}$ in Equation 17 was entered incorrectly into the computer program as $6^{1/3}/\pi$. As a result, the rate constant for differential sedimentation was underestimated by a factor of $(\pi)^{2/3}$ or 2.15. The numerical results presented in this chapter can be considered to correspond to the use of the collision efficiency factor for differential sedimentation of 1/2.15, or 0.47.

Correcting this error influences the results in that the flocculation rate achieved in the sedimentation tank is greater, which in turn allows removal by sedimentation to occur to a greater extent. Calculations have been made using this corrected constant. The trends reported in this chapter remain the same, and the conclusions are unchanged.

INDEX

INDEX

Jacket design by Carol Conway.
Editing and production by Susan Moses.

The book was composed by Service Composition Co., Baltimore, MD,
printed and bound by The Maple Press Co., York, PA.